BIOCHEMICAL CALCULATIONS

How to Solve Mathematical Problems in General Biochemistry

BIOCHEMICAL CALCULATIONS

How to Solve Mathematical Problems in General Biochemistry

IRWIN H. SEGEL
Department of Biochemistry
and Biophysics
University of California
Davis, California

John Wiley & Sons, Inc.   New York · London · Sydney · Toronto

*This book is dedicated with much love
to my sons Jonathan and Daniel*

PREFACE

Biochemistry is a quantitative science. Yet too often the subject is taught in a purely descriptive manner. This is understandable because many students in biological sciences have had only one course in general chemistry and one in introductory organic chemistry as prerequisites to general biochemistry. *Biochemical Calculations* was written to introduce students to some of the mathematical aspects of biochemistry that should be discussed in any introductory course. The problems discussed need no knowledge of mathematics beyond that required for general chemistry. In the very few problems where elementary calculus is employed, an alternate algebraic approximation is also given. In order for this book to be as useful as possible to the student, almost every problem is solved completely. There are no "rearranging terms and solving for X" statements. This book is meant to be used in conjunction with a standard textbook of general biochemistry. Nevertheless, a substantial amount of descriptive background (including mathematical derivations) is given in the appendices.

A mimeographed edition of *Biochemical Calculations* was used by more than 800 students on the Davis campus during 1966–1967. Most of the students were enrolled in our general biochemistry course. Many others were graduate students studying for their M.S. and Ph.D. examinations. It was the response of these students that prompted me to submit the book for publication and general distribution.

I wish to express my appreciation to Dr. Wayne W. Luchsinger and Dr. Ronald S. Watanabe for their advice and critical review of the original manuscript.

I thank Dr. Eric E. Conn and Dr. Paul K. Stumpf for their suggestions and encouragement during the writing of this book. I am especially grateful to Miss Leigh Denise Albizati for all her help.

October 1968
Davis, California

IRWIN H. SEGEL

CONTENTS

BIOCHEMICAL ENERGETICS   162
IV

ENZYME KINETICS   213
V

SPECTROPHOTOMETRY   259
VI

ISOTOPES IN BIOCHEMISTRY   279
VII

MISCELLANEOUS CALCULATIONS   303
VIII

Appendix: ACIDS AND BASES 317
I

Appendix: BIOCHEMICAL ENERGETICS 350
II

Appendix: ENZYME KINETICS 366
III

Appendix: SPECTROPHOTOMETRY 397
IV

BIOCHEMICAL CALCULATIONS

How to Solve Mathematical Problems in General Biochemistry

STRONG ACIDS AND BASES

I

A. pH AND ACIDITY OF STRONG ACIDS AND BASES

Problem I-1

What are the: (a) H^+ ion concentration, (b) pH, (c) OH^- ion concentration, and (d) pOH of a 0.001 M solution of HCl?

Solution

(a) HCl is a "strong" inorganic acid; that is, it is essentially 100% ionized in dilute solution. Consequently, when 0.001 mole of HCl is introduced into 1 liter of H_2O, it immediately dissociates into 0.001 M H^+ and 0.001 M Cl^-.

	HCl $\rightarrow$	H^+	$+ Cl^-$
Start:	0.001 M	0	0
Change:	-0.001 M	$+0.001$ M	$+ 0.001$ M
Final concentration:	0	0.001 M	0.001 M

The H^+ ion concentration, $[H^+]$, is 0.001 M, or more conveniently written, 10^{-3} M. Note that when we are dealing with strong acids, the H^+ contribution from the ionization of water is neglected.

(b) $\quad$ pH $= -\log [H^+]$ $\qquad$ or $\qquad$ pH $= +\log \dfrac{1}{[H^+]}$

$\qquad\qquad$ pH $= -\log 10^{-3}$ $\qquad\qquad\qquad$ pH $= \log \dfrac{1}{10^{-3}}$

$\qquad\qquad$ pH $= -(-3) = +3$ $\qquad\qquad\qquad$ pH $= \log 10^3$

$\qquad\qquad$ $\boxed{\text{pH} = 3}$ $\qquad\qquad\qquad\qquad\quad$ $\boxed{\text{pH} = 3}$

1

(c)

$$[H^+][OH^-] = K_w$$

$$[H^+][OH^-] = 1 \times 10^{-14}$$

$$[OH^-] = \frac{1 \times 10^{-14}}{[H^+]}$$

$$[OH^-] = \frac{1 \times 10^{-14}}{1 \times 10^{-3}}$$

$$\boxed{[OH^-] = 1 \times 10^{-11}}$$

(d)   We can define a shorthand notation for $[OH^-]$ as pOH where pOH is the negative logarithm of the hydroxyl ion concentration.

$$pOH = -\log[OH^-] \quad \text{or} \quad pOH = \log\frac{1}{[OH^-]}$$

$$pOH = -\log(10^{-11}) \qquad\qquad pOH = \log\frac{1}{10^{-11}}$$

$$pOH = -(-11) \qquad\qquad\qquad pOH = \log 10^{11}$$

$$\boxed{pOH = 11} \qquad\qquad\qquad\quad \boxed{pOH = 11}$$

Or, working with the expression for K_w, we can derive the following useful relationship:

$$[H^+][OH^-] = K_w$$

$$[H^+][OH^-] = 10^{-14}$$

$$\log[H^+] + \log[OH^-] = \log 10^{-14}$$

$$-\log[H^+] - \log[OH^-] = -\log 10^{-14}$$

$$-\log[H^+] = pH, \qquad -\log[OH^-] = pOH$$

$$\log 10^{-14} = -14, \qquad -\log 10^{-14} = 14$$

$$\therefore \quad \boxed{pH + pOH = 14}$$

$$pOH = 14 - 3 = 11$$

Problem I-2

What are the: (a) $[H^+]$, (b) $[OH^-]$, (c) pH, and (d) pOH of a 0.002 M solution of HNO_3?

Solution

(a) HNO_3 is a strong inorganic acid.

$$\boxed{[H^+] = 0.002\ M = 2 \times 10^{-3}\ M}$$

(b)
$$[H^+][OH^-] = 1 \times 10^{-14}$$
$$[OH^-] = \frac{1 \times 10^{-14}}{2 \times 10^{-3}}$$
$$[OH^-] = 0.5 \times 10^{-11}$$
$$\boxed{[OH^-] = 5 \times 10^{-12}\ M}$$

(c) $\text{pH} = \log \dfrac{1}{[H^+]}$

$\text{pH} = \log \dfrac{1}{2 \times 10^{-3}}$

$\text{pH} = \log 0.5 \times 10^3$ or $\text{pH} = \log 500$

$\text{pH} = \log 5 \times 10^2$ $\log 500 = 2.699$

$\text{pH} = \log 5 + \log 10^2$

$\text{pH} = 0.699 + 2$ $\boxed{\text{pH} = \mathbf{2.699}}$

$\boxed{\text{pH} = \mathbf{2.699}}$ where 2 = the number of places between the first significant figure and the decimal point and 0.699 = log of "5."

Check:
$$10^{-2}\ M\ [H^+] = \text{pH } 2$$
$$10^{-3}\ M\ [H^+] = \text{pH } 3$$
$$\therefore\ \ 2 \times 10^{-3}\ M\ [H^+] = \text{pH between 2 and 3}$$

(d) $\text{pH} + \text{pOH} = 14$ or $\text{pOH} = \log \dfrac{1}{[OH^-]}$

$\text{pOH} = 14.000 - 2.699$ $\text{pOH} = \log \dfrac{1}{5 \times 10^{-12}}$

$\boxed{\text{pOH} = \mathbf{11.301}}$ $\text{pOH} = \log 0.2 \times 10^{12}$

                                    $\text{pOH} = \log 2 \times 10^{11}$

                                    $\text{pOH} = \log 2 + \log 10^{11}$

                                    $\text{pOH} = 0.301 + 11$

                                    $\boxed{\text{pOH} = \mathbf{11.301}}$

Check:

$$10^{-11} M \text{ [OH}^-] = \text{pOH } 11$$

$$10^{-12} M \text{ [OH}^-] = \text{pOH } 12$$

$$\therefore \quad 5 \times 10^{-12} M \text{ [OH}^-] = \text{pOH between 11 and 12}$$

Part c could also be worked out using the original notation for pH (pH $= -\log$ [H$^+$]) and negative logarithms.

$$\text{pH} = -\log \text{ [H}^+]$$

$$\text{pH} = -\log 2 \times 10^{-3}$$

$$\text{pH} = -(\log 2 + \log 10^{-3})$$

$$\text{pH} = -\log 2 - \log 10^{-3}$$

$$\text{pH} = -\log 10^{-3} - \log 2$$

$$\text{pH} = +3 - \log 2$$

$$\text{pH} = 3 - 0.301$$

$$\boxed{\text{pH} = 2.699}$$

Similarly, part d could be solved using negative logarithms.

$$\text{pOH} = -\log \text{ [OH}^-]$$

$$\text{pOH} = -\log 5 \times 10^{-12}$$

$$\text{pOH} = -(\log 5 + \log 10^{-12})$$

$$\text{pOH} = -\log 5 - \log 10^{-12}$$

$$\text{pOH} = -\log 10^{-12} - \log 5$$

$$\text{pOH} = +12 - 0.699$$

$$\boxed{\text{pOH} = 11.301}$$

Note that in all of the above calculations the very small contribution of H$^+$ from H_2O has been neglected. Also note that values such as 0.5×10^3 were changed to 5×10^2 to simplify the determination of logarithms.

Problem I-3

What are the: (a) [H$^+$], (b) [OH$^-$], (c) pH, and (d) pOH of a $3 \times 10^{-4} M$ solution of H_2SO_4?

Solution

(a) H_2SO_4 is a strong inorganic acid. It completely ionizes in a dilute solution.

$$H_2SO_4 \rightarrow 2H^+ + SO_4^-$$

Note that every mole of H_2SO_4 that ionizes yields *2 moles* of H^+

$$\therefore \quad \boxed{[H^+] = 6 \times 10^{-4} \ M}$$

(b)
$$[OH^-] = \frac{1 \times 10^{-14}}{6 \times 10^{-4}}$$

$$= 0.167 \times 10^{-10} \ M$$

$$\boxed{[OH^-] = 1.67 \times 10^{-11} \ M}$$

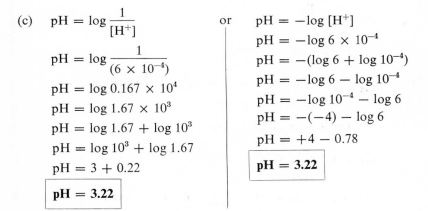

(c) $pH = \log \dfrac{1}{[H^+]}$ or $pH = -\log [H^+]$

$pH = \log \dfrac{1}{(6 \times 10^{-4})}$ $pH = -\log 6 \times 10^{-4}$

$pH = \log 0.167 \times 10^4$ $pH = -(\log 6 + \log 10^{-4})$

$pH = \log 1.67 \times 10^3$ $pH = -\log 6 - \log 10^{-4}$

$pH = \log 1.67 + \log 10^3$ $pH = -\log 10^{-4} - \log 6$

$pH = \log 10^3 + \log 1.67$ $pH = -(-4) - \log 6$

$pH = 3 + 0.22$ $pH = +4 - 0.78$

$\boxed{pH = 3.22}$ $\boxed{pH = 3.22}$

Check:

$$10^{-3} \ M \ [H^+] = pH \ 3$$

$$10^{-4} \ M \ [H^+] = pH \ 4$$

$$\therefore \quad 6 \times 10^{-4} \ M \ [H^+] = pH \ \text{between 3 and 4}$$

(d)
$$pH + pOH = 14$$

$$pOH = 14.00 - 3.22$$

$$\boxed{pOH = 10.78}$$

Problem I-4

What is the concentration of HNO_3 in a solution that has a pH of 3.4?

Solution

$$pH = \log \frac{1}{[H^+]}$$

$$3.4 = \log \frac{1}{[H^+]}$$

where 3 = number of places between first significant figure and the decimal point. Look up antilog of "4."

$$antilog = "251"$$

$$\therefore \quad \frac{1}{[H^+]} = 2510$$

$$[H^+] = \frac{1}{2510}$$

$$[H^+] = \frac{1}{2.51 \times 10^3}$$

$$[H^+] = 0.398 \times 10^{-3}$$

$$\boxed{[H^+] = 3.98 \times 10^{-4}}$$

or

$$[H^+] = 10^{-pH}$$

$$[H^+] = 10^{-3.4}$$

$$[H^+] = 10^{-4} \times 10^{+0.6}$$

Look up antilog of 0.6

$$antilog \ of \ 0.6 = "398"$$

$$= 3.98$$

$$\boxed{[H^+] = 3.98 \times 10^{-4}}$$

$$\therefore \quad HNO_3 = 3.98 \times 10^{-4} \ M \ \text{assuming } 100\% \text{ ionization}$$

Check:

$$pH \ 3 = 10^{-3} \ M \ [HNO_3]$$

$$pH \ 4 = 10^{-4} \ M \ [HNO_3]$$

$$\therefore \quad pH \ 3.4 = [HNO_3] \text{ between } 10^{-4} \text{ and } 10^{-3} \ M$$

Problem I-5

What are the: (a) molarity and (b) normality of an H_2SO_4 solution where the pH is 5.2?

Solution

(a)
$$pH = \log \frac{1}{[H^+]}$$

$$5.2 = \log \frac{1}{[H^+]}$$

$$\frac{1}{[H^+]} = \text{antilog of } 5.2$$

$$\frac{1}{[H^+]} = 1.585 \times 10^5$$

$$[H^+] = \frac{1}{1.585 \times 10^5}$$

$$[H^+] = 0.632 \times 10^{-5}$$

$$[H^+] = 6.32 \times 10^{-6} \ M$$

Every mole of H_2SO_4 yields 2 moles (g-ions) of H^+ upon ionization.

$$\therefore \ M_{H_2SO_4} = \frac{6.32 \times 10^{-6}}{2}$$

$$\boxed{H_2SO_4 = 3.16 \times 10^{-6} \ M}$$

(b)   One mole of $H_2SO_4 \equiv 2$ equivalents; i.e., a $1 \ M$ solution $\equiv 2 \ N$ solution or $N = M \times \#H^+$ ions produced upon ionization.

$$N = M \times 2$$

$$\boxed{\therefore \ \ H_2SO_4 = 6.32 \times 10^{-6} \ N}$$

Problem I-6

What are the: (a) $[OH^-]$, (b) pOH, (c) $[H^+]$, and (d) pH of a 0.004 M KOH solution?

Solution

(a)   KOH is a strong inorganic base. It is essentially 100% ionized in solution.

$$\boxed{\therefore \ \ [OH^-] = 4 \times 10^{-3} \ M}$$

(b)
$$pOH = \log \frac{1}{4 \times 10^{-3}}$$

$$= \log 0.25 \times 10^3$$

$$= \log 2.5 \times 10^2$$

$$= \log 2.5 + \log 10^2$$

$$= 0.398 + 2$$

$$\boxed{pOH = 2.398}$$

Check:

$$10^{-2} \, M \, [OH^-] = pOH \, 2$$

$$10^{-3} \, M \, [OH^-] = pOH \, 3$$

$$4 \times 10^{-3} \, M \, [OH^-] = pOH \text{ between 2 and 3}$$

(c)
$$[H^+][OH^-] = 1 \times 10^{-14}$$

$$[H^+] = \frac{1 \times 10^{-14}}{4 \times 10^{-3}}$$

$$[H^+] = 0.25 \times 10^{-11}$$

$$\boxed{[H^+] = 2.5 \times 10^{-12} \, M}$$

(d) pH + pOH = 14   or $pH = \log \dfrac{1}{2.5 \times 10^{-12}}$

   pH = 14 − 2.398

$$\boxed{pH = 11.602}$$

$$= \log 0.4 \times 10^{12}$$

$$= \log 4 \times 10^{11}$$

$$= \log 4 + \log 10^{11}$$

$$= 0.602 + 11$$

$$\boxed{pH = 11.602}$$

Problem I-7

What is the pH of a $10^{-8} \, M$ solution of HCl?

Solution

The first tendency of many students is to say "pH = 8." This is obviously incorrect. No matter how much one dilutes a strong acid, the solution will never become alkaline.

In this solution, the contribution of H^+ ions from H_2O is actually greater than the amount contributed by HCl. As a *first approximation*, therefore, the H^+ ions from the HCl may be neglected. The pH then is around 7.

As a *second approximation*, we can solve for pH while taking into account the H^+ ions from both sources.

$$\text{pH} = -\log[H^+]$$
$$[H^+] = 10^{-7} \text{ (from } H_2O) + 10^{-8} \text{ (from HCl)}$$
$$\text{pH} = -\log(1 \times 10^{-7} + 0.1 \times 10^{-7})$$

$$= \log\frac{1}{1.1 \times 10^{-7}}$$
$$= \log 0.909 \times 10^7$$
$$= \log 9.09 \times 10^6$$
$$= \log 9.09 + \log 10^6$$
$$= 0.953 + 6$$

$$\boxed{\text{pH} = 6.953}$$

The above solution is still not completely correct. It assumes that the contribution of H^+ ions from water is still $10^{-7}\ M$ in the presence of $10^{-8}\ M$ HCl. Actually, the slight increase in H^+ ions from HCl tends to depress the ionization of H_2O, i.e., shift the equilibrium of the HOH $\leftrightarrows$ $H^+ + OH^-$ reaction back to the left. An exact solution to the problem can be obtained in the following manner: Both HOH and HCl ionize to form H^+ ions.

$$\text{HOH} \leftrightarrows H^+ + OH^-$$

$$\text{HCl} \rightarrow H^+ + Cl^-$$

$$\text{Let} \quad X = [H^+] \text{ from } H_2O$$

$$\therefore \quad [OH^-] = X$$

$$[H^+] \text{ from HCl} = 10^{-8}\ M$$

$$[H^+] = X + 10^{-8}, [OH^-] = X$$

$$[H^+][OH^-] = 10^{-14}$$

$$(X + 10^{-8})(X) = 10^{-14}$$

$$X^2 + 10^{-8} X = 10^{-14}$$

$$X^2 + 10^{-8} X - 10^{-14} = 0$$

The above equation can be solved by substituting into the solution for the general quadratic equation:

$$X = \frac{-b \pm \sqrt{b^2 - 4ac}}{2a}$$

where:

$$a = 1$$
$$b = 10^{-8}$$
$$c = -10^{-14}$$

$$X = \frac{-10^{-8} \pm \sqrt{(10^{-8})^2 - 4(-10^{-14})}}{2}$$

$$= \frac{-10^{-8} \pm \sqrt{10^{-16} + 4 \times 10^{-14}}}{2}$$

$$= \frac{-10^{-8} \pm \sqrt{4.01 \times 10^{-14}}}{2}$$

$$= \frac{-10^{-8} \pm 2.0025 \times 10^{-7}}{2}$$

$$= \frac{-10^{-8} \pm 20.025 \times 10^{-8}}{2}$$

$$= \frac{19.025 \times 10^{-8}}{2} \quad \text{and} \quad \frac{-21.025 \times 10^{-8}}{2}$$

$$X = 9.5125 \times 10^{-8} \quad \text{(neglecting the negative value)}$$

$$[H^+] = X + 10^{-8}$$

$$= 9.5125 \times 10^{-8} + 10^{-8}$$

$$= 10.5125 \times 10^{-8}$$

$$pH = \log \frac{1}{10.5125 \times 10^{-8}}$$

$$= \log 0.09512 \times 10^8$$

$$= \log 9.512 \times 10^6$$

$$= \log 9.512 + \log 10^6$$

$$= 0.976 + 6$$

$$\boxed{pH = 6.976}$$

Problem I-8

What are the: (a) a_{H^+} and (b) γ_{H^+} in a 0.010 M solution of HNO_3 if the pH is 2.08?

Solution

In this problem we can no longer assume that $a_{H^+} = [H^+]$. It is obvious that if $a_{H^+} = [H^+]$ and pH $= -\log [H^+]$, the pH would be 2.0 and not 2.08. $\therefore \gamma \neq 1$.

(a)
$$pH = \log \frac{1}{a_{H^+}} \quad \text{or} \quad a_{H^+} = 10^{-pH}$$

$$2.08 = \log \frac{1}{a_{H^+}} \qquad a_{H^+} = 10^{-2.08}$$

$$\text{antilog of } 0.08 = \text{``120''} \qquad = 10^{-3} \times 10^{+0.92}$$

$$\frac{1}{a_{H^+}} = 120 \qquad = 10^{-3} \times 8.3$$

$$a_{H^+} = \frac{1}{120} \qquad \boxed{a_{H^+} = 8.3 \times 10^{-3}}$$

$$a_{H^+} = 0.0083$$

$$\boxed{a_{H^+} = 8.3 \times 10^{-3}}$$

(b)
$$a_{H^+} = \gamma_{H^+}[H^+]$$

$$(\gamma = \text{the activity coefficient})$$

$$0.0083 = \gamma_{H^+}(0.010)$$

$$\boxed{\gamma_{H^+} = 0.83}$$

Although the *actual* concentration of HNO_3 is 0.01 M, the solution behaves as if only 83% of the HNO_3 molecules are dissociated; the *effective* or *apparent* concentration (a_{H^+}) is 0.0083 M.

We could write an *apparent* K_i for the dissociation of HNO_3.

	HNO_3	$\rightleftharpoons$	H^+	+	NO_3^-
Start:	0.01 M		0		0
Change:	$-0.0083\ M$		$+0.0083\ M$		$+0.0083\ M$
Final concentration:	0.0017 M		0.0083 M		0.0083 M

$$K_{i(\text{app})} = \frac{[H^+][NO_3^-]}{[HNO_3]}$$

$$= \frac{(0.0083)(0.0083)}{(0.0017)} = \frac{(8.3 \times 10^{-3})^2}{(1.7 \times 10^{-3})}$$

$$= \frac{69 \times 10^{-6}}{1.7 \times 10^{-3}}$$

$$= 40.5 \times 10^{-3}$$

$$K_{i(\text{app})} = 4.05 \times 10^{-2}$$

Such $K_{i(\text{app})}$ are generally not calculated for strong inorganic acids. Although the "active acidity" of the solution suggests incomplete ionization of the HNO_3, the HNO_3 is actually 100% ionized. Interactions of ion-clouds, however, prevent full expression of the H^+ ions, i.e., the shielding effects of NO_3^- ions surrounding the H^+ ions make it seem that the H^+ ions are not all there.

Problem I-9

How many (a) H^+ ions and (b) OH^- ions are present in 250 ml of a solution of pH 3?

Solution

(a)
$$pH = 3$$

$$\therefore \quad [H^+] = 10^{-3} \, M \, (10^{-3} \text{ g-ions/liter})$$

$$1 \text{ g-ion/liter} = 6.023 \times 10^{23} \text{ ions/liter}$$

$$\therefore \quad 10^{-3} \text{ g-ions/liter} = 6.023 \times 10^{20} \text{ ions/liter}$$

$$\therefore \quad \frac{6.023 \times 10^{20}}{4} = \boxed{\mathbf{1.506 \times 10^{20} \text{ ions/250 ml}}}$$

(b)
$$pH + pOH = 14$$

$$pOH = 14 - 3$$

$$= 11$$

$$[OH^-] = 10^{-11} \, M \text{ or } 10^{-11} \text{ g-ions/liter}$$

$$10^{-11} \text{ g-ions/liter} \times 6.023 \times 10^{23} \text{ ions/g-ion} = 6.023 \times 10^{12} \text{ ions/liter}$$

$$\frac{6.023 \times 10^{12}}{4} = \boxed{\mathbf{1.506 \times 10^{12} \text{ ions/250 ml}}}$$

B. NEUTRALIZATION

Problem I-10

(a)   How many milliliters of 0.025 M H_2SO_4 are required to neutralize exactly 525 ml of 0.06 M KOH? (b) What is the pH of the "neutralized" solution?

Solution

(a)   number of moles (equivalents) of H^+ required =

$$\text{number of moles (equivalents) of } OH^- \text{ present}$$

$$\text{liters} \times N = \text{number of equivalents}$$

$$\text{liters}_{acid} \times N_{acid} = \text{liters}_{base} \times N_{base}$$

$$H_2SO_4 = 0.025 \, M$$

$$= 0.05 \, N$$

$$\text{liters}_{acid} \times 0.05 = 0.525 \times 0.06$$

$$\text{liters}_{acid} = \frac{0.525 \times 0.06}{0.05}$$

$$= 0.63 \text{ liter}$$

$$\boxed{\textbf{acid required} = \textbf{630 ml}}$$

(b)   The neutralized solution contains only K_2SO_4 which, being a salt of a strong acid and strong base, has no effect on pH.

$$\boxed{\therefore \quad \textbf{pH} = \textbf{7}}$$

Problem I-11

How many milliliters of 0.05 N HCl are required to neutralize exactly 8.0 g of NaOH?

Solution

At the equivalence point, the number of moles H^+ added equals the number of moles OH^- present.

$$\text{liters}_{\text{acid}} \times N_{\text{acid}} = \text{number of moles (equivalents) of } H^+ \text{ added}$$

$$\frac{\text{wt}_{g_{\text{NaOH}}}}{\text{MW}_{\text{NaOH}}} = \text{number of moles of NaOH (and OH}^-\text{) present}$$

$$\text{liters} \times N = \frac{\text{wt}_g}{\text{MW}}$$

$$\text{liters} \times 0.05 = \frac{8.0}{40}$$

$$\text{liters} = \frac{8.0}{40 \times 0.05}$$

$$= \frac{8.0}{2}$$

$$= 4.0 \text{ liters} = \boxed{\textbf{4000 ml}}$$

Problem I-12

What is the percent by weight of KOH in a solid mixture of KOH and KCl if 4.0 g of the solid mixture require 400 ml of 0.15 N HCl for complete neutralization?

Solution

$$\text{number of moles (equivalents) of } H^+ \text{ required} =$$

$$\text{number of moles (equivalents) of } OH^- \text{ present}$$

$$\text{liters}_{\text{acid}} \times N_{\text{acid}} = \text{number of moles (equivalents) of } H^+$$

$$0.40 \times 0.15 = 0.06 \text{ mole } H^+$$

$\therefore$ The solid mixture contains 0.06 mole (equivalents) of OH^-

$$\frac{\text{wt of pure KOH}}{\text{EW}} = \text{number of equivalents of } OH^- \text{ present}$$

KOH contains 1 OH^- per mole.

$\therefore$ 1 mole = 1 equivalent $\therefore$ MW = EW

$$\text{wt}_g \text{ of pure KOH} = 0.06 \times 56$$

$$= 3.36 \text{ g}$$

$$\frac{\text{wt of pure substance}}{\text{total wt of sample}} \times 100 = \% \text{ w/w pure substance}$$

$$\frac{3.36}{4.0} \times 100 \boxed{= \textbf{84.0}\%}$$

Problem I-13

What are the: (a) hydrogen ion concentration and (b) pH of a solution obtained by mixing 250 ml of 0.1 M KOH with 200 ml of 0.2 M HNO_3?

Solution

In this example, it can be calculated that the amount of H^+ ion added is more than enough required to neutralize exactly the OH^- ion present.

(a)

liters $\times$ M = number of moles

$0.250 \times 0.1 = 0.025$ mole OH^- ⎫ mixed together in 0.45 liter total
$0.200 \times 0.2 = 0.040$ mole H^+ ⎭ volume

$$H^+ + OH^- \rightarrow H_2O$$

The OH^- is the limiting ion. ∴ The 0.025 mole of OH^- reacts with 0.025 mole of H^+ to yield 0.025 mole of H_2O. And:

$0.040 - 0.025 = 0.015$ mole H^+ remain

$$[H^+] = \frac{0.015 \text{ mole } H^+}{0.45 \text{ liter}} = 0.0333 \text{ mole } H^+/\text{liter}$$

$$\boxed{[H^+] = 3.33 \times 10^{-2} M}$$

(b) $pH = \log \dfrac{1}{[H^+]}$

$= \log \dfrac{1}{3.33 \times 10^{-2}}$

$= \log 0.3 \times 10^2$

$= \log 3 \times 10^1$ or $pH = \log 30$

$= \log 3 + \log 10$

$= 0.477 + 1$ $\boxed{pH = 1.477}$

$\boxed{pH = 1.477}$

Check:

$0.1\ M\ [H^+] = $ pH 1

$0.01\ M\ [H^+] = $ pH 2

∴ $0.033\ M\ [H^+] = $ pH between 1 and 2

Problem I-14

What is the pH of the solution obtained when 0.16 g of solid NaOH is dissolved in 600 ml of 0.003 M H_2SO_4? (Assume that the volume of the solution remains 600 ml.)

Solution

$$\text{number of moles of NaOH} = \frac{\text{wt}_g}{\text{MW}}$$

$$\frac{0.16}{40} = 0.0040 \text{ mole NaOH}$$

$$\text{liters} \times M = \text{number of moles } H_2SO_4$$

$$0.6 \times 0.0030 = 0.0018 \text{ mole } H_2SO_4$$

$$= 0.0036 \text{ mole } H^+$$

The H^+ is the limiting ion.

$$\therefore \ 0.0036 \text{ mole } H^+ + 0.0040 \text{ mole } OH^- \rightarrow 0.0036 \text{ mole } H_2O +$$

$$0.0004 \text{ mole of } OH^- \text{ remain}$$

$$[OH^-] = \frac{4 \times 10^{-4} \text{ moles}}{0.6 \text{ liter}} = 6.67 \times 10^{-4} \ M$$

$$pOH = \log \frac{1}{6.67 \times 10^{-4}}$$

$$= \log 0.15 \times 10^4$$

$$= \log 1500$$

$$= 3.176$$

$$pH + pOH = 14$$

$$pH = 14 - 3.176$$

$$\boxed{pH = 10.824}$$

Check:

$$10^{-3} \ M \ [OH^-] = pH \ 11$$

$$10^{-4} \ M \ [OH^-] = pH \ 10$$

$$\therefore \ \ 6.67 \times 10^{-4} \ M \ [OH^-] = pH \text{ between 10 and 11}$$

C. TITRATION CURVES

Problem I-15

Draw the titration curve for the neutralization of 250 ml of 0.1 N HCl with 0.1 N KOH.

Solution

A titration curve is a plot of pH versus milliliters (or equivalents or moles) of standard titrant added. For the titration of a given amount of acid, the curve is a plot of pH versus milliliters (or equivalents) of base added, as shown in Figure AI-1.

First calculate (a) the number of moles of H^+ originally present and (b) the starting pH.

(a)   liters $\times$ N = number of equivalents

$0.250 \times 0.1 = 0.0250$ equivalents of HCl present

$= 0.0250$ mole H^+ ion present

(b) $$pH = -\log [H^+]$$

$$= -\log 10^{-1}$$

$$= -(-1)$$

$$\boxed{\textbf{pH} = \textbf{1 at start}}$$

Next calculate (c) the number of moles of H^+ ion remaining after the addition of a certain amount of OH^- and (d) the pH of the solution at that point.

(c)   After adding, for example, 10 ml of standard base:

liters $\times$ N = number of equivalents of KOH added

$0.010 \times 0.1 = 0.001$ equivalents of KOH added

$= 0.001$ mole OH^- added

0.001 mole $OH^- + 0.001$ mole $H^+ =$

0.001 mole H_2O formed (and also 0.001 mole KCl)

the *amount* of H^+ remaining $= 0.025 - 0.001$

$= 0.024$ mole

(d)   The *concentration* of H^+ remaining can be calculated:

$$M = \frac{\text{number of moles}}{\text{liter}}$$

$$= \frac{0.024}{0.26} = 0.0923 \ M \ [H^+] \text{ remains}$$

$$pH = \log \frac{1}{0.0923} = \log 10.83$$

$$\boxed{\textbf{pH} = \textbf{1.032}}$$

Note that in calculating the new molarity of the H^+ ion, the increase in volume of the solution has been taken into account.

Similarly, we can calculate the pH after adding another 10 ml, 20 ml, 30 ml, etc. For example, after adding 50 ml of base:

$$\text{liters} \times M = \text{number of moles}$$

$$0.050 \times 0.1 = 0.005 \text{ mole OH}^-$$

$$0.025 - 0.005 = 0.020 \text{ mole H}^+ \text{ remain}$$

$$M = \frac{0.020 \text{ mole}}{0.30 \text{ liter}}$$

$$= 0.0667 \ M \ [H^+] \text{ remains}$$

$$pH = \log \frac{1}{0.0667} = \log 15$$

$$\boxed{pH = 1.176}$$

The H^+ is completely neutralized when the number of moles of OH^- added equals the number of moles of H^+ originally present:

$$\text{number of moles H}^+ = \text{number of moles OH}^-$$

$$\text{liters}_{acid} \times M_{acid} = \text{liters}_{base} \times M_{base}$$

$$0.25 \times 0.1 = \text{liters}_{base} \times 0.1$$

$$\text{liters}_{base} = \frac{0.25 \times 0.1}{0.1}$$

$$= 0.25 \text{ liter}$$

Thus, after 250 ml of base have been added, the solution contains only KCl, the salt of the acid and base. Salts formed by the neutralization of strong acids with strong bases have no effect on pH. At neutrality, $[H^+] = [OH^-] = 10^{-7} \ M$. $\therefore$ pH = 7.

The titration curve should be drawn a little beyond the exact neutralization point. For example, calculate the pH after 275 ml of base have been added (i.e., 25 ml beyond the equivalence point).

$$\text{liters} \times M = \text{number of moles}$$

$$0.275 \times 0.1 = 0.0275 \text{ mole OH}^- \text{ added}$$

$$0.0275 - 0.0250 = 0.0025 \text{ mole OH}^- \text{ in excess}$$

or

$$0.025 \text{ liter excess} \times M = \text{number of moles of } OH^- \text{ in excess}$$

$$0.025 \times 0.1 = 0.0025 \text{ mole } OH^- \text{ in excess}$$

$$M_{OH^-} = \frac{\text{number of moles excess } OH^-}{\text{total liters of solution}}$$

$$M = \frac{2.5 \times 10^{-3}}{0.250 + 0.275} = \frac{2.5 \times 10^{-3}}{0.525} = 4.77 \times 10^{-3} M$$

$$pOH = \log \frac{1}{4.77 \times 10^{-3}} = \log 0.21 \times 10^3$$

$$= \log 2.1 \times 10^2 = \log 210$$

$$\boxed{pOH = 2.322}$$

$$pH = 14 - pOH$$

$$= 14 - 2.322$$

$$\boxed{pH = 11.678}$$

In the above calculation, we assumed that γ_{H^+} and γ_{OH^-} were equal to 1.0.

Problem I-16

(a) Draw the titration curve for the neutralization of 450 ml of 0.20 M NaOH with 0.075 M H_2SO_4. (b) How much H_2SO_4 must be added to neutralize completely the NaOH?

Solution

(a) This time the pH will start high and decrease as standard acid is added, as shown in Figure AI-2.

$$pH \text{ at start} = 14 - pOH \text{ at start}$$

$$pOH = \log \frac{1}{[OH^-]}$$

$$= \log \frac{1}{0.2}$$

$$= \log 5$$

$$pOH = 0.7$$

$$\therefore \quad pH = 14 - 0.7$$

$$\boxed{pH = 13.3 \text{ at start}} \qquad \text{(assuming } \gamma_{OH^-} = 1)$$

The solution contains $0.45 \times 0.20 = 0.090$ mole OH^- at the start. After adding, for example, 20 ml of acid, the resulting pH can be calculated as follows:

$$\text{liters}_{acid} \times N = \text{number of equivalents of } H^+ \text{ added}$$

H_2SO_4 is a diprotic acid.

$$\therefore \quad N = 2 \times M = 0.15 \, N$$

$$0.020 \times 0.15 = 0.003 \text{ equivalent of } H^+ \text{ added}$$

$$= 0.003 \text{ "mole" of } H^+ \text{ added}$$

$$\text{moles } OH^- \text{ remaining} = 0.090 - 0.003 = 0.087$$

$$M_{OH^-} = \frac{\text{number of moles}}{\text{liter}} = \frac{0.087}{0.45 + 0.02}$$

$$= \frac{0.087}{0.47}$$

$$= 0.185 \, M_{OH^-}$$

The resulting pH can be calculated by first calculating $pOH = \log 1/0.185$ and then $pH = 14 - pOH$ as in the preceding problem, or by first calculating $[H^+]$.

$$[H^+][OH^-] = 1 \times 10^{-14}$$

$$[H^+] = \frac{1 \times 10^{-14}}{0.185}$$

$$[H^+] = 5.41 \times 10^{-14}$$

$$pH = \log \frac{1}{5.41 \times 10^{-14}}$$

$$= \log 0.185 \times 10^{14}$$

$$= \log 1.85 \times 10^{13}$$

$$= \log 1.85 + \log 10^{13}$$

$$= 0.267 + 13$$

$$\boxed{pH = 13.267}$$

(b)   The base is completely neutralized when the number of moles of H^+ added equals the number of moles of OH^- present.

$$\text{liters}_{\text{acid}} \times N_{\text{acid}} = \text{liters}_{\text{base}} \times N_{\text{base}}$$

$$\text{liters}_{\text{acid}} = \frac{0.45 \times 0.20}{0.15} = \frac{0.09}{0.15}$$

$$= 0.600 \text{ liter}$$

When 600 ml of H_2SO_4 are added, the solution is neutral (pH = 7). As more acid is added, the pH decreases. Upon addition of acid beyond the equivalence point, the pH can be calculated as usual. For example, after 625 ml of H_2SO_4 are added:

$$\text{number of moles (equivalents) of } H^+ \text{ added} = \text{liters} \times N$$
$$= 0.625 \times 0.15$$
$$= 0.0936 \text{ mole}$$

$$\text{number of moles of } H^+ \text{ in excess} = 0.0936 - 0.0900$$
$$= 0.0036 \text{ mole}$$

$$\text{total volume of solution} = 0.450 + 0.625 \text{ liter}$$
$$= 1.075 \text{ liters}$$

$$M_{H^+} = \frac{0.0036}{1.075}$$
$$= 0.00335$$
$$= 3.35 \times 10^{-3} M$$

$$pH = \log \frac{1}{3.35 \times 10^{-3}}$$
$$= \log 0.299 \times 10^3$$
$$= \log 2.99 \times 10^2$$
$$= \log 2.99 + \log 10^2$$
$$= 0.476 + 2$$

$$\boxed{pH = 2.476}$$

D. PREPARATION OF STANDARD SOLUTIONS

Problem I-17

How many grams of solid NaOH are required to prepare (a) 500 ml of a 0.04 M solution and (b) 250 ml of a dilute NaOH solution of pH 9.7?

Solution

(a)   liters $\times M$ = number of moles NaOH required

$$0.5 \times 0.04 = 0.02 \text{ mole NaOH required}$$

$$\text{number of moles} = \frac{\text{wt}_g}{\text{MW}}$$

$$0.02 = \frac{\text{wt}_g}{40}$$

$$\boxed{\text{wt} = 0.8 \text{ g}}$$

$\therefore$ Weigh out 0.8 g, dissolve in water, and dilute to 500 ml.

(b)   pH = 9.7

$$\text{pH} = \log \frac{1}{[\text{H}^+]}$$

$$9.7 = \log \frac{1}{[\text{H}^+]}$$

$$\frac{1}{[\text{H}^+]} = \text{antilog of } 9.7$$

$$\frac{1}{[\text{H}^+]} = 5 \times 10^9$$

$$[\text{H}^+] = \frac{1}{5 \times 10^9}$$

$$= 0.2 \times 10^{-9}$$

$$= 2 \times 10^{-10} M$$

$$[\text{H}^+][\text{OH}^-] = 1 \times 10^{-14}$$

$$[\text{OH}^-] = \frac{1 \times 10^{-14}}{2 \times 10^{-10}} = 0.5 \times 10^{-4} = 5 \times 10^{-5} M$$

$$\boxed{\therefore \quad \textbf{NaOH} = \textbf{5} \times \textbf{10}^{-5} \textbf{\textit{M}}}$$

We need 250 ml of 5×10^{-5} M NaOH.

$$\text{liters} \times M = \text{number of moles}$$

$$0.25 \times 5 \times 10^{-5} = \text{number of moles}$$

$$= \boxed{\textbf{1.25} \times \textbf{10}^{-5} \textbf{ moles}}$$

$$\frac{wt_g}{MW} = \text{number of moles}$$

$$wt_g = \text{number of moles} \times MW$$

$$= (1.25 \times 10^{-5}) \times 40$$

$$= 50 \times 10^{-5}\,g$$

$$= 50 \times 10^{-2}\,mg$$

$$\boxed{wt = 0.50\ mg}$$

$\therefore$ Weigh out 0.50 mg and dissolve in sufficient water to make 250 ml of solution.

Problem I-18

How many milliliters of 5 M H_2SO_4 are required to make (a) 1500 ml of a 0.002 M H_2SO_4 solution and (b) 1500 ml of 0.002 N H_2SO_4?

Solution

(a) The number of moles of H_2SO_4 in the dilute solution equals the number of moles of H_2SO_4 taken from the concentrated solution.

$$\therefore \quad \text{liters} \times M = \text{number of moles}$$

$$\text{liters} \times M \text{ (dilute solution)} = \text{liters} \times M \text{ (concentrated solution)}$$

$$1.5 \times 0.002 = \text{liters} \times 5$$

$$\frac{1.5 \times 0.002}{5} = \text{liters concentrated solution required}$$

$$\frac{3 \times 10^{-3}}{5} = 0.6 \times 10^{-3}\ \text{liters} = \boxed{0.6\ ml}$$

$\therefore$ Take 0.6 ml. of the concentrated solution and dilute to 1.5 liters.
(b) Similarly, the number of equivalents of H_2SO_4 in the dilute solution equals the number of equivalents of H_2SO_4 taken from the concentrated solution.

$$\therefore \quad \text{liters} \times N = \text{number of equivalents}$$

$$\text{liters} \times N \text{ (dilute solution)} = \text{liters} \times N \text{ (concentrated solution)}$$
H_2SO_4 contains 2 hydrogens.

$\therefore$ N concentrated solution $= M \times 2$

concentrated solution $= 10\,N$

$1.5 \times 0.002 = \text{liters} \times 10$

$\dfrac{1.5 \times 0.002}{10} = \text{liters concentrated solution required}$

$\dfrac{3 \times 10^{-3}}{10} = 3 \times 10^{-4}\ \text{liters} = \boxed{\textbf{0. 3 ml}}$

$\therefore$ Take 0.3 ml of the concentrated solution and dilute to 1.5 liters.

Problem I-19

Describe the preparation of 2 liters of 0.4 M HCl starting with a concentrated HCl solution (28% w/w HCl, specific gravity $= 1.15$).

Solution

$$\text{liters} \times M = \text{number of moles}$$

$$\boxed{\textbf{2} \times \textbf{0.4} = \textbf{0.80 mole HCl needed}}$$

$$\frac{\text{wt}_{\text{g}}}{\text{MW}} = \text{number of moles}$$

$$\text{wt}_{\text{g}} = \text{number of moles} \times \text{MW}$$

$$\text{wt}_{\text{g}} = 0.80 \times 36.5$$

$$\boxed{\textbf{wt}_{\textbf{g}} = \textbf{29.2 g pure HCl needed}}$$

The stock solution is not pure HCl but only 28% HCl by weight.

$$\therefore\ \frac{29.2}{0.28} = \boxed{\textbf{104.2 g stock solution needed}}$$

Instead of weighing out 104.2 g of stock solution, we can calculate the *volume* required.

$$\text{vol} \times \text{den} = \text{wt}$$

$$\text{vol}_{\text{ml}} = \frac{\text{wt}_{\text{g}}}{\text{den}_{\text{g/ml}}}$$

$$\text{vol} = \frac{104.2}{1.15} = \boxed{\textbf{90.7 ml stock solution needed}}$$

$\therefore$ Measure out 90.7 ml of stock solution and dilute to 2 liters with water.

All of the above relationships (between weight, density, and percent w/w) can be combined into a single expression.

$$\boxed{\text{wt}_g = \text{vol}_{ml} \times \text{den}_{g/ml} \times \% \text{ (as decimal)}}$$

where:

wt_g = weight of *pure* substance required in grams

vol_{ml} = volume of dilute stock solution in ml

$\%$ = fraction of total weight that is pure substance

$$\therefore \quad \text{vol} = \frac{\text{wt}}{\text{den} \times \%} = \frac{29.2}{1.15 \times 0.28}$$

$$= 90.7 \text{ ml}$$

As an alternate method of solution, we can calculate the molarity of the stock solution. First calculate the weight of pure HCl in 1 liter of stock solution.

$$\text{wt} = \text{vol} \times \text{den} \times \%$$

$$\text{wt}_g = 1000 \text{ ml} \times 1.15 \text{ g/ml} \times 0.28$$

$$\text{wt} = 322 \text{ g}$$

In other words, 1000 ml (1 liter) of stock solution contains 322 g of pure HCl.

$$\text{number of moles} = \frac{\text{wt}_g}{\text{MW}} = \frac{322}{36.5} = 8.82$$

$\therefore$ The concentrated stock solution is 8.82 M.
We need 0.80 mole.

$$\text{liters} \times M = \text{number of moles}$$

$$\text{liters} = \frac{\text{number of moles}}{M}$$

$$= \frac{0.80}{8.82} = 0.0907 \text{ liter}$$

$\therefore$ Take 0.0907 liter (90.7 ml) of stock and dilute to 2 liters.

PRACTICE PROBLEM SET I

1. What is the molarity of pure ethanol—i.e., how many moles are present in 1 liter of pure ethanol? The density of ethanol is 0.8 g/ml.

2. Calculate the pH, pOH, and number of H^+ and OH^- ions per liter in each of the following solutions: (a) 0.01 M HCl, (b) 10^{-4} M HNO_3, (c) 0.0025 M H_2SO_4, (d) 3.7 × 10^{-5} M KOH, (e) 5 × 10^{-8} M HCl,

(f) $2.9 \times 10^{-3} M$ NaOH, (g) $1 M$ HCl, (h) $10 M$ HNO$_3$, and (i) $3 \times 10^{-5} N$ H$_2$SO$_4$.

3. Calculate the H$^+$ ion concentration (M), the OH$^-$ ion concentration, and the number of H$^+$ and OH$^-$ ions per liter in solutions having pH values of: (a) 2.73, (b) 5.29, (c) 6.78, (d) 8.65, (e) 9.52, (f) 11.41, and (g) 0.

4. Calculate the: (a) [H$^+$], (b) [OH$^-$], (c) pH, and (d) pOH of the final solution obtained after 100 ml of 0.2 M NaOH are added to 150 ml of 0.4 M H$_2$SO$_4$.

5. Calculate the pH and pOH of a solution obtained by adding 0.2 g of solid KOH to 1.5 liters of 0.002 M HCl.

6. The pH of a 0.10 M HCl solution was found to be 1.15. Calculate the (a) a_{H^+} and (b) γ_{H^+} in this solution.

7. The activity coefficient of the hydroxyl ion (γ_{OH^-}) is 0.72 in a 0.1 M solution of KOH. Calculate the pH and pOH of this solution.

8. Concentrated H$_2$SO$_4$ is 96% H$_2$SO$_4$ by weight and has a density of 1.84 g/ml. Calculate the volume of concentrated acid required to make: (a) 750 ml of 1 N H$_2$SO$_4$, (b) 600 ml of 1 M H$_2$SO$_4$, (c) 1000 grams of a dilute H$_2$SO$_4$ solution containing 12% H$_2$SO$_4$ by weight, (d) an H$_2$SO$_4$ solution containing 6.5 equivalents per liter, and (e) a dilute H$_2$SO$_4$ solution of pH 3.8.

9. Concentrated HCl is 37.5% HCl by weight and has a density of 1.19. (a) Calculate the molarity of the concentrated acid. (b) Describe the preparation of 500 ml of 0.2 M HCl. (c) Describe the preparation of 350 ml of 0.5 N HCl. (d) Describe the preparation of an HCl solution containing 25% HCl by weight. (e) Describe the preparation of a dilute HCl solution having a pH of 4.7.

10. Calculate the weight of solid NaOH required to prepare: (a) 5 liters of a 2 M solution, (b) 2 liters of a solution of pH 11.5, and (c) 500 ml of 62% w/w solution. The density of 62% NaOH solution is 1.15 g/ml.

11. How many milliliters of 0.12 M H$_2$SO$_4$ are required to neutralize exactly *half* of the OH$^-$ ions present in 540 ml of 0.18 N NaOH?

12. How many grams of solid Na$_2$CO$_3$ are required to neutralize exactly 2 liters of an HCl solution of pH 2.0?

13. How many milliliters of 0.15 M KOH are required to neutralize exactly 180 g of pure H$_2$SO$_4$?

14. Calculate the points and draw the titration curve for the neutralization and overtitration of 300 ml of 0.05 M NaOH with 0.02 M H$_2$SO$_4$.

WEAK ACIDS AND BASES

II

A. pH, ACIDITY, K_a, AND DEGREE OF IONIZATION

Problem II-1

The weak acid, HA, is 0.1% ionized (dissociated) in a 0.2 M solution.
(a) What is the equilibrium constant for the dissociation of the acid (K_a)?
(b) What is the pH of the solution? (c) How much "weaker" is the active
acidity of the HA solution compared to a 0.2 M solution of HCl? (d) How
many milliliters of 0.1 N KOH would be required to neutralize completely 500 ml of the 0.2 M HA solution?

Solution

(a)	HA	$\leftrightharpoons$	H^+	$+$	A^-
Start:	0.2 M		0		0
Change:	$-(0.1\%$ of 0.2 $M) =$				
	$-2 \times 10^{-4} M$		$+2 \times 10^{-4} M$		$+2 \times 10^{-4} M$
Equilibrium:	$0.2 - 2 \times 10^{-4} M$		$2 \times 10^{-4} M$		$2 \times 10^{-4} M$

$$K_a = \frac{[H^+][A^-]}{[HA]} = \frac{(2 \times 10^{-4})(2 \times 10^{-4})}{0.2 - 2 \times 10^{-4}}$$

When the amount of HA that has dissociated is small compared to the
original concentration of HA, the K_a expression may be simplified as shown

27

in the following equation

$$K_a = \frac{(2 \times 10^{-4})(2 \times 10^{-4})}{0.2}$$

$$= \frac{4 \times 10^{-8}}{2 \times 10^{-1}}$$

$$\boxed{K_a = 2 \times 10^{-7}}$$

As a general rule, the above simplification of the denominator term can be made if the acid is 10% or less dissociated.

Although HA is a "weak" acid, it is a much stronger acid than is water. Consequently, the H^+ ion contribution from H_2O is neglected.

(b)
$$pH = \log \frac{1}{[H^+]}$$

$$= \log \frac{1}{2 \times 10^{-4}}$$

$$= \log 0.5 \times 10^4$$

$$= \log 5 \times 10^3$$

$$= \log 5 + \log 10^3 = 0.7 + 3$$

$$\boxed{pH = 3.7}$$

(c)   A 0.2 M solution of HCl would be 100% ionized and yield 0.2 M H^+.

$$pH = \log \frac{1}{[H^+]}$$

$$= \log \frac{1}{2 \times 10^{-1}} = \log 0.5 \times 10^1 = \log 5$$

$$\boxed{pH = 0.7} \qquad \text{(assuming } \gamma = 1\text{)}$$

$\therefore$ The weak acid is 3 pH units less acid than a comparable HCl solution.

Remember the pH scale is a *logarithmic* scale, not a linear scale. The HA then is 10^3 or 1000 times less acid than the HCl (*not* 3 times).

(d)   Although the *active* acidity $[H^+]$ of the weak acid is 1000 times less than that of the HCl solution, the *total* acidity (free H^+ plus the undissociated hydrogen in HA) is the same. When OH^- is added, it reacts with the free H^+ to form H_2O. Some HA then immediately dissociates to H^+ and A^- to reestablish the equilibrium. This H^+ is also neutralized by further additions of OH^- and so on until all the HA is neutralized. Neutralization calculations for weak acids then may be conducted in the same manner as for strong acids.

number of moles of OH^- required = total number of moles H^+ present

$$\text{liters}_{\text{base}} \times N_{\text{base}} = \text{liters}_{\text{acid}} \times N_{\text{acid}}$$

HA is monoprotic.

$$\therefore \quad N = M$$

$$\text{liters}_{\text{base}} \times 0.1 = 0.5 \times 0.2$$

$$\text{liters}_{\text{base}} = \frac{0.5 \times 0.2}{0.1}$$

$$= \frac{0.1}{0.1}$$

$$= \boxed{\textbf{1 liter of base required}}$$

Problem II-2

The pH of a 0.1 M solution of a weak acid, HA, is 3.0. (a) What is the degree of ionization of HA in the 0.1 M solution? (b) What is K_a? (c) How many OH^- ions are present in 500 ml of the 0.1 M solution? (d) How many molecules of unionized HA are present in 250 ml of the 0.1 M solution?

Solution

(a) $$\text{pH} = 3$$

$$\text{pH} = -\log [H^+]$$

$$\therefore \quad [H^+] = 10^{-3} M$$

Each mole of H^+ comes from 1 mole of HA. $\therefore \quad 10^{-3}$ moles/liter of HA dissociated.

$$\% \text{ ionization} = \frac{[H^+]}{[HA]_{\text{orig}}} \times 100$$

$$= \frac{10^{-3}}{10^{-1}} \times 100 = \frac{10^{-1}}{10^{-1}}$$

$$= \boxed{1\%}$$

(b)   Each mole of HA that dissociates yields 1 mole of $[A^-]$ for each mole of $[H^+]$.

$$\therefore \quad [A^-] = 10^{-3} \, M$$

	HA	$\rightleftharpoons$	H^+	$+$	A^-
Start:	0.1 M		0		0
Change:	$-0.001 \, M$		$+0.001 \, M$		$+0.001 \, M$
Equilibrium:	$0.1 - 0.001 \, M$		$0.001 \, M$		$0.001 \, M$

$$K_a = \frac{(10^{-3})(10^{-3})}{(10^{-1} - 10^{-3})} = \frac{(10^{-3})(10^{-3})}{(10^{-1})}$$

The 10^{-3} term in the denominator is neglected because it is 100 times smaller than the 10^{-1} term.

$$K_a = \frac{10^{-6}}{10^{-1}}$$

$$\boxed{K_a = 10^{-5}}$$

(c)   The number of OH^- ions per liter equals $M \times$ Avogadro's number.

$$pH = 3 \quad \therefore \quad pOH = 11$$

$$\therefore \quad [OH^-] = 10^{-11} \, M$$

$$(10^{-11})(6.023 \times 10^{23}) = 6.023 \times 10^{12} \, OH^- \text{ ions/liter}$$

$$= \boxed{3.0115 \times 10^{12} \, OH^- \text{ ions/500 ml}}$$

(d)   The number of molecules of HA per liter equals $M \times$ Avogadro's number.

$$(10^{-1})(6.023 \times 10^{23}) = 6.023 \times 10^{22} \text{ molecules/liter}$$

$$= 6.023 \times 0.25 \times 10^{22} \text{ molecules/250 ml}$$

$$= \boxed{1.505 \times 10^{22} \text{ molecules/250 ml}}$$

Problem II-3

The K_a for a weak acid, HA, is 1.6×10^{-6}. What are the: (a) pH and (b) degree of ionization of the acid in a $10^{-3} \, M$ solution? (c) Calculate pK_a.

Solution

(a) Let $x = M$ of HA that dissociates. ∴ $x = M$ of H^+ and also M of A^- produced.

	HA	⇌	H^+	+	A^-
Start:	$10^{-3}\ M$		0		0
Change:	$-x\ M$		$+x\ M$		$+x\ M$
Equilibrium:	$10^{-3} - x\ M$		$x\ M$		$x\ M$

$$K_a = \frac{[H^+][A^-]}{[HA]} = \frac{(x)(x)}{10^{-3} - x} = 1.6 \times 10^{-6}$$

First calculate x assuming that x is very much smaller than the concentration of unionized acid, i.e., assuming that the acid is less than 10% ionized. The denominator of the K_a expression may then be simplified.

$$1.6 \times 10^{-6} = \frac{x^2}{10^{-3}}$$

$$x^2 = 1.6 \times 10^{-9} = 16 \times 10^{-10}$$

$$x = \sqrt{16 \times 10^{-10}} = \sqrt{16} \times \sqrt{10^{-10}} = 4.0 \times 10^{-5}$$

$$\boxed{[H^+] = 4 \times 10^{-5}\ M}$$

$$pH = \log \frac{1}{[H^+]}$$

$$= \log \frac{1}{4 \times 10^{-5}} = \log 0.25 \times 10^5 = \log 2.5 \times 10^4$$

$$= \log 2.5 + \log 10^4 = 0.398 + 4$$

$$\boxed{pH = 4.398}$$

(b)   degree of ionization $= \dfrac{[H^+]}{[HA]_{orig}} \times 100$

$$= \frac{4 \times 10^{-5}}{10^{-3}} \times 100 = \frac{4 \times 10^{-3}}{10^{-3}}$$

$$= \boxed{4\%}$$

The acid is indeed less than 10% ionized. Therefore, the simplification of the denominator term in the expression for K_a is reasonably valid.

(c)   The pK_a is the negative logarithm of K_a.

$$pK_a = -\log K_a = \log \frac{1}{K_a}$$

$$pK_a = \log \frac{1}{1.6 \times 10^{-6}} = \log 6.25 \times 10^5$$

$$pK_a = \log 6.25 + \log 10^5$$

$$pK_a = 0.796 + 5$$

$$\boxed{pK_a = 5.796}$$

Problem II-4

What are the: (a) concentration of a weak acid, HA, where $K_a = 6 \times 10^{-5}$, and (b) H^+ concentration in a solution in which the acid is 2% dissociated?

Solution

(a)   Let $y = M$ of $[HA]_{orig}$.

	HA	$\rightleftharpoons$	H^+	+	A^-
Start:	$y\ M$		0		0
Change:	-2% of $y\ M$		$+2\%$ of $y\ M$		$+2\%$ of $y\ M$
Equilibrium:	$y - 2\%$ of $y\ M =$				
	$y - 0.02\,y$		$0.02\,y$		$0.02\,y$

$$K_a = \frac{[H^+][A^-]}{[HA]} = \frac{(0.02y)(0.02y)}{y - 0.02y} = 6 \times 10^{-5}$$

Since the acid is only 2% dissociated, the denominator may be simplified.

$$6 \times 10^{-5} = \frac{(2 \times 10^{-2}y)^2}{y} = \frac{(4 \times 10^{-4})y^2}{y}$$

$$6 \times 10^{-5}y = (4 \times 10^{-4})y^2$$

$$\frac{6 \times 10^{-5}}{4 \times 10^{-4}} = \frac{y^2}{y}$$

$$1.5 \times 10^{-1} = y$$

$$\boxed{[HA] = 0.15\ M}$$

(b) $$[H^+] = 0.02y$$

$$= (0.02)(0.15) = (2 \times 10^{-2})(1.5 \times 10^{-1})$$

$$\boxed{[H^+] = 3.0 \times 10^{-3}\ M}$$

Problem II-5

The K_a for a weak acid, HA, is 3×10^{-5}. The pH of a dilute solution of the acid is 2.4. (a) What is the concentration of HA in the solution? (b) What is the degree of ionization of the solution?

Solution

(a)
$$pH = \log \frac{1}{[H^+]}$$

$$2.4 = \log \frac{1}{[H^+]}$$

$$\frac{1}{[H^+]} = \text{antilog of } 2.4$$

$$= 2.51 \times 10^2$$

$$[H^+] = \frac{1}{2.51 \times 10^2}$$

$$= 0.398 \times 10^{-2}$$

$$[H^+] = 3.98 \times 10^{-3} \, M$$

Let $y = [HA]_{orig}$.

	HA	$\leftrightharpoons$	H⁺	+	A⁻
Start:	$y \, M$		0		0
Change:	$-3.98 \times 10^{-3} \, M$		$+3.98 \times 10^{-3} \, M$		$+3.98 \times 10^{-3} \, M$
Equilibrium:					
	$y - 3.98 \times 10^{-3} \, M$		$3.98 \times 10^{-3} \, M$		$3.98 \times 10^{-3} \, M$

$$K_a = \frac{[H^+][A^-]}{[HA]} = 3 \times 10^{-5}$$

$$\frac{(3.98 \times 10^{-3})^2}{y - 3.98 \times 10^{-3}} = 3 \times 10^{-5}$$

First assume that y is much larger than 3.98×10^{-3}.

$$\frac{(3.98 \times 10^{-3})^2}{y} = 3 \times 10^{-5}$$

$$3 \times 10^{-5} y = (3.98 \times 10^{-3})^2$$

$$y = \frac{15.86 \times 10^{-6}}{3 \times 10^{-5}}$$

$$y = 5.29 \times 10^{-1}$$

$$\boxed{[HA]_{orig} = 0.529 \, M}$$

(b) $$\text{degree of ionization} = \frac{[H^+]}{[HA]_{orig}} \times 100\%$$

$$= \frac{3.98 \times 10^{-3}}{5.29 \times 10^{-1}} \times 100\%$$

$$= \frac{3.98 \times 10^{-1}}{5.29 \times 10^{-1}} \%$$

$$= \boxed{0.753\%}$$

Thus, the assumption that y is much greater than 3.98×10^{-3} is valid.

Problem II-6

A weak acid, HA, is 2% dissociated in a solution of pH 4.2. Calculate: (a) the concentration of undissociated acid in the solution and (b) K_a.

Solution

(a) $$pH = \log \frac{1}{[H^+]}$$

$$4.2 = \log \frac{1}{[H^+]}$$

$$\frac{1}{[H^+]} = \text{antilog of } 4.2 = 1.58 \times 10^4$$

$$[H^+] = \frac{1}{1.58 \times 10^4} = 0.633 \times 10^{-4}$$

$$\boxed{[H^+] = 6.33 \times 10^{-5}\ M}$$

$$\text{degree of dissociation} = \frac{[H^+]}{[HA]_{orig}} \times 100\%$$

Let $Z = [HA]_{orig}$.

$$6.33 \times 10^{-5} = 2\% \text{ of } Z$$

$$2 = \frac{6.33 \times 10^{-5}}{Z} \times 100 = \frac{6.33 \times 10^{-3}}{Z}$$

$$2Z = 6.33 \times 10^{-3}$$

$$Z = 3.165 \times 10^{-3}$$

$$\boxed{[HA] = 3.165 \times 10^{-3}\ M}$$

(b)
$$K_a = \frac{[H^+][A^-]}{[HA]} = \frac{(6.33 \times 10^{-5})(6.33 \times 10^{-5})}{3.165 \times 10^{-3}}$$

$$= \frac{40 \times 10^{-10}}{3.165 \times 10^{-3}} = 12.62 \times 10^{-7}$$

$$\boxed{K_a = 1.262 \times 10^{-6}}$$

Problem II-7

(a) Calculate K_b for the dissociation of the weak inorganic base "NH_4OH" if it is 1% dissociated in a 0.25 M solution. (b) What is the pH of the 0.25 M solution?

Solution

(a) "NH_4OH" is actually a solution of NH_3 in H_2O. The production of OH^- ions can be thought of as occurring in two steps:

$$NH_3 + HOH \rightleftharpoons NH_4OH \rightleftharpoons NH_4^+ + OH^-$$

Start:
0.25 M 0   0

Change: -1% of 0.25 M =

$-0.0025\ M$ $+0.0025\ M$ $+0.0025\ M$

Equilibrium:
$0.25 - 0.0025\ M$ $0.0025\ M$ $0.0025\ M$

$$K_b = \frac{[NH_4^+][OH^-]}{["NH_4OH"]} = \frac{(0.0025)(0.0025)}{(0.25 - 0.0025)}$$

where ["NH_4OH"] refers to the sum of NH_3 plus the small amount of actual NH_4OH actually present.

Because the base is less than 10% dissociated, the K_b expression may be simplified.

$$K_b = \frac{(2.5 \times 10^{-3})^2}{0.25}$$

$$K_b = \frac{6.25 \times 10^{-6}}{2.5 \times 10^{-1}}$$

$$\boxed{K_b = 2.5 \times 10^{-5}}$$

(b)

$$pOH = \log \frac{1}{[OH^-]}$$

$$pOH = \log \frac{1}{2.5 \times 10^{-3}} = \log 0.4 \times 10^3 = \log 400$$

$$\boxed{pOH = 2.6}$$

$$pH = 14.0 - 2.6$$

$$\boxed{pH = 11.4}$$

Problem II-8

(a) Calculate the apparent degree of dissociation of an amine in a 0.3 M aqueous solution if the pH of the solution is 10.6. (b) Calculate K_b.

Solution

Amines are organic bases. They yield OH^- ions in aqueous solution by the following reaction:

$$R\text{-}NH_2 + HOH \rightleftharpoons R\text{-}NH_3^+ + OH^-$$

The situation is analogous to the solution of NH_3 in water. If you wish, you can consider that the amine reacts first with H_2O to yield $R\text{-}NH_3OH$ which then dissociates to $R\text{-}NH_3^+$ and OH^-.

(a)

$$pH = \log \frac{1}{[H^+]}$$

$$10.6 = \log \frac{1}{[H^+]}$$

$$\frac{1}{[H^+]} = \text{antilog of } 10.6$$

$$= 3.98 \times 10^{10}$$

$$[H^+] = \frac{1}{3.98 \times 10^{10}} = 0.251 \times 10^{-10} = 2.51 \times 10^{-11} M$$

$$[H^+][OH^-] = 1 \times 10^{-14}$$

$$[OH^-] = \frac{1 \times 10^{-14}}{2.51 \times 10^{-11}} = 0.398 \times 10^{-3} = 3.98 \times 10^{-4} M$$

$$\text{apparent degree of dissociation} = \frac{[\text{OH}^-]}{[\text{``R-NH}_3\text{OH''}]} \times 100$$

$$= \frac{3.98 \times 10^{-4}}{3 \times 10^{-1}} \times 100$$

$$= \frac{3.98 \times 10^{-2}}{3 \times 10^{-1}}$$

$$= 1.325 \times 10^{-1}$$

$$= \boxed{\mathbf{0.1325\%}}$$

(b)
$$K_b = \frac{[\text{R-NH}_3^+][\text{OH}^-]}{[\text{``R-NH}_3\text{OH''}]}$$

where "R-NH$_3$OH" is the sum of R-NH$_2$ plus any small amount of R-NH$_3$OH that might be present.

$$K_b = \frac{(3.98 \times 10^{-4})(3.98 \times 10^{-4})}{3 \times 10^{-1}}$$

$$= \frac{15.9 \times 10^{-8}}{3 \times 10^{-1}}$$

$$\boxed{K_b = \mathbf{5.3 \times 10^{-7}}}$$

Problem II-9

Calculate the: (a) H^+ ion concentration in a $0.02\,M$ solution of a moderately strong acid, HA, where $K_a = 3 \times 10^{-2}\,M$, and (b) degree of dissociation of the acid.

Solution

(a) Let:

$$x = M \text{ of HA dissociated}$$

$$= M \text{ of H}^+ \text{ produced}$$

	HA	$\leftrightharpoons$	H^+	+	A^-
Start:	0.02 M		0		0
Change:	$-x\,M$		$+x\,M$		$+x\,M$
Equilibrium:	$0.02 - x\,M$		$x\,M$		$x\,M$

$$K_a = \frac{[\text{H}^+][\text{A}^-]}{[\text{HA}]}$$

$$3 \times 10^{-2} = \frac{(x)(x)}{0.02 - x}$$

The phrase "moderately strong acid" suggests that the acid is more than 10% dissociated. Therefore, the denominator term in the K_a expression should not be simplified.

$$(3 \times 10^{-2})(0.02 - x) = x^2$$

$$6 \times 10^{-4} - 3 \times 10^{-2}x = x^2$$

$$x^2 + 3 \times 10^{-2}x - 6 \times 10^{-4} = 0$$

Solve for x using the general solution for quadratic equations.

$$x = \frac{-b \pm \sqrt{b^2 - 4ac}}{2a}$$

where:

$$a = 1$$

$$b = 3 \times 10^{-2}$$

$$c = -6 \times 10^{-4}$$

$$x = \frac{-3 \times 10^{-2} \pm \sqrt{(3 \times 10^{-2})^2 - 4(-6 \times 10^{-4})}}{2}$$

$$= \frac{-3 \times 10^{-2} \pm \sqrt{(9 \times 10^{-4}) + 24 \times 10^{-4}}}{2}$$

$$= \frac{-3 \times 10^{-2} \pm \sqrt{33 \times 10^{-4}}}{2}$$

$$= \frac{-3 \times 10^{-2} \pm 5.74 \times 10^{-2}}{2}$$

$$= \frac{+2.74 \times 10^{-2}}{2} \qquad \text{(neglecting the negative answer)}$$

$$\boxed{[H^+] = 1.37 \times 10^{-2}\ M}$$

(b) $$\text{degree of dissociation} = \frac{[H^+]}{[HA]_{orig}} \times 100$$

$$= \frac{0.0137}{0.0200} \times 100$$

$$\boxed{= 68.5\%}$$

Problem II-10

A moderately strong acid, HA, is 25% ionized in a 0.1 M solution. (a) What is the K_a? (b) What is the pH of the solution?

Solution

(a)

	HA	$\leftrightharpoons$	H^+	+	A^-
Start:	0.1 M		0		0
Change:	-25% of 0.1 $M =$				
	-0.025 M		$+0.025$ M		$+0.025$ M
Equilibrium:	$0.1 - 0.025 =$				
	0.075 M		0.025 M		0.025 M

$$K_a = \frac{(0.025)(0.025)}{(0.075)} = \frac{(2.5 \times 10^{-2})^2}{7.5 \times 10^{-2}}$$

$$= \frac{6.25 \times 10^{-4}}{7.5 \times 10^{-2}}$$

$$= 0.833 \times 10^{-2}$$

$$\boxed{K_a = 8.33 \times 10^{-3}}$$

(b)

$$pH = \log \frac{1}{[H^+]}$$

$$= \log \frac{1}{0.025}$$

$$= \log 40$$

$$\boxed{pH = 1.60}$$

Problem II-11

What would be the K_a value that divides "weak" acids from "moderately strong" acids if we define a "weak" acid as one that is less than 10% dissociated in a 0.1 M solution and a "moderately strong" acid as one that is more than 10% dissociated in a 0.1 M solution?

Solution

	HA	$\leftrightharpoons$	H^+	$+$	A^-
Start:	0.1 M		0		0
Change:	-10% of 0.1 $M =$				
	$-0.01\ M$		$+0.01\ M$		$+0.01\ M$
Equilibrium:	$0.1 - 0.01\ M =$				
	0.09 M		0.01 M		0.01 M

$$K_a = \frac{[H^+][A^-]}{[HA]}$$

$$= \frac{(10^{-2})^2}{(9 \times 10^{-2})}$$

$$= \frac{1 \times 10^{-4}}{9 \times 10^{-2}}$$

$$= 0.111 \times 10^{-2}$$

$$\boxed{K_a = 1.11 \times 10^{-3}}$$

Problem II-12

(a)   Calculate the concentration of unionized acetic acid and acetate ions in a solution prepared by mixing 250 ml of 0.1 M HNO_3 with 250 ml of 0.3 M acetic acid.   (b) What is the degree of dissociation of the acetic acid in the above solution?   The $K_a = 1.8 \times 10^{-5}$ for acetic acid.

Solution

(a) HNO_3 is a strong acid.   It will be 100% dissociated.

$$\text{number of moles of } HNO_3 \text{ added} = 0.25 \text{ liter} \times 0.1\ M$$
$$= 0.025 \text{ mole}$$

$$M \text{ of } HNO_3 \text{ in final solution} = \frac{0.025 \text{ mole}}{0.50 \text{ liter}} = 0.05\ M$$

$$\therefore \quad M \text{ of } H^+ \text{ from } HNO_3 = 0.05\ M$$

Let

$$y = M \text{ of HA that dissociates}$$
$$y = M \text{ of } A^-$$

and

$$y = M \text{ of } H^+ \text{ from HA}$$

number of moles of HA present $= 0.25$ liter $\times 0.3$ $M = 0.075$ mole

$$\therefore \quad M \text{ of HA} = \frac{0.075 \text{ mole}}{0.50 \text{ liter}} - y = 0.15 - y \ M$$

$$K_a = \frac{[H^+][A^-]}{[HA]} = \frac{(0.05 + y)(y)}{(0.15 - y)} = 1.8 \times 10^{-5}$$

We may determine y exactly. However, because HA is a much weaker acid than HNO_3, and the amounts of HA and HNO_3 present are of the same order of magnitude, we may assume that y is small compared to 0.05 and 0.15. Simplifying:

$$1.8 \times 10^{-5} = \frac{(0.05)(y)}{(0.15)}$$

$$y = \frac{(0.15)(1.8 \times 10^{-5})}{(0.05)} = 5.4 \times 10^{-5}$$

$$\boxed{\begin{array}{l} \therefore \quad [A^-] = 5.4 \times 10^{-5} \ M \\ [HA] = 0.15 \ M \end{array}}$$

Thus our assumption that $[A^-]$ is much smaller than 0.05 and 0.15 is valid.

(b) degree of dissociation $= \dfrac{[A^-]}{[HA]} \times 100\% = \dfrac{5.4 \times 10^{-5}}{0.15} \times 100\%$

$$= \frac{5.4 \times 10^{-3}}{0.15} \% = 3.6 \times 10^{-2}\%$$

$$= \boxed{0.036\%}$$

In the absence of HNO_3, we could calculate that $[A^-]$ in a 0.15 M HA solution would be 5.2×10^{-3} M and that the HA would be 1.1% dissociated. We can see that the addition of a strong acid to a solution of a weak acid suppresses the dissociation of the weak acid.

Problem II-13

(a) Calculate the concentrations of unionized acetic acid [HA], acetate ions $[A^-]$, and hydrogen ions $[H^+]$ in a solution prepared by adding 1 ml of 1 M HCl to 500 ml of 0.15 M acetic acid. (b) What is the degree of dissociation of the acetic acid in the solution?

Solution

(a) number of moles of HCl added $= 10^{-3}$ liters $\times 1 \, M = 10^{-3}$ moles

$$\therefore \quad [\text{H}^+] \text{ from HCl} = \frac{10^{-3} \text{ moles}}{0.501 \text{ liter}} = 2 \times 10^{-3} \, M$$

We can estimate quickly the H^+ ion concentration that would be present in 0.15 M HA alone.

$$K_a = \frac{[\text{H}^+][\text{A}^-]}{[\text{HA}]} = \frac{(y)(y)}{0.15} = 1.8 \times 10^{-5}$$

$$y^2 = 0.27 \times 10^{-5} = 2.7 \times 10^{-6}$$

$$y = 1.64 \times 10^{-3}$$

$$[\text{H}^+] = 1.64 \times 10^{-3} \, M$$

Thus, the contributions of H^+ ions from the HCl and the HA are of the same order of magnitude. Consequently, both sources must be considered in any calculations. The situation is similar to that described in Problem I-7.

Let

$$y = M \text{ of } \text{H}^+ \text{ produced by the dissociation of HA}$$

$$\therefore \quad y = M \text{ of } \text{A}^-$$

and

$$0.15 - y = M \text{ of HA remaining}$$

$$M \text{ of } \text{H}^+ \text{ from HCl} = 2 \times 10^{-3}$$

$$\therefore \quad [\text{H}^+] = y + 2 \times 10^{-3} \, M$$

$$K_a = \frac{[\text{H}^+][\text{A}^-]}{[\text{HA}]} = \frac{(y + 2 \times 10^{-3})(y)}{(0.15 - y)} = 1.8 \times 10^{-5}$$

We can solve for y exactly. However, we know that in the absence of HCl, y will be small compared to 0.15 M. In the presence of excess H^+ from the HCl, the dissociation of HA will be depressed. Therefore, y is even smaller than it would be in the absence of HCl. Consequently, we can eliminate y in the denominator. Cross-multiplying:

$$0.27 \times 10^{-5} = y^2 + 2 \times 10^{-3}y$$

$$y^2 + 2 \times 10^{-3}y - 0.27 \times 10^{-5} = 0$$

(If we carried the y term in the denominator, our expression at this point would be: $y^2 + 2.0185 \times 10^{-3}y - 0.27 \times 10^{-5} = 0$.)

$$y = \frac{-b \pm \sqrt{b^2 - 4ac}}{2a}$$

where:

$$a = 1$$
$$b = 2 \times 10^{-3}$$
$$c = -0.27 \times 10^{-5}$$

$$y = \frac{-2 \times 10^{-3} \pm \sqrt{(2 \times 10^{-3})^2 + 1.08 \times 10^{-5}}}{2}$$

$$y = \frac{-2 \times 10^{-3} \pm \sqrt{4 \times 10^{-6} + 10.8 \times 10^{-6}}}{2}$$

$$y = \frac{-2 \times 10^{-3} \pm \sqrt{14.8 \times 10^{-6}}}{2}$$

$$y = \frac{-2 \times 10^{-3} \pm 3.85 \times 10^{-3}}{2}$$

$$y = \frac{1.85 \times 10^{-3}}{2} = 0.925 \times 10^{-3}$$

$$[H^+] = (0.925 \times 10^{-3}) + (2 \times 10^{-3}) = 2.925 \times 10^{-3} \ M$$
$$[A^-] = 9.25 \times 10^{-4} \ M$$
$$[HA] = 0.15 - 9.25 \times 10^{-4} \ M = 0.1491 \ M$$

(b) $\qquad$ degree of dissociation $= \dfrac{[A^-]}{[HA]} \times 100\%$

$$= \frac{9.25 \times 10^{-4}}{1.5 \times 10^{-1}} \times 100\%$$

$$= \frac{9.25 \times 10^{-2}}{1.5 \times 10^{-1}} \%$$

$$= 6.17 \times 10^{-1}\%$$

$$= \boxed{0.617\%}$$

In the absence of HCl, the HA would be 1.1% dissociated. Thus, the small amount of HCl suppresses the dissociation significantly.

Problem II-14

An acid is 10% ionized in a 0.30 M solution. (a) Calculate the error in the determination of K_a that results from simplifying the denominator term of the K_a expression as described in the previous problems. (b) Calculate the error involved if the acid were more dilute, e.g., 0.003 M.

Solution

(a) $\qquad\qquad$ HA $\quad\rightleftharpoons\quad$ H$^+$ $\quad+\quad$ A$^-$

Start: $\qquad\qquad$ 0.30 M $\qquad\qquad$ 0 $\qquad\qquad$ 0

Change: $\qquad\qquad$ -10% of 0.30 M =

$\qquad\qquad\qquad$ $-0.030\ M$ $\qquad$ $+0.03\ M$ $\quad$ $+0.03\ M$

Equilibrium: $\qquad\quad$ $0.300 - 0.030$ =

$\qquad\qquad\qquad$ 0.270 M $\qquad\quad$ 0.03 M $\qquad$ 0.03 M

Using the simplified expression for K_a:

$$K_a = \frac{[\text{H}^+][\text{A}^-]}{[\text{HA}]} = \frac{(0.03)(0.03)}{(0.30)}$$

$$= \frac{(3 \times 10^{-2})^2}{3 \times 10^{-1}} = \frac{9 \times 10^{-4}}{3 \times 10^{-1}}$$

$$\boxed{K_a = 3.00 \times 10^{-3}}$$

Using the actual value of [HA]:

$$K_a = \frac{(0.03)(0.03)}{(0.27)} = \frac{9 \times 10^{-4}}{2.7 \times 10^{-1}}$$

$$\boxed{K_a = 3.33 \times 10^{-3}}$$

In the above example, the simplification results in an error of about 10% in the value of K_a calculated.

(b) If the acid were 0.003 M:

$\qquad\qquad\qquad$ HA $\quad\rightleftharpoons\quad$ H$^+$ $\quad+\quad$ A$^-$

Start: $\qquad\qquad$ 0.003 M $\qquad\qquad$ 0 $\qquad\qquad$ 0

Change: $\qquad\qquad$ -10% of 0.003 M =

$\qquad\qquad\qquad$ $-3 \times 10^{-4}\ M$ $\quad$ $+3 \times 10^{-4}\ M$ $\quad$ $+3 \times 10^{-4}\ M$

Equilibrium: $\qquad\quad$ $0.003 - 0.0003$ =

$\qquad\qquad\qquad$ 0.0027 =

$\qquad\qquad\qquad$ $2.7 \times 10^{-3}\ M$ $\quad$ $3 \times 10^{-4}\ M$ $\qquad$ $3 \times 10^{-4}\ M$

$$K_a = \frac{(3 \times 10^{-4})^2}{(3 \times 10^{-3})} \qquad\qquad K_a = \frac{(3 \times 10^{-4})^2}{(2.7 \times 10^{-3})}$$

$$= \frac{9 \times 10^{-8}}{3 \times 10^{-3}} \qquad\qquad\qquad = \frac{9 \times 10^{-8}}{2.7 \times 10^{-3}}$$

$$\boxed{K_a = 3.0 \times 10^{-5}} \qquad\qquad \boxed{K_a = 3.33 \times 10^{-5}}$$

Thus, the percent error in K_a is independent of the concentration of HA for a *given degree of dissociation*.

Problem II-15

Show how the degree of dissociation of a weak acid, HA, varies with the concentration.

Solution

$$HA \rightleftharpoons H^+ + A^-$$

$$K_a = \frac{[H^+][A^-]}{[HA]}$$

Assume

$$K_a = 10^{-5}$$

First calculate the degree of dissociation for, e.g., a 0.1 *M* solution. Let:

$$x = M \text{ of HA dissociated}$$
$$= M \text{ of } H^+ \text{ produced}$$

$$10^{-5} = \frac{(x)(x)}{0.1}$$

$$x^2 = 10^{-6}$$

$$x = 10^{-3}$$

$$[H^+] = 10^{-3}$$

$$\text{degree of dissociation} = \frac{[H^+]}{[HA]_{orig}} \times 100\%$$

$$= \frac{10^{-3}}{10^{-1}} \times 100\% = \boxed{1\%}$$

Next calculate the degree of dissociation for, e.g., a 10^{-3} *M* solution.

$$10^{-5} = \frac{x^2}{10^{-3}}$$

$$x^2 = 10^{-8}$$

$$x = 10^{-4}$$

$$[H^+] = 10^{-4} M$$

$$\text{degree of dissociation} = \frac{10^{-4}}{10^{-3}} \times 100\% = \boxed{10\%}$$

Thus, as a weak acid is diluted, the degree of dissociation increases. However, if the $[HA]_{orig}$ is $100 \times K_a$ or greater, the degree of dissociation is only 10% or less. Under this condition, the $[HA]_{orig}$ may be substituted for the [HA] term in the K_a expression without a significant error. As a general rule, the relationship between the degree of dissociation, the K_a value, and the concentration of a weak monoprotic acid is that shown below

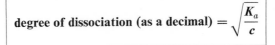

$$\text{degree of dissociation (as a decimal)} = \sqrt{\frac{K_a}{c}}$$

where $c = $ the concentration (M).

Problem II-16

At what concentration (in terms of K_a) of a weak acid, HA, will the acid be 50% dissociated?

Solution

Let $c = $ initial concentration of HA. At equilibrium, when the acid is 50% dissociated:

$$[HA] = \frac{c}{2}$$

$$[H^+] = \frac{c}{2}$$

$$[A^-] = \frac{c}{2}$$

$$K_a = \frac{(c/2)(c/2)}{(c/2)} = \frac{c^2/4}{c/2}$$

$$K_a \frac{c}{2} = \frac{c^2}{4}$$

$$\frac{K_a c}{2} = \frac{c^2}{4}$$

$$2c^2 = 4K_a c$$

$$\frac{2c^2}{c} = 4K_a$$

$$2c = 4K_a$$

$$\boxed{c = 2K_a}$$

When the original concentration of a weak acid, HA, is equal to twice its dissociation constant, the acid is 50% ionized.

Problem II-17

What is the H^+ ion concentration in a 0.12 M solution of a diprotic weak acid, H_2A, if both hydrogens of the acid dissociate simultaneously according to the equation shown below, and K_a for the overall ionization is 2×10^{-8}?

Solution

$$H_2A \rightleftharpoons 2H^+ + A^=$$

Let

$$y = M \text{ of } H_2A \text{ that dissociates}$$
$$\therefore \quad 2y = M \text{ of } H^+ \text{ produced}$$

and

$$y = M \text{ of } A^= \text{ produced}$$

Start:	0.12 M	0	0
Change:	$-y\ M$	$+2y\ M$	$+y\ M$
Equilibrium:	$0.12 - y\ M$	$2\,y\ M$	$y\ M$

$$K_a = \frac{[H^+]^2[A^=]}{[H_2A]} = 2 \times 10^{-8}$$

$$\frac{(2y)^2(y)}{0.12 - y} = 2 \times 10^{-8}$$

If y is small compared to 0.12, the K_a expression may be simplified.

$$\frac{(4y^2)y}{0.12} = 2 \times 10^{-8}$$
$$4y^3 = 0.24 \times 10^{-8}$$
$$y^3 = 0.06 \times 10^{-8}$$
$$y^3 = 0.6 \times 10^{-9}$$
$$y = \sqrt[3]{0.6} \times \sqrt[3]{10^{-9}}$$
$$y = 0.84 \times 10^{-3}$$

$$\boxed{[H^+] = 16.8 \times 10^{-4}\ M}$$

Note: This problem is given as an exercise only. There are probably no weak diprotic acids that ionize in a single step to $2H^+$ and $A^=$. In general $K_{a_1} \gg K_{a_2}$, and $[H^+] \gg [A^=]$. In other words, $[H^+] \neq 2[A^=]$.

Problem II-18

The weak diprotic inorganic acid, H_2S, dissociates in two steps, each with a specific K_a as shown below.

$$H_2S \rightleftharpoons H^+ + HS^- \qquad K_{a_1} = 1.1 \times 10^{-7}$$
$$HS^- \rightleftharpoons H^+ + S^= \qquad K_{a_2} = 1 \times 10^{-15}$$

What are the: (a) H^+ ion concentration, (b) HS^- ion concentration, and (c) $S^=$ ion concentration in a 0.15 M solution of H_2S?

Solution

(a) We can see from the values of K_{a_1} and K_{a_2} that H_2S is a much stronger acid than HS^-. Consequently, almost all of the H^+ ions produced will come from the first ionization of H_2S.

$$K_{a_1} = \frac{[H^+][HS^-]}{[H_2S]} = 1.1 \times 10^{-7}$$

Let

$$y = M \text{ of } H_2S \text{ that dissociates to } H^+ \text{ and } HS^-$$
$$\therefore \quad y = M \text{ of } H^+ \text{ produced}$$

and

$$y = M \text{ of } HS^- \text{ produced}$$

Actually, $[HS^-]$ is somewhat smaller than y because of the dissociation of HS^-. However, again we can see from the value of K_{a_2} that the amount of HS^- that dissociates will be extremely small compared to $[HS^-]$.

$$\frac{(y)(y)}{0.15 - y} = 1.1 \times 10^{-7}$$

The y is small compared to 0.15. (The low value of K_{a_1} tells us that H_2S is a "weak" acid.)

$$\frac{y^2}{1.5 \times 10^{-1}} = 1.1 \times 10^{-7}$$

$$y^2 = 1.65 \times 10^{-8}$$
$$y = \sqrt{1.65 \times 10^{-8}}$$
$$= \sqrt{1.65} \times \sqrt{10^{-8}}$$
$$= 1.285 \times 10^{-4}$$

$$\boxed{[H^+] = 1.285 \times 10^{-4} \ M}$$

(b)   For every H^+ ion produced, a HS^- ion is also produced.

$$\therefore \quad [HS^-] = 1.285 \times 10^{-4} \, M$$

(c)   The concentration of $S^=$ in the solution can be calculated from the K_{a_2} value. Similarly, we can also calculate the amount of H^+ contributed by the ionization of HS^-. Let:

$$Z = M \text{ of } S^= \text{ produced by the ionization of } HS^-$$
$$Z = M \text{ of } H^+ \text{ produced in the second ionization}$$
$$[H^+] = 1.285 \times 10^{-4} + Z \, M$$
$$[HS^-] = 1.285 \times 10^{-4} - Z \, M$$

$$K_a = \frac{[H^+][S^=]}{[HS^-]} = 1 \times 10^{-15}$$

$$\frac{(1.285 \times 10^{-4} + Z)(Z)}{(1.285 \times 10^{-4} - Z)} = 1 \times 10^{-15}$$

Since Z is very small compared to $[HS^-]$ and $[H^+]$, as described above, the K_{a_2} expression may be simplified.

$$\frac{(1.285 \times 10^{-4})(Z)}{(1.285 \times 10^{-4})} = 1 \times 10^{-15}$$

$$\frac{1.285 \times 10^{-19}}{1.285 \times 10^{-4}} = Z$$

$$Z = 1 \times 10^{-15}$$

$$[S^=] = 1 \times 10^{-15} \, M$$

We can also see that the H^+ produced in the second ionization ($1 \times 10^{-15} \, M$) is extremely small compared to the amount produced in the first ionization. In fact, HS^- is a much weaker acid than H_2O.

Alternate Solution for $[S^=]$

For the stepwise dissociation of any polyprotic acid such as H_2S, we can write an overall $K_{a_{1,2}}$, where $K_{a_{1,2}} = K_{a_1} \times K_{a_2}$.

$$H_2S \rightleftharpoons 2H^+ + S^=$$

$$K_a = \frac{[H^+]^2[S^=]}{[H_2S]}$$

The value for $[H^+]$ is fixed by the first dissociation. Note that $[H^+]$ is *not* $2[S^=]$. $[H^+]$ would equal $2[S^=]$ only if all of the H_2S that dissociated went to $2H^+$ and $S^=$.

$$K_{a_{1,2}} = K_{a_1} \times K_{a_2}$$

$$K_{a_{1,2}} = (1.1 \times 10^{-7})(1 \times 10^{-15})$$

$$K_{a_{1,2}} = 1.1 \times 10^{-22}$$

Let

$$y = M \text{ of } S^=$$

$$1.1 \times 10^{-22} = \frac{(1.285 \times 10^{-4})^2(y)}{0.15 - 1.285 \times 10^{-4}} = \frac{(1.65 \times 10^{-8})y}{1.5 \times 10^{-1}}$$

$$1.65 \times 10^{-23} = (1.65 \times 10^{-8})y$$

$$y = \frac{1.65 \times 10^{-23}}{1.65 \times 10^{-8}}$$

$$y = 1 \times 10^{-15}$$

$$\boxed{[S^=] = 1 \times 10^{-15} M}$$

Problem II-19

Consider a dicarboxylic acid that dissociates in two consecutive steps. Suppose that K_{a_1} and K_{a_2} are only about one order of magnitude apart so that the second ionization *might* have to be considered in very accurate calculations of the H^+ ion concentration of the solution. (a) Set up the equations necessary to calculate the pH of a 0.1 M solution of H_2A.
(b) Estimate the H^+ ion concentration of the solution, first assuming that the secondary dissociation contributes negligible H^+, and then (c) assuming that 50% of the HA^- produced in the first ionization dissociates further. (d) What would be the value of K_{a_2} if the second dissociation proceeded to the extent of 50%? (e) Approximately what is the actual degree of secondary dissociation?

Solution

(a)
$$H_2A \rightleftharpoons HA^- + H^+ \qquad K_{a_1} = 6.4 \times 10^{-5}$$
$$\downarrow$$
$$A^= + H^+ \qquad K_{a_2} = 6.4 \times 10^{-6}$$

Let

$$X = M \text{ of } HA^- \text{ present at equilibrium}$$

$$Z = M \text{ of } A^= \text{ present at equilibrium}$$

Every mole of H_2A that disappeared is present *either* as HA^- *or* $A^=$.

$$\therefore \quad (X + Z) = M \text{ of } H_2A \text{ that dissociated}$$

$$\therefore \quad [H_2A] = 0.1 - (X + Z)$$

The solution must be neutral. That is, the sum of all positive charges must equal the sum of all negative charges. The total $[H^+]$ present must be sufficient to neutralize the total $[A^=]$ and $[HA^-]$ present. Furthermore, it takes 2 moles of H^+ to neutralize 1 mole of $A^=$.

$$\therefore \quad [H^+] = 2[A^=] + [HA^-]$$

$$[H^+] = 2Z + X$$

Check: Assume that 10 μmoles of H_2A disappeared of which 6 μmoles remain as HA^- and 4 μmoles converted to $A^=$.

10 μmoles $H_2A \rightarrow$ 10 μmoles $HA^- + $ 10 μmoles H^+

4 μmoles $HA^- \rightarrow$ 4 μmoles $H^+ + $ 4 μmoles $A^=$

Sum: 10 μmoles $H_2A \rightarrow$ 6 μmoles $HA^- + $ 14 μmoles $H^+ + $ 4 μmoles $A^=$

$$[H^+] = 2[A^=] + [HA^-]$$

$$14 = 2(4) + 6$$

$$14 = 8 + 6$$

$$14 = 14$$

We can now set up two equations:

$$K_{a_1} = \frac{[HA^-][H^+]}{[H_2A]} \quad \text{and} \quad K_{a_2} = \frac{[A^=][H^+]}{[HA^-]}$$

$$6.4 \times 10^{-5} = \frac{(X)(2Z + X)}{0.1 - (X + Z)} \quad \text{and} \quad 6.4 \times 10^{-6} = \frac{(Z)(2Z + X)}{(X)}$$

Two equations with two unknowns can be solved although those shown above are quite messy.

The calculation described below will allow us to *estimate* the H^+ ion concentration and, at the same time, point out some interesting relationships between K_{a_1}, K_{a_2}, and the degrees of primary and secondary dissociation.

(b) *First* let us calculate the $[H^+]$ assuming that the secondary dissociation is negligible, i.e., assuming that virtually all the H^+ in the solution is derived from the $H_2A \rightarrow HA^- + H^+$ ionization. Let:

$$y = M \text{ of } H_2A \text{ dissociated}$$
$$\therefore \quad y = M \text{ of } HA^- \text{ and } M \text{ of } H^+$$
$$K_{a_1} = \frac{[H^+][HA^-]}{[H_2A]} = 6.4 \times 10^{-5}$$
$$\frac{(y)(y)}{0.1 - y} = 6.4 \times 10^{-5}$$
$$y^2 = 6.4 \times 10^{-6}$$
$$y = 2.53 \times 10^{-3}$$

$$\boxed{\therefore \quad [H^+] = 2.53 \times 10^{-3} \, M}$$

$$pH = \log \frac{1}{2.53 \times 10^{-3}}$$
$$= \log 0.412 \times 10^3$$
$$= \log 4.12 \times 10^2$$
$$= 0.614 + 2$$

$$\boxed{pH = 2.614}$$

(c)   Next, let us ignore the true value of K_{a_2} and assume that, for example, 50% of the HA^- produced in the first dissociation dissociates further to H^+ and $A^=$. What would the $[H^+]$ and pH of the solution be under these conditions?

$$H_2A \rightleftharpoons HA^- + H^+$$
$$HA^- \rightleftharpoons H^+ + A^=$$

Let $y = M$ of HA^- present at equilibrium. Since 50% of the HA^- *originally* produced ionized further, $[HA^-] = [A^=]$.

$$\therefore \quad y = M \text{ of } [A^=] \text{ present at equilibrium}$$

The total number of positive charges in the solution must equal the total number of negative charges.

$$\therefore \quad [H^+] = 2[A^=] + [HA^-]$$
$$[H^+] = 2y + y$$
$$[H^+] = 3y$$

The H_2A that disappeared is present as either HA^- or $A^=$.

$$\therefore \quad [H_2A] \doteq 0.1 - ([HA^-] + [A^=])$$
$$[H_2A] = 0.1 - 2y$$
$$K_{a_1} = \frac{[HA^-][H^+]}{[H_2A]} = 6.4 \times 10^{-5}$$
$$\frac{(y)(3y)}{0.1 - 2y} = 6.4 \times 10^{-5}$$

Because of the small value of K_{a_1}, we know that 0.1 is much larger than the amount of H_2A that disappeared ($2y$).

$$\therefore \quad 3y^2 = 6.4 \times 10^{-6}$$
$$y^2 = 2.13 \times 10^{-6}$$
$$y = \sqrt{2.13 \times 10^{-6}}$$
$$y = 1.46 \times 10^{-3}$$
$$[H^+] = 3y$$

$$\boxed{[H^+] = 4.38 \times 10^{-3} \, M}$$

$$pH = \log \frac{1}{4.38 \times 10^{-3}}$$
$$= \log 0.228 \times 10^3$$
$$= \log 2.28 \times 10^2$$
$$= 0.358 + 2$$

$$\boxed{pH = 2.358}$$

Thus, if the degree of the second dissociation were as high as 50%, the H^+ ion concentration would be less than double that calculated, ignoring the second dissociation completely. The pH would be only slightly lower.

The pH would be lower for two obvious reasons: (1) the removal of a product (HA^-) of the first dissociation causes more H_2A to dissociate (yielding more H^+ than would normally accumulate if HA^- were not removed) and (2) in the process of removing HA^-, still more H^+ is produced by ionization.

(d)   Now let us calculate a value for K_{a_2}, assuming that the degree of secondary ionization is indeed 50%.

$$K_{a_2} = \frac{[H^+][A^=]}{[HA^-]}$$

$$= \frac{(4.38 \times 10^{-3})(1.46 \times 10^{-3})}{(1.46 \times 10^{-3})}$$

$$\boxed{K_{a_2} = 4.38 \times 10^{-3}}$$

The K_{a_2} could also be calculated from the dissociation constant for the overall dissociation.

$$K_{a_{1,2}} = K_{a_1} \times K_{a_2} = \frac{[H^+]^2[A^=]}{[H_2A]}$$

$$= \frac{(4.38 \times 10^{-3})^2(1.46 \times 10^{-3})}{0.1 - 2.92 \times 10^{-3}}$$

$$= \frac{(19.2 \times 10^{-6})(1.46 \times 10^{-3})}{10^{-1}}$$

$$K_{a_{1,2}} = 28.1 \times 10^{-8}$$

$$K_{a_2} = \frac{K_{a_{1,2}}}{K_{a_1}}$$

$$K_{a_2} = \frac{28.1 \times 10^{-8}}{6.4 \times 10^{-5}}$$

$$\boxed{K_{a_2} = 4.38 \times 10^{-3}}$$

We can see immediately that our assumption of 50% ionization is incorrect. The calculated K_{a_2} value is far too large. The K_{a_2} for weak acids is never greater than K_{a_1}. Thus, the extent to which the secondary dissociation contributes to the H^+ ion concentration of the solution is negligible.

(e)   We can estimate the degree to which HA^- dissociates by employing the actual value for K_{a_2}.

$$HA^- \rightleftharpoons H^+ + A^= \qquad K_{a_2} = 6.4 \times 10^{-6}$$

Let

$$y = M \text{ of } A^= \text{ produced}$$
$$[HA^-] = 2.43 \times 10^{-3} M - y$$
$$[H^+] = 2.43 \times 10^{-3} M + y$$
$$K_{a_2} = \frac{[H^+][A^=]}{[HA^-]} = 6.4 \times 10^{-6}$$

Simplifying:

$$\frac{(2.53 \times 10^{-3})(y)}{(2.53 \times 10^{-3})} = 6.4 \times 10^{-6}$$

$$y = 6.4 \times 10^{-6}$$

$$\boxed{[A^=] = 6.4 \times 10^{-6} \ M}$$

$$\text{degree of dissociation} = \frac{[A^=]}{[HA^-]} \times 100\%$$

$$= \frac{6.4 \times 10^{-6}}{2.53 \times 10^{-3}} \times 100$$

$$= \frac{6.4 \times 10^{-4}}{2.53 \times 10^{-3}} = 2.53 \times 10^{-1}$$

$$\boxed{\textbf{degree of dissociation} = \textbf{0.253}\%}$$

B. HYDROLYSIS REACTIONS, IONIZATION OF INTERMEDIATE IONS OF POLYPROTIC ACIDS ("AMPHOLYTES")

Problem II-20

(a)   Calculate the pH of a 0.1 M solution of NH_4Cl. The K_b for NH_4OH is 1.8×10^{-5}. (b) What is the degree of hydrolysis of the salt?

Solution

(a) NH_4Cl is a salt of a weak base and a strong acid. Therefore a solution of NH_4Cl will be acidic because of hydrolysis of the NH_4^+ ion.

$$NH_4^+ + HOH \leftrightharpoons NH_4OH + H^+$$

$$K_h = K_a = \frac{K_w}{K_b} = \frac{1 \times 10^{-14}}{1.8 \times 10^{-5}} = 5.55 \times 10^{-10}$$

$$K_h = \frac{[NH_4OH][H^+]}{[NH_4^+]} = 5.55 \times 10^{-10}$$

Let

$$y = M \text{ of } NH_4OH \text{ produced upon hydrolysis}$$
$$\therefore \quad y = M \text{ of } H^+ \text{ produced upon hydrolysis}$$
$$\frac{(y)(y)}{(0.1 - y)} = 5.55 \times 10^{-10}$$

Simplifying:

$$y^2 = 5.55 \times 10^{-11}$$
$$y = \sqrt{55.5} \times \sqrt{10^{-12}}$$
$$[H^+] = 7.42 \times 10^{-6}$$
$$pH = \log \frac{1}{7.42 \times 10^{-6}}$$

$$\boxed{pH = 5.13}$$

As described in Appendix I-J, $K_h = K_a$ for a weak base. Therefore, we could also solve the problem considering the ionization of NH_4^+ as a conjugate acid.

$$NH_4^+ \rightleftharpoons H^+ + NH_3$$

$$K_a = \frac{[H^+][NH_3]}{[NH_4^+]} = 5.55 \times 10^{-10}$$

(b)   degree of hydrolysis $= \dfrac{[H^+]}{[NH_4^+]} \times 100\%$

$$= \frac{7.42 \times 10^{-6}}{10^{-1}} \times 100$$

$$= \frac{7.42 \times 10^{-4}}{10^{-1}}$$

$$= \boxed{7.42 \times 10^{-3}\%}$$

Because the NH_4^+ is less than 10% hydrolyzed (ionized), our substitution of 0.1 M for $0.1 - y$ M is reasonably valid.

Problem II-21

(a)   Calculate the pH of a 0.2 M KCN solution. The K_a for HCN is 4.9×10^{-10}. (b) What is the degree of hydrolysis of the salt?

Solution

(a) KCN is a salt of a strong base and weak acid. Therefore, a solution of KCN will be basic because of hydrolysis of the CN^- ion; that is, because of the ionization of the CN^- ion as a base.

$$CN^- + HOH \rightleftharpoons HCN + OH^-$$

$$K_h = K_b = \frac{K_w}{K_a} = \frac{1 \times 10^{-14}}{4.9 \times 10^{-10}} = 2.04 \times 10^{-5}$$

$$K_h = \frac{[HCN][OH^-]}{[CN^-]} = 2.04 \times 10^{-5}$$

Let

$$y = M \text{ of HCN produced upon hydrolysis}$$
$$\therefore \quad y = M \text{ of OH}^- \text{ produced upon hydrolysis}$$
$$\frac{(y)(y)}{(0.2 - y)} = 2.04 \times 10^{-5}$$

Simplifying:

$$y^2 = 4.08 \times 10^{-6}$$
$$y = \sqrt{4.08} \times \sqrt{10^{-6}}$$
$$[OH^-] = 2.04 \times 10^{-3}$$
$$pOH = \log \frac{1}{2.04 \times 10^{-3}}$$
$$pOH = 2.69$$
$$pH = 14 - 2.69$$

$$\boxed{pH = 11.31}$$

(b) degree of hydrolysis $= \dfrac{[OH^-]}{[CN^-]} \times 100\%$

$$= \frac{2.04 \times 10^{-3}}{2 \times 10^{-1}} \times 100 = \boxed{1.02\%}$$

Thus, our assumption that y is small compared to 0.2 is valid.

Problem II-22

Calculate: (a) the OH^-, HS^-, and $S^=$ ion concentrations in a 0.1 M solution of Na_2S, and (b) the degree of hydrolysis of the Na_2S. Assume K_{a_2} for H_2S is 10^{-14}.

Solution

(a) S$^=$ is the conjugate base of a very weak acid, HS$^-$. Consequently, we would expect the degree of hydrolysis to be quite extensive.

$$S^= + HOH \rightleftharpoons HS^- + OH^-$$

$$K_{eq} = K_{h_2} = \frac{K_w}{K_{a_2}}$$

$$K_{h_2} = \frac{K_w}{K_{a_2}} = \frac{1 \times 10^{-14}}{1 \times 10^{-14}} = 1$$

The high value of K_{h_2} shows us at a glance that S$^=$ is indeed extensively hydrolyzed. (Compare the value of K_{h_2} with those in the previous problems.)

We can estimate the degree of hydrolysis quite easily. For example, if the salt were 90% hydrolyzed, [HS$^-$] and [OH$^-$] would be 0.09 M each and [S$^=$] would be 0.01 M.

$$K_{h_2} = \frac{(9 \times 10^{-2})(9 \times 10^{-2})}{(1 \times 10^{-2})} = \frac{81 \times 10^{-4}}{1 \times 10^{-2}} = 0.81$$

But K_{h_2} is actually larger than 0.81. Therefore, the salt is more than 90% hydrolyzed. If the salt were 99% hydrolyzed, [HS$^-$] and [OH$^-$] would be 0.099 M each and [S$^=$] would be 0.001 M.

$$K_{h_2} = \frac{(9.9 \times 10^{-2})(9.9 \times 10^{-2})}{1 \times 10^{-3}} = \frac{98 \times 10^{-2}}{1 \times 10^{-3}} = 9.8$$

But K_{h_2} is actually smaller than 9.8. Therefore, the degree of hydrolysis lies between 90 and 99%.

We can solve the problem in several ways. Exact solutions involving quadratic equations can be obtained.

Method 1

Let

$y = M$ of S$^=$ that reacted

$\therefore$ $y = M$ of HS$^-$ produced

and

$y = M$ of OH$^-$ produced

and

$0.1 - y = M$ of S$^=$ remaining

$$K_{h_2} = \frac{(y)(y)}{(0.1 - y)} = 1$$

$$y^2 = 0.1 - y$$

$$y^2 + y - 0.1 = 0$$

Method 2

Let

$y = M$ of S$^=$ remaining

$\therefore$ $0.1 - y = M$ of HS$^-$ produced

and

$0.1 - y = M$ of OH$^-$ produced

$$K_{h_2} = \frac{(0.1 - y)(0.1 - y)}{(y)} = 1$$

$$y = y^2 - 0.2y + 0.01$$

$$y^2 - 1.2y + 0.01 = 0$$

Method 1—*continued*	**Method 2**—*continued*

$$y = \frac{-b \pm \sqrt{b^2 - 4ac}}{2a}$$

where

$$a = 1$$
$$b = 1$$
$$c = -0.1$$

$$y = \frac{-1 \pm \sqrt{1 \pm 0.4}}{2} = \frac{-1 \pm \sqrt{1.4}}{2}$$

$$y = \frac{-1 \pm 1.183}{2} = \frac{+0.183}{2}$$

$$y = 0.0915$$

$$[OH^-] = 0.0915\ M$$

$$[HS^-] = 0.0915\ M$$

$$\therefore\ [S^-]_{remain} = 0.0085\ M$$

$$y = \frac{-b \pm \sqrt{b^2 - 4ac}}{2a}$$

where

$$a = 1$$
$$b = -1.2$$
$$c = 0.01$$

$$y = \frac{+1.2 \pm \sqrt{+1.44 - 0.04}}{2}$$

$$y = \frac{1.2 \pm \sqrt{1.4}}{2} = \frac{1.2 \pm 1.183}{2}$$

$$y = \frac{2.383}{2} \quad \text{and} \quad \frac{0.017}{2}$$

$$y = 1.19 \text{ and } 0.0085$$

Obviously the 1.19 answer is impossible because the original $[S^=]$ was only 0.1 M

$$\therefore\ [S^=]_{remain} = \mathbf{0.0085\ M}$$
$$[OH^-] = \mathbf{0.0915\ M}$$
$$[HS^-] = \mathbf{0.0915\ M}$$

(b)   degree of hydrolysis $= \dfrac{[OH^-]}{[S^=]_{orig}} \times 100\%$

$$= \frac{0.0915}{0.10} \times 100 = \boxed{\mathbf{91.5\%}}$$

We could simplify our calculations by assuming that the amount of $S^=$ remaining is very small compared to the amount that reacted. The K_{h_2} expression in Method 2 can then be simplified:

$$K_{h_2} = \frac{(0.1)(0.1)}{y} = 1 \qquad y = 0.01$$

$$\therefore\ [HS^-] = 0.09\ M \qquad [OH^-] = 0.09\ M \qquad [S^=] = 0.01\ M$$

The answers are almost the same as that obtained by the exact method.

Problem II-23

Calculate the pH of a 0.3 M solution of $KHCO_3$. For carbonic acid, $K_{a_1} = 4.3 \times 10^{-7}$ and $K_{a_2} = 5.6 \times 10^{-11}$.

Solution

$KHCO_3$ is an intermediate salt of a polyprotic acid, H_2CO_3. Consequently, it will undergo both hydrolysis (1) and further dissociation (2).

(1) $\quad HCO_3^- + HOH \rightleftharpoons H_2CO_3 + OH^- \qquad K_{eq_1} = K_{h_1} = \dfrac{K_w}{K_{a_1}}$

(2) $\qquad\qquad HCO_3^- \rightleftharpoons H^+ + CO_3^= \qquad K_{eq_2} = K_{a_2}$

or

$$HCO_3^- + HOH \rightleftharpoons H_3O^+ + CO_3^=$$

$$K_{h_1} = \frac{1 \times 10^{-14}}{4.3 \times 10^{-7}} = 2.33 \times 10^{-8} \qquad K_{a_2} = 5.6 \times 10^{-11}$$

The greater value of K_{h_1} indicates that the solution is basic. However, the solution is not as alkaline as you might expect. The strong base, OH^-, produced in reaction 1 reacts further with the weak acid, HCO_3^-, according to reaction 3.

(3) $\qquad OH^- + HCO_3^- \rightleftharpoons HOH + CO_3^= \qquad K_{eq_3} = \dfrac{1}{K_{h_2}}$

Because OH^- is a much stronger base than H_2O, reaction 3 proceeds to a much greater extent than reaction 2. The removal of some OH^- ions disturbs the equilibrium of reaction 1, causing more HCO_3^- to hydrolyze. The newly formed OH^- reacts further with HCO_3^- and so on until equilibrium is attained. Because of reaction 3, the amount of OH^- ions present at equilibrium is much less than we would calculate solely on the basis of hydrolysis (1) and ionization (2).

The major reactions occurring in the solution, and ultimately responsible for establishing the pH, are reactions 1 and 3

(1) $HCO_3^- + HOH \rightleftharpoons OH^- + H_2CO_3 \qquad K_{eq_1} = K_{h_1} = \dfrac{K_w}{K_{a_1}}$

(3) $HCO_3^- + OH^- \rightleftharpoons HOH + CO_3^= \qquad K_{eq_3} = \dfrac{1}{K_{h_2}}$

Sum: (4) $\quad 2HCO_3^- \rightleftharpoons H_2CO_3 + CO_3^= \qquad K_{eq_4} = K_{dis} = K_{eq_1} \times K_{eq_3}$

The net effect, reaction 4, is a disproportionation reaction. The extent to which it occurs for any intermediate ion of a polyprotic acid can be calculated from the ratio of the next to the preceding K_a values.

$$K_{dis} = K_{eq_1} \times K_{eq_3} = \frac{K_w}{K_{a_1}} \times \frac{1}{K_{h_2}}$$

$$K_{h_2} = \frac{K_w}{K_{a_2}}$$

$$K_{dis} = \frac{K_w}{K_{a_1}} \times \frac{K_{a_2}}{K_w}$$

$$\boxed{K_{dis} = \frac{K_{a_2}}{K_{a_1}}}$$

We can arrive at the same disproportionation reaction by another line of reasoning. Assume that the H^+ ion released in reaction 2 reacts with a different molecule of HCO_3^-, as in reaction 5.

(2) $\qquad\qquad HCO_3^- \rightleftharpoons H^+ + CO_3^= \qquad K_{eq_2} = K_{a_2}$

(5) $\qquad HCO_3^- + H^+ \rightleftharpoons H_2CO_3 \qquad K_{eq_5} = \frac{1}{K_{a_1}}$

Sum: (4) $\qquad 2HCO_3^- \rightleftharpoons H_2CO_3 + CO_3^=$

$$K_{eq_4} = K_{dis} = K_{eq_2} \times K_{eq_5} = \frac{K_{a_2}}{K_{a_1}}$$

For HCO_3^- we can see that K_{dis} is about 5500 times greater than K_h and 10^7 times greater than K_{a_2}. Thus, the disproportionation is the most important reaction establishing the H_2CO_3 and $CO_3^=$ concentrations in the solution.

It is obvious that reaction 4 does not contain any H^+ or OH^- terms. Nevertheless, the H^+ and OH^- concentrations of the solution are *fixed* by the H_2CO_3, HCO_3^-, and $CO_3^=$ concentrations because all equilibria containing these components must be satisfied.

In writing reaction 4, we assumed that reactions 1 and 3 proceed to the same extent (we cancelled HOH and OH^-). Actually, reaction 1 proceeds slightly further to the right than does reaction 3; that is why the solution is basic. However, the actual amounts of OH^- produced and utilized are much greater than the *difference between* the amounts produced and utilized. Therefore, we can safely use reaction 4 as a basis for calculating $[HCO_3^-]$, $[CO_3^=]$, and $[H_2CO_3]$.

For example, suppose 10,000 μmoles of HCO_3^- react with 10,000 μmoles of HOH to produce 10,000 μmoles of OH^- and 10,000 μmoles of H_2CO_3. Then 9998 μmoles of the OH^- produced react with 9998 μmoles more of HCO_3^- to produce 9998 μmoles of $CO_3^=$ and 9998 μmoles of HOH.

(1) $10,000\ HCO_3^- + 10,000\ HOH \rightleftharpoons 10,000\ OH^- + 10,000\ H_2CO_3$

(3) $9998\ HCO_3^- + 9998\ OH^- \rightleftharpoons 9998\ CO_3^= + 9998\ HOH$

Sum: (4) $19,998\ HCO_3^- + 2HOH \rightleftharpoons 2\ OH^- + 9998\ CO_3^= + 10,000\ H_2CO_3$

or:

$1.9998\ HCO_3^- + 0.0002\ HOH \rightleftharpoons 0.0002\ OH^- + 0.9998\ CO_3^= + 1\ H_2CO_3$

The above example shows that the assumption that reaction 4 is $2HCO_3^- \rightleftharpoons H_2CO_3 + CO_3^=$ is valid.

Calculation

$$K_{dis} = \frac{[H_2CO_3][CO_3^=]}{[HCO_3^-]^2} = \frac{K_{a_2}}{K_{a_1}} = \frac{5.6 \times 10^{-11}}{4.3 \times 10^{-7}} = 1.301 \times 10^{-4}$$

To calculate the concentrations of each component of the equilibrium, let

$$y = M \text{ of } HCO_3^- \text{ that disappeared}$$

$$\therefore \frac{y}{2} = M \text{ of } H_2CO_3 \text{ produced}$$

and

$$\frac{y}{2} = M \text{ of } CO_3^= \text{ produced}$$

$$[HCO_3^-] = 0.3 - y$$

$$K_{dis} = \frac{(y/2)(y/2)}{(0.3 - y)^2} = \frac{y^2/4}{(0.3 - y)^2} = 1.3 \times 10^{-4}$$

$$\frac{y^2}{4(0.3 - y)^2} = 1.3 \times 10^{-4} \qquad \frac{y^2}{(0.3 - y)^2} = 5.2 \times 10^{-4}$$

Taking square roots of both sides:

$$\frac{y}{(0.3 - y)} = \sqrt{5.21} \times \sqrt{10^{-4}} = 2.28 \times 10^{-2}$$

$$y = 0.683 \times 10^{-2} - 2.28 \times 10^{-2}y$$

$$y + 0.0228y = 6.83 \times 10^{-3}$$

$$1.0228y = 6.83 \times 10^{-3}$$

$$\boxed{y = 6.67 \times 10^{-3}}$$

$$\therefore \quad [H_2CO_3] = 3.34 \times 10^{-3} \, M$$

$$[CO_3^=] = 3.34 \times 10^{-3} \, M$$

$$[HCO_3^-] = 0.300 - 0.007 = 0.293 \, M$$

The H^+ ion concentration and, consequently, the pH of the solution may be calculated from any K_{eq} expression containing the above components. For example,

$$K_{a_1} = \frac{[H^+][HCO_3^-]}{[H_2CO_3]} = 4.3 \times 10^{-7}$$

or

$$K_{a_2} = \frac{[H^+][CO_3^=]}{[HCO_3^-]} = 5.6 \times 10^{-11}$$

or

$$K_{h_1} = \frac{[H_2CO_3][OH^-]}{[HCO_3^-]} = \frac{K_w}{K_{a_1}}$$

or

$$K_{h_2} = \frac{[HCO_3^-][OH^-]}{[CO_3^=]} = \frac{K_w}{K_{a_2}}$$

where everything except $[H^+]$ or $[OH^-]$ is known.

The H^+ ion concentration and pH may also be calculated using the expressions derived in Appendix I.

$$pH = \frac{pK_{a_1} + pK_{a_2}}{2}$$

$$[H^+] = \sqrt{K_{a_2}K_{a_1}}$$

Using the above expressions, you will find that $[H^+] = 5 \times 10^{-9} \, M$ and $[OH^-] = 2 \times 10^{-6} \, M$. If $[OH^-]$ were calculated solely from the hydrolysis equation, reaction 1, $[OH^-]$ would be $83.5 \times 10^{-6} \, M$. The $[H^+]$ calculated solely from the ionization equation, reaction 2, would be $4.1 \times 10^{-6} \, M$. Thus, if we calculate $[H^+]$ and $[OH^-]$ solely from reactions 1 and 2, and then subtract the $[H^+]$ from $[OH^-]$ to obtain excess $[OH^-]$, our answer would still be significantly in error.

Problem II-24

Calculate the pH of a 0.1 M solution of KH_2PO_4. For H_3PO_4, $K_{a_1} = 7.5 \times 10^{-3}$ and $K_{a_2} = 6.2 \times 10^{-8}$.

Solution

KH_2PO_4 is a salt of a polyprotic acid. Consequently, it will undergo further dissociation (ionization as an acid) (1), hydrolysis (ionization as a base) (2), and disproportionation (3).

(1) $$H_2PO_4^- \leftrightharpoons H^+ + HPO_4^= \qquad K_{eq_1} = K_{a_2}$$

(2) $$H_2PO_4^- + HOH \leftrightharpoons H_3PO_4 + OH^- \qquad K_{eq_2} = K_{h_1} = \frac{K_w}{K_{a_1}}$$

(3) $$2H_2PO_4^- \leftrightharpoons HPO_4^= + H_3PO_4 \qquad K_{eq_3} = K_{dis} = \frac{K_{a_2}}{K_{a_1}}$$

Reaction 3 may be thought of either as the sum of the hydrolysis reaction plus the reaction between the H^+ produced and more $H_2PO_4^-$, or as the reaction between 2 molecules of $H_2PO_4^-$ where 1 molecule donates a H^+ ion (acts as an acid) and the other accepts the H^+ ion (acts as a base). See the previous problem involving HCO_3^- for the derivation of K_{dis} by both assumptions.

$$K_{a_2} = 6.2 \times 10^{-8}$$

$$K_{h_1} = 1.33 \times 10^{-12}$$

$$K_{dis} = 8.27 \times 10^{-6}$$

The relative values of K_{a_2} and K_{h_1} indicate that the solution will be acidic.

The pH is calculated below using: (a) the disproportionation reaction to extablish the $HPO_4^=$ concentration, and (b) the expression derived in Appendix I-K.

(a) $$2H_2PO_4^- \rightleftharpoons HPO_4^= + H_3PO_4$$

Let

$$y = M \text{ of } H_2PO_4^- \text{ reacting}$$

$$\therefore \frac{y}{2} = M \text{ of } HPO_4^= \text{ produced}$$

and

$$\frac{y}{2} = M \text{ of } H_3PO_4 \text{ produced}$$

$$K_{dis} = \frac{[HPO_4^=][H_3PO_4]}{[H_2PO_4^-]^2}$$

$$8.27 \times 10^{-6} = \frac{(y/2)(y/2)}{(0.1 - y)^2} = \frac{y^2}{4(0.1 - y)^2}$$

$$\sqrt{8.27 \times 10^{-6}} = \frac{y}{2(0.1 - y)} = 2.875 \times 10^{-3}$$

$$\frac{y}{0.1 - y} = 5.75 \times 10^{-3}$$

$$y = 5.75 \times 10^{-4} - 5.75 \times 10^{-3}y$$

$$1.00575y = 5.75 \times 10^{-4}$$

$$y = 5.75 \times 10^{-4}$$

$$[HPO_4^=] = \frac{y}{2} = \frac{5.75 \times 10^{-4}}{2}$$

$$[HPO_4^=] = 2.875 \times 10^{-4} \, M$$

$$[H_2PO_4^-] = 0.1000 - 0.000575 = 0.0994 \, M$$

The $[H^+]$ may be calculated from K_{a_2} expression.

$$K_{a_2} = \frac{[H^+][HPO_4^=]}{[H_2PO_4^-]}$$

$$[H^+] = \frac{K_{a_2}[H_2PO_4^-]}{[HPO_4^=]}$$

$$[H^+] = \frac{(6.2 \times 10^{-8})(9.94 \times 10^{-2})}{(2.875 \times 10^{-4})}$$

$$\boxed{[H^+] = 2.145 \times 10^{-5}}$$

$$pH = \log \frac{1}{2.145 \times 10^{-5}}$$

$$= \log 4.66 \times 10^4$$

$$\boxed{pH = 4.67}$$

(b)

$$pH = \frac{pK_{a_1} + pK_{a_2}}{2}$$

$$= \frac{2.125 + 7.208}{2}$$

$$= \frac{9.333}{2}$$

$$\boxed{pH = 4.67}$$

The actual pH of a 0.1 M KH_2PO_4 solution measured in the author's laboratory was 4.46. The discrepancy between calculated and experimental values results from our use of concentrations rather than activities. If we consult a chemical handbook, we will find that the activity coefficients for the $H_2PO_4^-$ and $HPO_4^=$ ions are 0.744 and 0.445, respectively, in 0.1 M solutions. Recalculation of a_{H^+} from the K_{a_2} expression using the values for $[H_2PO_4^-]$ and $[HPO_4^=]$ obtained from K_{dis} yields: $a_{H^+} = 3.5 \times 10^{-5}$ M and pH = 4.456.

Comparison of the above problem with the previous one involving HCO_3^- shows that the extent to which the disproportionation reaction affects pH depends on the relative values of K_{dis} and K_a (or K_h, whichever is predominant). In the above problem, K_{dis} is only about 100 times greater than K_{a_2}. The $[H^+]$ calculated solely from the dissociation reaction is about 4 times greater than the true value. In the HCO_3^- problem, K_{dis} is 5500 times greater than K_h. In this problem, the $[OH^-]$ calculated solely from the hydrolysis reaction is about 40 times greater than the true value.

Problem II-25

(a) What is the pH of a 0.1 M solution of monosodium succinate?
(b) What are the concentrations of unionized succinic acid, Hsuccinate^{-1} and succinate^{-2} in the solution ($pK_{a_1} = 4.19$ and $pK_{a_2} = 5.57$)?

Solution

Monosodium succinate [Hsuccinate^{-1}] is an intermediate ion of a polyprotic acid. The pH of a 0.1 M solution and the concentrations of all three ionic forms of succinic acid may be calculated as shown in the preceding problems:

(a) $$pH = \frac{pK_{a_1} + pK_{a_2}}{2} = \frac{4.19 + 5.57}{2} = \frac{9.76}{2}$$

$$\boxed{pH = 4.88}$$

(b) $$2\text{Hsuccinate}^{-1} \rightarrow \text{H}_2\text{succinate} + \text{succinate}^{-2}$$

$$K_{\text{dis}} = \frac{[\text{H}_2\text{succinate}][\text{succinate}^{-2}]}{[\text{Hsuccinate}^{-1}]^2} = \frac{K_{a_2}}{K_{a_1}} = \frac{2.7 \times 10^{-6}}{6.4 \times 10^{-5}} = 4.22 \times 10^{-2}$$

Let

$$y = M \text{ of Hsuccinate}^{-1} \text{ that disappears}$$

$$\therefore \quad \frac{y}{2} = M \text{ of H}_2\text{succinate produced}$$

and

$$\frac{y}{2} = M \text{ of succinate}^{-2} \text{ produced}$$

$$\frac{(y/2)(y/2)}{(0.1 - y)^2} = 4.22 \times 10^{-2}$$

$$\sqrt{\frac{y^2}{4(0.1 - y)^2}} = \sqrt{4.22 \times 10^{-2}}$$

$$\frac{y}{2(0.1 - y)} = 2.055 \times 10^{-1}$$

$$\frac{y}{(0.2 - 2y)} = 0.2055$$

$$y = 0.0411 - 0.411y$$

$$1.411y = 0.0411$$

$$y = \frac{0.0411}{1.411}$$

$$y = 0.0291$$

$$\frac{y}{2} = 0.0146$$

H$_2$succinate = 0.0146 *M*

succinate^{-2} = 0.0146 *M*

Hsuccinate^{-1} = 0.1000 − 0.0291 = 0.0709 *M*

The concentrations of the various ionic species may also be calculated in another manner. First calculate the ratio of Hsuccinate^{-1}/H$_2$succinate using the Henderson-Hasselbalch equation.

$$pH = pK_{a_1} + \log \frac{[\text{Hsuccinate}^{-1}]}{[\text{H}_2\text{succinate}]}$$

$$4.88 = 4.19 + \log \frac{[\text{Hsuccinate}^{-1}]}{[\text{H}_2\text{succinate}]}$$

$$0.69 = \log \frac{[\text{Hsuccinate}^{-1}]}{[\text{H}_2\text{succinate}]}$$

$$\frac{[\text{Hsuccinate}^{-1}]}{[\text{H}_2\text{succinate}]} = \text{antilog of } 0.69 = 5.0$$

$\therefore$ The ratio of Hsuccinate^{-1}/H$_2$succinate is 5/1.
Next calculate the ratio of Hsuccinate^{-1}/succinate^{-2}.

$$pH = pK_{a_2} + \log \frac{[\text{succinate}^{-2}]}{[\text{Hsuccinate}^{-1}]}$$

$$4.88 = 5.57 + \log \frac{[\text{succinate}^{-2}]}{[\text{Hsuccinate}^{-1}]}$$

$$-0.69 = \log \frac{[\text{succinate}^{-2}]}{[\text{Hsuccinate}^{-1}]}$$

$$+0.69 = \log \frac{[\text{Hsuccinate}^{-1}]}{[\text{succinate}^{-2}]}$$

$$\frac{[\text{Hsuccinate}^{-1}]}{[\text{succinate}^{-2}]} = \text{antilog of } 0.69 = 5.0$$

$\therefore$ The ratio of Hsuccinate^{-1}/succinate^{-2} is also 5/1.   The concentrations of all three ionic species are in the ratio of 1:5:1.

$$\text{H}_2\text{succinate} \rightleftharpoons \text{Hsuccinate}^{-1} \rightleftharpoons \text{succinate}^{-2}$$
$$1 \qquad\qquad 5 \qquad\qquad 1$$

The *total* concentration of succinate is 0.1 M.

$$\tfrac{1}{7} \times 0.1\ M = \text{H}_2\text{succinate concentration}$$
$$\tfrac{5}{7} \times 0.1\ M = \text{Hsuccinate}^{-1} \text{ concentration}$$
$$\tfrac{1}{7} \times 0.1\ M = \text{Hsuccinate}^{-1} \text{ concentration}$$

$$\boxed{\begin{array}{l} \therefore \quad [\text{H}_2\text{succinate}] = 0.0143\ M \\ [\text{Hsuccinate}^{-1}] = 0.0714\ M \\ [\text{succinate}^{-2}] = 0.0143\ M \end{array}}$$

The small difference between the values obtained by the two methods results from rounding off the pK_a and antilog values.

Problem II-26

Calculate the concentration of all ionic species of succinate present in a solution (buffer) of 0.1 M succinate, pH 4.59.

Solution

We can see from the titration curve sketched in Figure II-1 that the major ionic species present at pH 4.59 are H_2succinate and Hsuccinate^{-1} with the latter predominating. However, because pK_{a_2} is close to pK_{a_1}, an appreciable amount of succinate^{-2} is also present.

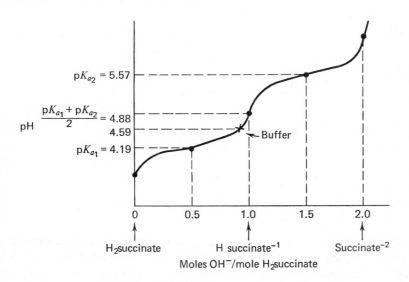

Figure II-1 Titration curve of succinic acid

First calculate the ratio of Hsuccinate^{-1}/H_2succinate using the Henderson-Hasselbalch equation.

$$pH = pK_{a_1} + \log \frac{[\text{Hsuccinate}^{-1}]}{[\text{H}_2\text{succinate}]}$$

$$4.59 = 4.19 + \log \frac{[\text{Hsuccinate}^{-1}]}{[\text{H}_2\text{succinate}]}$$

$$0.40 = \log \frac{[\text{Hsuccinate}^{-1}]}{[\text{H}_2\text{succinate}]}$$

$$\frac{[\text{Hsuccinate}^{-1}]}{[\text{H}_2\text{succinate}]} = \text{antilog of } 0.40 = 2.5$$

∴ The Hsuccinate^{-1}/H_2succinate ratio is 2.5/1.

Next calculate the Hsuccinate^{-1}/succinate^{-2} ratio at pH 4.59.

$$pH = pK_{a_2} + \log \frac{[\text{succinate}^{-2}]}{[\text{Hsuccinate}^{-1}]}$$

$$4.59 = 5.57 + \log \frac{[\text{succinate}^{-2}]}{[\text{Hsuccinate}^{-1}]}$$

$$-0.98 = \log \frac{[\text{succinate}^{-2}]}{[\text{Hsuccinate}^{-1}]}$$

$$+0.98 = \frac{[\text{Hsuccinate}^{-1}]}{[\text{succinate}^{-2}]}$$

$$\frac{[\text{Hsuccinate}^{-1}]}{[\text{succinate}^{-2}]} = \text{antilog of } 0.98 = 9.5$$

$\therefore$ The Hsuccinate^{-1}/succinate^{-2} ratio is 9.5/1.
The three ionic species are in the following proportions:

H$_2$succinate		Hsuccinate^{-1}		succinate^{-2}
1	:	2.5		
		9.5	:	1

The two ratios must be expressed relative to a common component such as Hsuccinate^{-1}. So, if 1 part succinate^{-2} is present for every 9.5 parts of Hsuccinate^{-1}, calculate how much succinate^{-2} is present for every 2.5 parts of Hsuccinate^{-1}.

$$\frac{1}{9.5} = \frac{y}{2.5}$$

$$9.5y = 2.5$$

$$y = \frac{2.5}{9.5} = 0.263$$

That is, the ratio of the three ionic species is:

H$_2$succinate		Hsuccinate^{-1}		succinate^{-2}	total
1	:	2.5	:	0.263	3.763 parts

The *total* succinate concentration is 0.1 M.

$$\frac{1}{3.763} \times 0.1 \ M = \boxed{\textbf{0.0266 } \textit{\textbf{M}} \textbf{ H}_2\textbf{succinate}}$$

$$\frac{2.5}{3.763} \times 0.1 \ M = \boxed{\textbf{0.0664 } \textit{\textbf{M}} \textbf{ Hsuccinate}^{-1}}$$

$$\frac{0.263}{3.763} \times 0.1 \ M = \boxed{\textbf{0.00698 } \textit{\textbf{M}} \textbf{ succinate}^{-2}}$$

Problem II-27

What is the pH of a 0.1 M solution of NH_4CN?

Solution

NH_4CN is a salt of a weak acid and a weak base. Both ions will hydrolyze.

(1) $NH_4^+ + HOH \rightleftharpoons NH_4OH + H^+$ $\qquad K_{eq_1} = K_{h_{NH_4^+}} = K_{a_{NH_4^+}}$
$$= 5.55 \times 10^{-10}$$

or

$$NH_4^+ \rightleftharpoons NH_3 + H^+$$

or

$$NH_4^+ + H_2O \rightleftharpoons NH_3 + H_3O^+$$

(2) $CN^- + HOH \rightleftharpoons HCN + OH^-$ $\qquad K_{eq_2} = K_{h_{CN^-}} = K_{b_{CN^-}}$
$$= 2.04 \times 10^{-5}$$

The greater value for the K_h of the CN^- hydrolysis suggests that CN^- will hydrolyze to a greater extent than NH_4^+. Thus, the solution will be basic. However, the actual OH^- ion concentration is not as great as you might expect from the relative K_h values. The newly formed OH^- reacts with some more NH_4^+ to form NH_3 and HOH. The removal of some OH^- disturbs the equilibrium of reaction 2, causing more CN^- to hydrolyze. The newly formed OH^- reacts with some more NH_4^+ and so on until an equilibrium is attained. The reaction of OH^- with NH_4^+ causes more NH_4^+ to disappear than would normally disappear if no weak conjugate base were present. That is, the NH_4^+ hydrolyzes to a greater extent than it would in a solution of, for example, NH_4Cl of equal concentration. Similarly, because OH^- is removed, the CN^- hydrolyzes to a much greater extent than it would in a solution of, for example, KCN or NaCN.

The reaction of NH_4^+ with the strong base, OH^-, proceeds to a much greater extent than the reaction of NH_4^+ with the weak base, H_2O. Consequently, the major reactions taking place in solution, and ultimately responsible for the final pH are:

(2) $CN^- + HOH \rightleftharpoons HCN + OH^-$ $K_{eq_2} = K_{h_{CN^-}}$

(3) $OH^- + NH_4^+ \rightleftharpoons NH_3 + HOH$ $K_{eq_3} = \dfrac{1}{K_{b_{NH_3}}}$

Sum:

(4) $CN^- + NH_4^+ \rightleftharpoons HCN + NH_3$ $K_{eq_4} = K_{dis} = K_{eq_2} \times K_{eq_3}$

The sum of reactions 2 and 3, reaction 4, is known as a displacement reaction. (The conjugate base of a weak acid displaces the conjugate base of a stronger acid.)

$$K_{dis} = K_{eq_2} \times K_{eq_3} = K_{h_{CN^-}} \times \dfrac{1}{K_{b_{NH_3}}} = \dfrac{K_{h_{CN^-}}}{K_{b_{NH_3}}}$$

or because

$$K_{h_{CN^-}} = \dfrac{K_w}{K_{a_{HCN}}} \quad \text{and} \quad \dfrac{1}{K_{b_{NH_3}}} = \dfrac{K_{a_{NH_4^+}}}{K_w}$$

$$K_{dis} = \dfrac{K_w}{K_{a_{HCN}}} \times \dfrac{K_{a_{NH_4^+}}}{K_w}$$

$$\boxed{K_{dis} = \dfrac{K_{a_{NH_4^+}}}{K_{a_{HCN}}}}$$

Thus, the extent to which displacement reactions occur (i.e., K_{dis}) can be calculated by the ratio of K_a for the stronger acid/K_a for the weaker acid. For reaction 4:

$$K_{dis} = \dfrac{5.55 \times 10^{-10}}{4.9 \times 10^{-10}} = 1.132$$

The extremely large value of K_{dis} compared to K_h indicates that the displacement reaction will be the most significant reaction in the solution. In writing reaction 4, we assumed that reactions 2 and 3 proceed to the same extent (that is, we cancelled HOH and OH^-). Actually, reaction 2 proceeds to a slightly greater extent than reaction 3 because, as our calculations will show, the solution is basic. However, the *difference* between the extent of reactions 2 and 3 is extremely small compared to the *total number* of moles of OH^- produced and utilized in these reactions. (See the HCO_3^- problem for a similar example.) Consequently, we can assume that reaction 4 is the only reaction occurring in the solution. From reaction 4 and its K_{dis}, we can calculate the equilibrium concentrations of HCN, NH_3, CN^-, and NH_4^+. Because *all* equilibria involving these components must be satisfied, we can then calculate $[H^+]$ or $[OH^-]$.

$$K_{dis} = \frac{[HCN][NH_3]}{[CN^-][NH_4^+]} = 1.132$$

Let

$$y = M \text{ of } CN^- \text{ that reacted} \qquad \text{(that disappeared)}$$

$$\therefore \quad y = M \text{ of } NH_4^+ \text{ that reacted}$$

(The much smaller amount that reacted in reaction 1 is ignored.)

$$[CN^-] = 0.1 - y \qquad [HCN] = y$$

$$[NH_4^+] = 0.1 - y \qquad [NH_3] = y$$

$$K_{dis} = \frac{(y)(y)}{(0.1 - y)(0.1 - y)} = \frac{y^2}{(0.1 - y)^2} = 1.132$$

$$\sqrt{\frac{y^2}{(0.1 - y)^2}} = \sqrt{1.132}$$

$$\frac{y}{0.1 - y} = 1.065 \qquad y = 0.1065 - 1.065y$$

$$2.065y = 0.1065 \qquad y = \frac{0.1065}{2.065} \qquad y = 0.0515$$

$$\therefore \quad [CN^-] = 0.0485 \, M \qquad [HCN] = 0.0515 \, M$$

$$[NH_4^+] = 0.0485 \, M \qquad [NH_3] = 0.0515 \, M$$

The degree of hydrolysis may be calculated.

$$\text{degree of hydrolysis} = \frac{[HCN]}{[CN^-]} \times 100\% \qquad \text{or} \qquad \frac{[NH_3]}{[NH_4^+]} \times 100\%$$

$$= \frac{0.0515}{0.1} \times 100\% = \boxed{51.5\%}$$

The pH of the solution may be calculated from any appropriate K expression.

$$K_h = \frac{[NH_3][H^+]}{[NH_4^+]}$$

$$5.55 \times 10^{-10} = \frac{(0.0515)[H^+]}{(0.0485)}$$

$$[H^+] = \frac{(0.0485)5.55 \times 10^{-10}}{(0.0515)}$$

$$\boxed{[H^+] = 5.23 \times 10^{-10}\ M}$$

$$pH = \log \frac{1}{5.23 \times 10^{-10}}$$

$$= \log 1.91 \times 10^9$$

$$\boxed{pH = 9.28}$$

$$K_h = \frac{[HCN][OH^-]}{[CN^-]}$$

$$2.04 \times 10^{-5} = \frac{(0.0515)[OH^-]}{(0.0485)}$$

$$[OH^-] = \frac{(0.0485)2.04 \times 10^{-5}}{(0.0515)}$$

$$[OH^-] = 1.92 \times 10^{-5}$$

$$[H^+] = \frac{K_w}{[OH^-]}$$

$$= \frac{1 \times 10^{-14}}{(1.92 \times 10^{-5})}$$

$$\boxed{[H^+] = 5.23 \times 10^{-10}\ M}$$

or

$$K_a = \frac{[H^+][CN^-]}{[HCN]}$$

$$4.9 \times 10^{-10} = \frac{[H^+](0.0485)}{(0.0515)}$$

$$[H^+] = \frac{(0.0515)4.9 \times 10^{-10}}{(0.0485)}$$

$$\boxed{[H^+] = 5.23 \times 10^{-10}\ M}$$

Note that the actual OH^- ion concentration is about 100 times less than that which we would calculate solely from the hydrolysis of CN^- (reaction 2).

We can derive expressions from which the H^+ ion concentration and the pH may be determined directly from the K_{a_1} and K_{a_2} values:

$$K_{dis} = \frac{[HCN][NH_3]}{[CN^-][NH_4^+]} = \frac{K_{a_{NH_4^+}}}{K_{a_{HCN}}} \quad \text{and} \quad \begin{array}{c} [HCN] = [NH_3] \\ [CN^-] = [NH_4^+] \end{array}$$

$$\frac{[HCN]^2}{[CN^-]^2} = \frac{K_{a_{NH_4^+}}}{K_{a_{HCN}}} \qquad \frac{[HCN]}{[CN^-]} = \frac{\sqrt{K_{a_{NH_4^+}}}}{\sqrt{K_{a_{HCN}}}}$$

Substituting the $[HCN]/[CN^-]$ ratio into the $K_{a_{HCN}}$ expression:

$$K_{a_{HCN}} = \frac{[H^+][CN^-]}{[HCN]}$$

$$[H^+] = \frac{K_{a_{HCN}}[HCN]}{[CN^-]}$$

$$[H^+] = \frac{K_{a_{HCN}}\sqrt{K_{a_{NH_4^+}}}}{\sqrt{K_{a_{HCN}}}}$$

$$\frac{K_{a_{HCN}}}{\sqrt{K_{a_{HCN}}}} = \sqrt{K_{a_{HCN}}}$$

$$[H^+] = \sqrt{K_{a_{NH_4^+}}}\sqrt{K_{a_{HCN}}}$$

$$\boxed{[H^+] = \sqrt{K_{a_{NH_4^+}}K_{a_{HCN}}}}$$

or, in general,

$$\boxed{[H^+] = \sqrt{K_{a_1}K_{a_2}}}$$

and

$$\boxed{pH = \frac{pK_{a_1} + pK_{a_2}}{2}}$$

C. TITRATIONS OF WEAK ACIDS AND BASES*

A weak acid dissociates in an aqueous solution to yield a small amount of H^+ ions.

(1) $$HA \leftrightharpoons H^+ + A^-$$

When OH^- ions are added, they are neutralized by the H^+ ions to form H_2O.

(2) $$OH^- + H^+ \rightarrow H_2O$$

The removal of H^+ ions disturbs the equilibrium between the weak acid and its ions. Consequently, more HA ionizes to reestablish the equilibrium. The newly produced H^+ ions can then be neutralized by more OH^- and so on until all of the hydrogen originally present is neutralized. The

* See Appendix I–L.

overall result, the sum of equations 1 and 2, is the titration of HA with OH⁻.

$$(3) \qquad\qquad HA + OH^- \rightleftharpoons H_2O + A^-$$

The number of equivalents of OH⁻ required equals the total number of equivalents of hydrogen present (as H⁺ plus HA). Neutralization calculations are conducted in exactly the same manner as for strong acids.

The pH at the exact end (equivalence) point of the titration is not 7 but higher because of the hydrolysis of the A⁻ ion; that is, because reaction 3 itself is an equilibrium reaction. In the absence of any remaining HA, the A⁻ ion reacts with H_2O to produce OH⁻ ions and the undissociated weak acid, HA. Hydrolysis of salts was discussed in Section B. In this section, we concern ourselves with calculations of [H⁺] and pH in partially neutralized solutions of weak acids and bases. Because equilibrium conditions must always be satisfied in solutions of weak acids and bases, the H⁺ ion concentration and pH during the titration can be calculated from the K_a expression or from the Henderson-Hasselbalch equation, provided the concentration of conjugate acid and conjugate base (or the ratio of their concentrations) is known. When calculating the values for [HA] and [A⁻] during a titration, it is safe to assume that moles HA_{remain} = moles HA_{orig} − moles $HA_{titrated}$, and moles A⁻ = moles $HA_{titrated}$ throughout *most* of the titration curve. Significant errors (resulting from hydrolysis of the salt) arise only when the last equivalence point is approached. The weaker the acid (in terms of K_a as well as original concentration), the sooner (in terms of percent of the original acid titrated) anomalous answers result from ignoring hydrolysis.

Problem II-28

Calculate the amount of 0.1 N KOH required to titrate completely 250 ml of 0.2 M acetic acid.

Solution

number of equivalents of OH⁻ required =
        number of equivalents of hydrogen (H⁺ plus HA) present

$$number\ of\ equivalents\ OH^- = liters_{base} \times N_{base}$$
$$number\ of\ equivalents\ of\ H = liters_{acid} \times N_{acid}$$
$$liters_{base} \times N_{base} = liters_{acid} \times N_{acid}$$

Let X = $liters_{base}$.

$(X)\ 0.1 = 0.25 \times 0.2$

$$X = \frac{0.05}{0.1} = 0.5 \text{ liter}$$

$\therefore$ **500 ml of KOH required**

Problem II-29

(a) Write the reactions showing the successive titrations of the carboxyl groups in citric acid. (b) Calculate the amount of 0.2 M KOH required to titrate completely all the citric acid present in 300 ml of a 0.1 M solution. (c) How much KOH would be required to convert all the citric acid present to dipotassium citrate?

Solution

(a)

$$
\begin{array}{l}
\text{H}_2\text{C—COOH} \\
\quad | \\
\text{HO—C—COOH} + \text{OH}^- \rightarrow \\
\quad | \\
\text{H}_2\text{C—COOH}
\end{array}
\quad
\begin{array}{l}
\text{H}_2\text{C—COO}^- \\
\quad | \\
\text{HO—C—COOH} + \text{H}_2\text{O} \\
\quad | \\
\text{H}_2\text{C—COOH}
\end{array}
$$

$$
\begin{array}{l}
\text{H}_2\text{C—COO}^- \\
\quad | \\
\text{HO—C—COOH} + \text{OH}^- \rightarrow \\
\quad | \\
\text{H}_2\text{C—COOH}
\end{array}
\quad
\begin{array}{l}
\text{H}_2\text{C—COO}^- \\
\quad | \\
\text{HO—C—COO}^- + \text{H}_2\text{O} \\
\quad | \\
\text{H}_2\text{C—COOH}
\end{array}
$$

$$
\begin{array}{l}
\text{H}_2\text{C—COO}^- \\
\quad | \\
\text{HO—C—COO}^- + \text{OH}^- \rightarrow \\
\quad | \\
\text{H}_2\text{C—COOH}
\end{array}
\quad
\begin{array}{l}
\text{H}_2\text{C—COO}^- \\
\quad | \\
\text{HO—C—COO}^- + \text{H}_2\text{O} \\
\quad | \\
\text{H}_2\text{C—COO}^-
\end{array}
$$

(b) number of equivalents of OH^- required =
number of equivalents of H present

$$0.1\ M\ \text{H}_3\text{A} = 0.3\ N\ \text{H}_3\text{A}$$

$$\text{liters}_{\text{base}} \times N_{\text{base}} = \text{liters}_{\text{acid}} \times N_{\text{acid}}$$

$$\text{liters}_{\text{base}} = \frac{0.3 \times 0.3}{0.2} = \frac{0.09}{0.2} = 0.45$$

$\therefore$ **450 ml of KOH required**

(c)   To convert H_3A to K_2HA (i.e., $HA^=$), only 2 out of the 3 equivalents in citric acid have to be titrated.

$\frac{2}{3} \times 450$ ml =

300 ml of KOH required

liters $\times M$ = number of moles of H_3A

$0.3 \times 0.1 = 0.03$ mole of H_3A

$\therefore$ 0.03 mole of OH^- titrates the H_3A to H_2A^-; 0.03 mole more of OH^- converts the H_2A^- to $HA^=$.

$\therefore$ 0.06 moles of KOH required

$$\text{liters}_{\text{base}} = \frac{\text{number of moles}}{M}$$

$$= \frac{0.06}{0.2} = 0.3 = \boxed{\textbf{300 ml}}$$

Problem II-30

What are the H^+ ion concentration and pH of a solution obtained by mixing 200 ml of 0.1 M KOH with 300 ml of 0.15 M acetic acid? The $K_a = 1.7 \times 10^{-5}$ and $pK_a = 4.77$.

Solution

$$HA + OH^- \rightleftharpoons HOH + A^-$$

number of moles HA originally present $= 0.3$ liter $\times 0.15$ M
$$= 0.045 \text{ mole}$$

number of moles HA titrated $= 0.2$ liter $\times 0.1$ M
$$= 0.020 \text{ mole}$$

$\therefore$ number of moles HA remaining $= 0.045 - 0.020$
$$= 0.025 \text{ mole}$$

number of moles A^- produced $= 0.020$ mole

final volume $= 0.2 + 0.3$ liter
$$= 0.5 \text{ liter}$$

In the final solution:

$$[HA] = \frac{0.025 \text{ mole}}{0.5 \text{ liter}} = 0.05 \ M \qquad [A^-] = \frac{0.020 \text{ mole}}{0.5 \text{ liter}} = 0.04 \ M$$

$$K_a = \frac{[H^+][A^-]}{[HA]}$$

$$1.7 \times 10^{-5} = \frac{[H^+](0.04)}{(0.05)}$$

$$8.5 \times 10^{-7} = [H^+](4 \times 10^{-2})$$

$$\frac{8.5 \times 10^{-7}}{4 \times 10^{-2}} = [H^+]$$

$$\boxed{[H^+] = 2.12 \times 10^{-5} \, M}$$

$$pH = \log \frac{1}{2.12 \times 10^{-5}}$$

$$= \log 4.7 \times 10^4$$

$$\boxed{pH = 4.67}$$

$$pH = pK_a + \log \frac{[A^-]}{[HA]}$$

$$pH = 4.77 + \log \frac{(0.04)}{(0.05)}$$

$$= 4.77 - \log \frac{(0.05)}{(0.04)}$$

$$= 4.77 - \log 1.25$$

$$= 4.77 - 0.1$$

$$\boxed{pH = 4.67}$$

$$\frac{1}{[H^+]} = \text{antilog of } 4.67 = 4.7 \times 10^4$$

$$[H^+] = \frac{1}{4.7 \times 10^4} = 0.212 \times 10^{-4}$$

$$\boxed{[H^+] = 2.12 \times 10^{-5} \, M}$$

Problem II-31

Calculate the pH of a solution prepared by mixing 500 ml of 0.25 M sodium acetate with 250 ml of 0.10 M HCl.

Solution

Strong acids react with salts (conjugate bases) of weak acids to produce the unionized weak parent acid (conjugate acid).

$$H^+ + A^- \rightleftharpoons HA$$

number of moles acetate present = 0.5 liter × 0.25 M = 0.125 mole

number of moles H^+ added = 0.25 liter × 0.10 M = 0.025 mole

The 0.025 mole of H^+ reacts with the 0.025 mole of acetate to produce 0.025 mole of HA.

$$[A^-] = \frac{0.125 - 0.025 \text{ mole}}{0.75 \text{ liter}} = \frac{0.10 \text{ mole}}{0.75 \text{ liter}} = 0.133 \, M$$

$$[HA] = \frac{0.025 \text{ mole}}{0.75 \text{ liter}} = 0.033 \, M$$

A small amount of HA dissociates and a small amount of A^- hydrolyzes, but the amounts undergoing these reactions are small compared to the amounts present. The pH of the solution may be calculated from the Henderson-Hasselbalch equation as shown in the above problem.

Problem II-32

(a) Write the reactions for the stepwise titration of H_3PO_4 with NaOH. (b) Calculate the H^+ ion concentration and pH of a solution prepared by mixing 125 ml of 0.1 M H_3PO_4 with 125 ml of 0.1 M NaOH and (c) 175 ml of 0.1 M NaOH.

Solution

(a)
$$H_3PO_4 + OH^- \rightleftharpoons H_2PO_4^- + H_2O$$

$$H_2PO_4^- + OH^- \rightleftharpoons HPO_4^= + H_2O$$

$$HPO_4^= + OH^- \rightleftharpoons PO_4^{\equiv} + H_2O$$

(b)   0.125 liter $\times$ 0.1 M H_3PO_4 = 0.0125 mole of H_3PO_4

The addition of 0.125 liter $\times$ 0.1 M NaOH = 0.0125 mole of OH^- is sufficient to convert all the H_3PO_4 present to $H_2PO_4^-$. The final solution will contain 0.0125 mole $H_2PO_4^-$/0.25 liter = 0.05 M. The exact $[H^+]$ and pH of a 0.05 M $H_2PO_4^-$ solution can be calculated, taking into account the disproportionation reaction or from the pK_{a_1} and pK_{a_2} values as shown in Appendix I-K.

$$pH = \frac{pK_{a1} + pK_{a2}}{2}$$

$$pH = \frac{2.12 + 7.21}{2} = \boxed{4.67}$$

(c)   0.175 liter $\times$ 0.1 M NaOH = 0.0175 mole OH^-

The 0.0175 mole of OH^- is sufficient to convert all the H_3PO_4 present to $H_2PO_4^-$ and partially titrate some $H_2PO_4^-$ to $HPO_4^=$. The final solution contains $H_2PO_4^-$ (conjugate acid) and $HPO_4^=$ (conjugate base). The pH of the solution may be calculated from the K_{a_2} expression or the Henderson-Hasselbalch equation. First calculate the M of the $H_2PO_4^-$ and $HPO_4^=$ present.

0.0125 mole H_3PO_4 + 0.0175 mole OH^-
$$\rightarrow 0.0125 \text{ mole } H_2PO_4^- + 0.0050 \text{ mole } OH^-$$
0.0125 mole $H_2PO_4^-$ + 0.0050 mole OH^-
$$\rightarrow 0.0050 \text{ mole } HPO_4^= + 0.0075 \text{ mole } H_2PO_4^-$$

$$[H_2PO_4^-] = \frac{0.0075 \text{ mole}}{0.300 \text{ liter}} = 0.025 \ M$$

$$[HPO_4^=] = \frac{0.005 \text{ mole}}{0.300 \text{ liter}} = 0.0167 \ M$$

$$pH = pK_a + \log \frac{[HPO_4^=]}{[H_2PO_4^-]}$$

$$pH = 7.21 + \log \frac{(0.0167)}{(0.025)}$$

$$= 7.21 - \log \frac{(0.025)}{(0.0167)}$$

$$= 7.21 - \log 1.496 = 7.21 - 0.175$$

$$\boxed{pH = 7.035}$$

Check: The pH is less than pK_{a_2} $\therefore$ $H_2PO_4^-$ must be less than half titrated to $HPO_4^=$. The 0.005 mole (titrated) is less than half of 0.0125 mole.

Problem II-33

Calculate the appropriate values and plot the curve for the titration of 250 ml of 0.2 M propionic acid with 0.1 M KOH ($K_a = 1.34 \times 10^{-5}$ and $pK_a = 4.87$).

Solution (see Figure AI-3.)

(a) First calculate the starting pH using the K_a expression and the appropriate assumptions.

$$K_a = \frac{[H^+][A^-]}{[HA]} = \frac{(y)(y)}{0.2 - y} = 1.34 \times 10^{-5}$$

$$y^2 = (1.34 \times 10^{-5})(0.2)$$

$$y = \sqrt{2.68 \times 10^{-6}}$$

$$\therefore \quad \boxed{[H^+] = 1.64 \times 10^{-3} \ M}$$

$$pH = \log \frac{1}{1.64 \times 10^{-3}}$$

$$\boxed{pH = 2.785}$$

(b) Next calculate the concentration of propionate (salt) and subsequently the pH at the equivalence point, using the expression for the hydrolysis of the propionate anion.

$$A^- + HOH \rightleftharpoons HA + OH^-$$

$$K_h = \frac{K_w}{K_a} = \frac{1 \times 10^{-14}}{1.34 \times 10^{-5}} = 7.46 \times 10^{-10}$$

$$K_h = \frac{[HA][OH^-]}{[A^-]} = 7.46 \times 10^{-10}$$

$$\text{liters}_{base} \times N_{base} = \text{liters}_{acid} \times N_{acid}$$

$$\text{liters}_{base} = \frac{0.25 \times 0.2}{0.1} = 0.50$$

$$\therefore \quad \text{final volume} = 0.50 + 0.25 = 0.75 \text{ liter}$$

and

$$[A^-] = \frac{0.05 \text{ mole}}{0.75 \text{ liter}} = 0.066 \ M$$

Let

$$y = M \text{ of HA produced upon hydrolysis}$$
$$\therefore \quad y = M \text{ of OH}^- \text{ produced upon hydrolysis}$$

$$\frac{(y)(y)}{(0.066 - y)} = 7.46 \times 10^{-10}$$

Simplifying:

$$y^2 = 0.493 \times 10^{-10} = 49.3 \times 10^{-12}$$

$$y = \sqrt{49.3} \times \sqrt{10^{-12}}$$

$$[OH^-] = 7.03 \times 10^{-6} \ M$$

$$[H^+] = \frac{1 \times 10^{-14}}{[OH^-]} = \frac{1 \times 10^{-14}}{7.03 \times 10^{-6}} = 1.42 \times 10^{-9}$$

$$pH = \log \frac{1}{1.42 \times 10^{-9}} = \log 7.05 \times 10^8$$

$$\boxed{pH = 8.848}$$

(c) Next calculate values for the pH of the solution between the starting pH and the pH at the end point; that is, after various additions of base to the acid solution (10 ml, 50 ml, 100 ml, 200 ml, 300 ml, 400 ml, etc.). Use the Henderson-Hasselbalch equation and be sure to correct for the changing volume. You have already seen that the pH at the equivalence point is 8.848. Consequently, if the Henderson-Hasselbalch equation gives a pH greater than 8.848, you have approached too close to the equivalence point and the assumptions that moles $A^- = $ moles $HA_{titrated}$ and moles $HA_{remain} = $ moles $HA_{orig} - $ moles $HA_{titrated}$ are not valid. In other words, the extent of hydrolysis is appreciable.

For example, after 200 ml of 0.1 M KOH are added, 0.20 liter $\times$ 0.1 M KOH $= 0.020$ mole of OH^- added and 0.020 mole of HA is converted to A^-.

At the start we had 0.25 liter $\times$ 0.2 $M = 0.050$ mole of HA.

$\therefore$ After adding 200 ml of 0.1 M KOH, $0.05 - 0.02 = 0.03$ mole of HA remains and 0.02 mole of A^- is produced.

$$[HA] = \frac{0.03 \text{ mole}}{0.45 \text{ liter}} = 0.0667 \ M$$

$$[A^-] = \frac{0.02}{0.45} = 0.0445 \ M$$

$$pH = pK_a + \log \frac{[A^-]}{[HA]}$$

$$pH = 4.87 + \log \frac{(0.0445)}{(0.0667)}$$

$$pH = 4.87 - \log \frac{(0.0667)}{(0.0445)}$$

$$pH = 4.87 - \log 1.5$$

$$pH = 4.87 - 0.17$$

$$\boxed{pH = 4.70}$$

Check: The pH is less than pK_a $\therefore$ HA is less than half titrated. The 0.02 mole is less than 0.05/2 mole.

(d)   Finally, calculate the pH for the overtitration of the HA after the addition of, for example, 510 ml, 525 ml, 550 ml, and 600 ml of base. For these calculations it is sufficient to assume that $[OH^-] = [OH^-]_{excess}$ and $pOH = -\log [OH^-]_{excess}$.

Problem II-34

Calculate the values and plot the curve for the titration of 200 ml of 0.3 M sodium acetate with 0.2 M HCl.

Solution (see Figure AI-4.)

The pH starts high and decreases as HCl is added.

(a)   At the start the solution contains only OAc⁻. The starting pH is basic and can be calculated from the K_h expression.

(b)   At the equivalence point, all the OAc⁻ is converted to HOAc. The pH at the equivalence point can be calculated from the K_a expression.

(c)   The pH values between the starting point and the equivalence point can be calculated from the Henderson-Hasselbalch equation.

(d)   Beyond the equivalence point the pH may be calculated from $pH = -\log [H^+]_{excess}$.

Problem II-35

Calculate the appropriate values and plot the curve for the titration of 100 ml of 0.02 M H_3PO_4 with KOH. For simplicity, assume that the KOH solution is very concentrated so that the volume remains constant during the titration. Plot pH versus moles OH⁻ added.

Solution (see Figure II-2.)

The calculations are made in the same way as those described previously. The starting pH is calculated from the K_{a_1} expression. The pH between equivalence points is calculated from the Henderson-Hasselbalch equation by using the appropriate pK_a value. For example, between H_3PO_4 and $H_2PO_4^-$, pK_{a_1} is used; between $H_2PO_4^-$ and $HPO_4^=$, pK_{a_2} is used; between $HPO_4^=$ and $PO_4^\equiv$, pK_{a_3} is used. The pH at the first equivalence point is calculated from the pK_{a_1} and pK_{a_2} values (as described in Problem II-24). The pH at the second equivalence point is similarly calculated from the pK_{a_2} and pK_{a_3} values. The pH at the third equivalence point is calculated solely from the $K_{h_3}(K_{b_1})$ expression. ($PO_4^\equiv$ is the final, not an intermediate, ion of the polyprotic acid. It undergoes only hydrolysis, that is, ionization as a base.)

D. BUFFERS

A "buffer" is something that resists change. In common chemical usage, a pH buffer is a substance, or mixture of substances, that permits solutions to resist *large changes* in pH upon the addition of small amounts of H⁺ or OH⁻ ions. To put it another way, a buffer helps maintain a *near constant* pH upon the addition of small amounts of H⁺ or OH⁻ ions to a solution.

Common buffer mixtures contain two substances, a conjugate acid and a conjugate base. An "acidic" buffer contains a weak acid and a salt of the weak acid (conjugate base). A "basic" buffer contains a weak base and a salt of the weak base (conjugate acid). Together the two species (conjugate acid plus conjugate base) resist large changes in pH by *partially* absorbing additions of H⁺ or OH⁻ ions to the system. If H⁺ ions are added to the buffered solution, they react partially with the conjugate base present to form the conjugate acid. Thus, some H⁺ ions are taken out of circulation. If OH⁻ ions are added to the buffered solution, they react partially with the conjugate acid present to form water and the conjugate base. Thus, some OH⁻ ions are taken out of circulation. Buffered solutions *do* change in pH upon the addition of H⁺ or OH⁻ ions. However, the change is much less than that which would occur if no buffer were present. The amount of change depends on the strength of the buffer and the $[A^-]/[HA]$ ratio.

Problem II-36

(a) Describe the components of an "acetate" buffer. (b) Show the reactions by which an acetate buffer resists changes in pH upon the addition of OH⁻ and H⁺ ions.

Solution

(a) An "acetate" buffer contains unionized acetic acid [HOAc] as the conjugate acid and acetate ions [OAc⁻] as the conjugate base. The OAc⁻ may be provided directly by NaOAc, KOAc, etc., or by neutralizing a portion of the HOAc with KOH or NaOH.

(b) In a solution containing a weak acid such as HOAc, a certain condition must be met—namely, the product of [H⁺][OAc⁻] divided by [HOAc] must be constant:

$$K_a = \frac{[H^+][OAc^-]}{[HOAc]}$$

A change in the concentration of any one of the three components of the K_a expression causes the concentrations of the other two to alter appropriately so that [H⁺][OAc⁻] divided by [HOAc] is still the same constant value (K_a).

For example, if OH⁻ ions are added to the system, they react with the H⁺ ions present to form H_2O.

(1) $$\boxed{OH^- + H^+ \rightarrow H_2O}$$

The reduction in [H⁺] disturbs the equilibrium momentarily. Consequently, more HOAc dissociates to reestablish the equilibrium condition.

(2)
$$\text{HOAc} \rightleftharpoons \text{H}^+ + \text{OAc}^-$$

The net result (as well as the sum of the above two equations) is as if the OH⁻ ions react directly with the conjugate acid of the acetate buffer to yield H_2O plus more conjugate base [OAc⁻].

(3)
$$\text{OH}^- + \text{HOAc} \rightleftharpoons \text{H}_2\text{O} + \text{OAc}^-$$

All of this, of course, happens almost instantaneously.

Similarly, if H⁺ ions are added to the system, the equilibrium again shifts. This time the conjugate base [OAc⁻] reacts with some of the excess H⁺ ions to form unionized HOAc.

(4)
$$\text{H}^+ + \text{OAc}^- \rightleftharpoons \text{HOAc}$$

It should be emphasized that the excess H⁺ or OH⁻ ions are not *completely* neutralized by the buffer; that is, the pH does not remain absolutely constant upon addition of H⁺ or OH⁻ ions to a buffer. The reactions by which H⁺ and OH⁻ ions are absorbed are themselves equilibrium reactions and do not go to completion.

Problem II-37

Show mathematically why an acetate buffer cannot maintain an absolutely constant pH upon the addition of H⁺.

Solution

Suppose we have a buffer containing 0.01 M HA and 0.01 M A⁻. Assume that the K_a of the weak acid is also 10^{-5}. Consequently, the H⁺ ion concentration must also be 10^{-5} M.

$$\frac{[\text{H}^+][\text{A}^-]}{[\text{HA}]} = K_a$$

$$\frac{(10^{-5})(10^{-2})}{(10^{-2})} = 10^{-5}$$

Now suppose 10^{-3} M H$^+$ is added to the buffer. If *all* of the H$^+$ reacts with A$^-$ to yield HA (thus maintaining [H$^+$] at 10^{-5} M), the new concentration of [HA] would be 1.1×10^{-2} M and the new concentration of [A$^-$] would be 0.9×10^{-2} M. Substituting these values into the K_a expression, we can see that the [H$^+$][A$^-$] divided by [HA] is *not* constant and equal to 10^{-5}.

$$\frac{(10^{-5})(0.9 \times 10^{-2})}{(1.1 \times 10^{-2})} \neq 10^{-5}$$

Problem II-38

Consider a 0.002 M acidic buffer containing 10^{-3} M HA and 10^{-3} M A$^-$. The pH $=$ p$K_a = 5$ ($K_a = 10^{-5}$). Suppose that 5×10^{-4} moles of H$^+$ are added to 1 liter of the buffer (assume that the volume remains at 1 liter). (a) Calculate the exact concentrations of A$^-$ and HA and the pH of the solution after addition of the HCl. (b) Calculate the concentrations of A$^-$ and HA and the pH of the solution assuming that the increase in the amount of HA (and the decrease in the amount of A$^-$) is equal to the amount of H$^+$ added.

Solution

The added H$^+$ is partially utilized by reacting with A$^-$ to form unionized HA.

$$H^+ + A^- \rightleftharpoons HA$$

(a)   Let

$$y = M \text{ of H}^+ \text{ utilized by the buffer}$$

$$[A^-] = 10^{-3} - y \ M$$

$$[HA] = 10^{-3} + y \ M$$

$$[H^+] = 10^{-5} + 5 \times 10^{-4} - y = 51 \times 10^{-5} - y$$

$$K_a = \frac{[H^+][A^-]}{[HA]} = \frac{(51 \times 10^{-5} - y)(10^{-3} - y)}{(10^{-3} + y)} = 10^{-5}$$

Cross-multiplying:

$$10^{-8} + 10^{-5}y = 51 \times 10^{-8} - 51 \times 10^{-5}y - 10^{-3}y + y^2$$

Rearranging and collecting terms:

$$y^2 - 10^{-3}y - 51 \times 10^{-5}y - 10^{-5}y + 51 \times 10^{-8} - 10^{-8} = 0$$

$$y^2 - 100 \times 10^{-5}y - 51 \times 10^{-5}y - 1 \times 10^{-5}y + 50 \times 10^{-8} = 0$$

$$y^2 - 152 \times 10^{-5}y + 50 \times 10^{-8} = 0$$

$$y = \frac{-b \pm \sqrt{b^2 - 4ac}}{2a}$$

where

$$a = 1$$

$$b = -152 \times 10^{-5}$$

$$c = 50 \times 10^{-8}$$

$$y = \frac{+152 \times 10^{-5} \pm \sqrt{(-15.2 \times 10^{-4})^2 - 4(50 \times 10^{-8})}}{2}$$

$$y = \frac{15.2 \times 10^{-4} \pm \sqrt{231 \times 10^{-8} - 200 \times 10^{-8}}}{2}$$

$$y = \frac{15.2 \times 10^{-4} \pm \sqrt{31 \times 10^{-8}}}{2} = \frac{15.20 \times 10^{-4} \pm 5.57 \times 10^{-4}}{2}$$

$$y = \frac{20.77 \times 10^{-4}}{2} \quad \text{and} \quad \frac{9.63 \times 10^{-4}}{2}$$

$$y = 10.38 \times 10^{-4} \quad \text{and} \quad 4.815 \times 10^{-4} \, M$$

The higher value is obviously incorrect because only $5 \times 10^{-4} \, M \, H^+$ was added.

$$\therefore \quad y = 4.815 \times 10^{-4} \, M$$

Thus, of the $5 \times 10^{-4} \, M \, H^+$ originally added, $4.815 \times 10^{-4} \, M$ was utilized by the buffer. The final H^+ ion concentration was increased by $0.185 \times 10^{-4} \, M$.

$$[H^+]_{final} = (1 \times 10^{-5}) + (1.85 \times 10^{-5}) = \boxed{\mathbf{2.85 \times 10^{-5} \, M}}$$

$$[A^-]_{final} = (10 \times 10^{-4}) - (4.82 \times 10^{-4}) = \boxed{\mathbf{5.18 \times 10^{-4} \, M}}$$

$$[HA]_{final} = (10 \times 10^{-4}) + (4.82 \times 10^{-4}) = \boxed{\mathbf{14.82 \times 10^{-4}\ M}}$$

$$pH_{final} = \log \frac{1}{[H^+]_{final}}$$

$$= \log \frac{1}{2.85 \times 10^{-5}}$$

$$= \log 0.351 \times 10^5 = \log 3.51 \times 10^4$$

$$\boxed{\mathbf{pH = 4.546}}$$

In other words, the pH decreased by 0.454 unit.
(b) If we assume that virtually *all* the H^+ reacts with A^- to form HA, in order to simplify the calculations:

$$[A^-] = (10 \times 10^{-4}) - (5 \times 10^{-4}) = \boxed{\mathbf{5.0 \times 10^{-4}\ M}}$$

$$[HA] = (10 \times 10^{-4}) + (5 \times 10^{-4}) = \boxed{\mathbf{15 \times 10^{-4}\ M}}$$

The estimated new $[H^+]$ is that which is in equilibrium with $15 \times 10^{-4}\ M$ HA and $5 \times 10^{-4}\ M\ A^-$. The estimated value will be slightly high because, as shown in part (a), the $[HA]/[A^-]$ ratio is actually a little less than 3/1.

$$K_a = \frac{[H^+][A^-]}{[HA]}$$

Let $y = M$ of H^+ present

$$10^{-5} = \frac{(y)(5 \times 10^{-4})}{(15 \times 10^{-4})}$$

$$15 \times 10^{-9} = (5 \times 10^{-4})y$$

$$y = \frac{15 \times 10^{-9}}{5 \times 10^{-4}}$$

$$y = 3 \times 10^{-5}$$

$$\boxed{\mathbf{H^+ = 3 \times 10^{-5}\ M}}$$

$$pH = pK_a + \log \frac{[A^-]}{[HA]}$$

$$pH = 5.00 + \log \frac{(5 \times 10^{-4})}{(15 \times 10^{-4})}$$

$$pH = 5.00 - \log \frac{(15 \times 10^{-4})}{(5 \times 10^{-4})}$$

$$pH = 5.00 - \log 3$$

$$pH = 5.00 - 0.477$$

$$\boxed{\mathbf{pH = 4.523}}$$

The calculated H^+ ion concentration increase is $2 \times 10^{-5}\ M$ (compared to the true value of $1.85 \times 10^{-5}\ M$).

The calculated pH decrease is 0.477 unit (compared to the true value of 0.454).

We can see that the error introduced by assuming that the buffer reacts *completely* with the added H^+ is small. In the above problem, the buffer is relatively weak and the amount of H^+ ion added is of the same order of magnitude as the original A^- concentration. In practice, the concentration of the buffer employed would be high compared to the expected change in H^+ (or OH^-) ion concentration. Consequently, buffer calculations may be simplified greatly, as shown in part (b) above, without undue error.

Problem II-39

(a) Describe the components of a 0.25 M "ethylamine" buffer. (b) Show how the buffer resists change upon addition of H^+ and OH^- ions.

Solution

(a) An "ethylamine" buffer contains the conjugate acid of ethylamine plus the conjugate base of ethylamine; that is, $R\text{-}NH_3^+$ (conjugate acid) and $R\text{-}NH_2$ (conjugate base). The $R\text{-}NH_3^+$ may be provided directly by ethylamine hydrochloride [$R\text{-}NH_3Cl$] or by neutralizing a portion of the $R\text{-}NH_2$ with, for example, HCl. The *total* ethylamine present (conjugate acid plus conjugate base) equals 0.25 M. For example, if the pH = pK_a, [$R\text{-}NH_2$] and [$R\text{-}NH_3^+$] are equal at 0.125 M each.

(b) The conjugate acid portion reacts with excess OH^- ions.

$$R\text{-}NH_3^+ + OH^- \rightleftharpoons H_2O + R\text{-}NH_2$$

The conjugate base portion reacts with excess H^+ ions.

$$R\text{-}NH_2 + H^+ \rightleftharpoons R\text{-}NH_3^+$$

As in the previous example with the acetate buffer, the partial neutralization of the OH^- ions by the conjugate acid results in the production of the conjugate base; the partial neutralization of the H^+ ions by the conjugate base results in the production of the conjugate acid.

Problem II-40

What are the concentrations of HOAc and OAc^- in a 0.2 M "acetate" buffer, pH 5.00? The K_a for acetic acid is 1.70×10^{-5} ($pK_a = 4.77$).

Solution

A "0.2 M acetate" buffer contains a *total* of 0.2 mole of "acetate" per liter. Some of the total acetate is in the conjugate acid form, HOAc, and some is in the conjugate base form, OAc⁻. The proportions (and hence the concentrations) of each form may be solved by using either the K_a expression or the Henderson-Hasselbalch equation.

$$K_a = \frac{[H^+][OAc^-]}{[HOAc]}$$

$$pH = pK_a + \log\frac{[OAc^-]}{[HOAc]}$$

Let

$$y = M \text{ of } OAc^-$$

$$\therefore \quad 0.2 - y = M \text{ of HOAc}$$

$$pH = 5$$

$$\therefore \quad [H^+] = 10^{-5}$$

$$1.7 \times 10^{-5} = \frac{(10^{-5})(y)}{(0.2 - y)}$$

$$3.4 \times 10^{-6} - 1.7 \times 10^{-5}y$$

$$= 1 \times 10^{-5}y$$

$$3.4 \times 10^{-6} = 2.7 \times 10^{-5}y$$

$$y = \frac{3.4 \times 10^{-6}}{27 \times 10^{-6}}$$

$$y = 0.126$$

$$\boxed{[OAc^-] = 0.126 \ M}$$

$$[HOAc] = 0.200 - 0.126$$

$$\boxed{[HOAc] = 0.074 \ M}$$

Let

$$y = M \text{ of } OAc^-$$

$$\therefore \quad 0.2 - y = M \text{ of HOAc}$$

$$5.00 = 4.77 + \log\frac{y}{0.2 - y}$$

$$0.23 = \log\frac{y}{0.2 - y}$$

$$\frac{y}{0.2 - y} = \text{antilog of } 0.23$$

$$\frac{y}{0.2 - y} = 1.70$$

$$0.34 - 1.70y = y$$

$$0.34 = 2.7y$$

$$y = 0.126$$

$$\boxed{[OAc^-] = 0.126 \ M}$$

$$[HOAc] = 0.2 - y \ M$$

$$[HOAc] = 0.2 - 0.126$$

$$\boxed{[HOAc] = 0.074 \ M}$$

Or

$$pH = pK_a + \log \frac{[OAc^-]}{[HOAc]}$$

$$5 = 4.77 + \log \frac{[OAc^-]}{[HOAc]}$$

$$0.23 = \log \frac{[OAc^-]}{[HOAc]}$$

$$\frac{[OAc^-]}{[HOAc]} = \text{antilog of } 0.23$$

$$\frac{[OAc^-]}{[HOAc]} = 1.70$$

$$\frac{[OAc^-]}{[HOAc]} = \frac{1.70}{1}$$

$$\therefore \quad \frac{1.70}{2.70} \text{ of total} = OAc^-$$

$$\frac{1.00}{2.70} \text{ of total} = HOAc$$

$$\frac{1.70}{2.7} \times 0.2\ M = 0.126$$

$$\boxed{[OAc^-] = 0.126\ M}$$

$$\frac{1.00}{2.70} \times 0.2\ M = 0.074$$

$$\boxed{[HOAc] = 0.074\ M}$$

Check: The pH is higher than the pK_a. $\therefore$ The solution should contain more conjugate base than conjugate acid. Conjugate base = 0.126 M. Conjugate acid = 0.074 M.

Problem II-41

What is the pH of a solution that contains 0.15 M potassium acetate and 0.30 M acetic acid?

Solution

KOAc is a salt. It is completely dissociated in an aqueous solution.

	KOAc	$\rightarrow$	K$^+$	+	OAc$^-$
Start:	0.15 M		0		0
Change:	$-0.15\ M$		$+0.15\ M$		$+0.15\ M$
Final:	0		0.15 M		0.15 M

Acetic acid is only partially dissociated.

	HOAc	$\leftrightharpoons$	H$^+$	+	OAc$^-$
Start:	0.30 M		0		0
Change:	$-y\ M$		$+y\ M$		$+y\ M$
Equilibrium:	$0.30 - y\ M$		$y\ M$		$y\ M$

The OAc$^-$ also undergoes hydrolysis.

	OAc$^-$ + HOH	$\leftrightharpoons$	HOAc	+	OH$^-$
Start:	0.15 M		0		0
Change:	$-z\ M$		$+z\ M$		$+z\ M$
Equilibrium:	$0.15 - z\ M$		$z\ M$		$z\ M$

The OH$^-$ produced tends to decrease the amount of H$^+$ in solution.

$$[\text{OAc}^-] = 0.15 + y - z\ M$$
$$[\text{HOAc}] = 0.30 - y + z\ M$$
$$[\text{H}^+] = y - z\ M$$

$$K_a = \frac{[\text{H}^+][\text{OAc}^-]}{[\text{HOAc}]} \qquad \text{pH} = \text{p}K_a + \log\frac{[\text{OAc}^-]}{[\text{HOAc}]}$$

$$1.70 \times 10^{-5} = \frac{(y - z)(0.15 + y - z)}{(0.30 - y + z)} \qquad \text{pH} = 4.77 + \log\frac{(0.15 + y - z)}{(0.30 - y + z)}$$

The amount of OAc$^-$ ions contributed by the HOAc ($y\ M$) is very small compared to the amount contributed by the potassium acetate (0.15 M). Also, the amount of HOAc that dissociates ($y\ M$) is small compared to the original HOAc concentration (0.30 M). Similarly, we can show that the extent of hydrolysis of the OAc$^-$ in the presence of 0.3 M HOAc is extremely small. Therefore, the K_a and pK_a expressions may be simplified.

$$1.70 \times 10^{-5} = \frac{(y)(0.15)}{(0.30)}$$

$$0.51 \times 10^{-5} = 0.15y$$

$$y = 3.4 \times 10^{-5} \, M$$

$$\boxed{[H^+] = 3.4 \times 10^{-5} \, M}$$

$$pH = \log \frac{1}{3.4 \times 10^{-5}}$$

$$= \log 0.294 \times 10^5$$

$$= \log 2.94 \times 10^4$$

$$= 0.47 + 4$$

$$\boxed{pH = 4.47}$$

$$pH = 4.77 + \log \frac{(0.15)}{(0.30)}$$

$$pH = 4.77 - \log \frac{(0.30)}{(0.15)}$$

$$= 4.77 - \log 2$$

$$= 4.77 - 0.30$$

$$\boxed{pH = 4.47}$$

Notice that the expression $+\log (0.15)/(0.30)$ was changed to $-\log (0.30)/(0.15)$ in order to simplify the determination of the logarithm.

Check: The solution contains more conjugate acid than conjugate base. ∴ The pH should be lower than the pK_a. The $pK_a = 4.77$; pH = 4.47.

Problem II-42

What is the pH of a solution that contains 0.1 M NH$_4$Cl and 0.2 M NH$_3$?

Solution

NH$_4$Cl is a salt. It is completely dissociated in an aqueous solution.

$$\text{NH}_4\text{Cl} \rightarrow \text{NH}_4^+ + \text{Cl}^-$$

	NH$_4$Cl	NH$_4^+$	Cl$^-$
Start:	0.10 M	0	0
Change:	-0.10 M	$+0.10$ M	$+0.10$ M
Final:	0	0.10 M	0.10 M

NH$_3$ is a weak base. It partially ionizes in aqueous solution.

$$\text{NH}_3 + \text{HOH} \rightleftharpoons \text{NH}_4^+ + \text{OH}^-$$

$$K_{eq} = K_b = 1.8 \times 10^{-5}$$

	NH$_3$	NH$_4^+$	OH$^-$
Start:	0.20 M	0	0
Change:	$-y$ M	$+y$ M	$+y$ M
Equilibrium:	$0.20 - y$ M	y M	y M

The NH_4^+ undergoes hydrolysis (i.e., ionizes as an acid).

$$NH_4^+ \ \leftrightarrows \ NH_3 + H^+$$
$$K_{eq} = K_a = 5.55 \times 10^{-10}$$

Start:	0.10 M	0	0
Change:	$-z\,M$	$+z\,M$	$+z\,M$
Equilibrium:	0.10 $- z\,M$	$z\,M$	$z\,M$

The H^+ produced tends to decrease the amount of OH^- in solution. The final concentrations of the buffer components are:

$$[NH_3] = 0.20 - y + z M$$
$$[NH_4^+] = 0.10 + y - z M$$
$$[OH^-] = y - z M$$

Because NH_3 is a weak base, the amount of NH_4^+ formed by ionization ($y\,M$) is small compared to the amount contributed by the NH_4Cl (0.10 M). In fact, the high concentration of NH_4^+ tends to suppress the ionization of NH_3 so that the amount of NH_4^+ formed by ionization of the NH_3 is even smaller than that which would be there if no NH_4Cl were present. The amount of NH_3 that ionizes ($y\,M$) is also small compared to the amount of NH_3 originally present (0.2 M). Thus the y term may be dropped from the $[NH_4^+]$ and $[NH_3]$ expressions.

Similarly, because NH_4^+ is a weak acid, the amount of NH_3 formed by dissociation (hydrolysis) ($z\,M$) is small compared to the amount of NH_3 already present (and even less than the amount that would form if no NH_3 were present). Also z is small compared to the amount of NH_4^+ present. The relative values of K_b and K_a show us that the OH^--producing reaction proceeds to a much greater extent than the H^+-producing reaction. Therefore, z is small compared to y and may be neglected in all terms.

Thus the problem may be solved assuming that:

$$[NH_3] = [NH_3] \text{ originally added} = 0.20 \ M$$
$$[NH_4^+] = [NH_4^+] \text{ originally added} = 0.10 \ M$$
$$[OH^-] = y$$

$$pH = pK_a + \log \frac{\text{base}}{\text{acid}}$$

$$pH = 9.26 + \log \frac{0.2}{0.1}$$

$$pH = 9.26 + \log 2$$

$$pH = 9.26 + 0.3$$

$$\boxed{\mathbf{pH = 9.56}}$$

General Rule

As a general rule for solving buffer problems, we can assume that the concentrations of conjugate acid and conjugate base present in solution are the same as the concentrations of each originally added to the solution (or produced by partial titration of one or the other). This general rule does not hold when the buffer is extremely dilute (when the concentrations of buffer components are in the region of the K_a value). In such dilute buffers, the changes (y and z) are large compared to the original concentrations.

Problem II-43

An enzyme-catalyzed reaction was carried out in a 0.2 M "tris" buffer, pH 7.8. As a result of the reaction, 0.03 mole/liter of H^+ was produced. (a) What was the ratio of $tris^+$ (conjugate acid)/$tris^0$ (conjugate base) at the start of the reaction? (b) What are the concentrations of $tris^+$ and $tris^0$ at the start of the reaction? (c) Show the reaction by which the buffer maintained a near constant pH. (d) What were the concentrations of $tris^0$ and $tris^+$ at the end of the reaction? (e) What was the pH at the end of the reaction?

Solution

(a)
$$pH = pK_a + \log \frac{tris^0}{tris^+}$$

$$7.8 = 8.1 + \log \frac{tris^0}{tris^+}$$

$$-0.3 = \log \frac{tris^0}{tris^+}$$

$$+0.3 = \log \frac{tris^+}{tris^0}$$

$$\frac{tris^+}{tris^0} = \text{antilog of } 0.3 = 2$$

$$\boxed{\frac{tris^+}{tris^0} = \frac{2}{1}}$$

(b)

$\frac{2}{3} \times 0.2\ M =$ $\boxed{\textbf{0.133 }\textit{M}\textbf{ tris}^+}$ and $\frac{1}{3} \times 0.2\ M =$ $\boxed{\textbf{0.067 }\textit{M}\textbf{ tris}^0}$

Check: The pH is less than the pK_a; $\therefore$ [conjugate acid] $>$ [conjugate base]; 0.133 $M > 0.067$ M.

(c) The conjugate base reacts with the excess H^+.

$$\text{tris}^0 + H^+ \longrightarrow \text{tris}^+$$

(d) As a result of the reaction, the amounts of tris^+ and tris^0 change as shown below.

$$[\text{tris}^+] = 0.133 + 0.030 = \boxed{0.163 \ M}$$

$$[\text{tris}^0] = 0.067 - 0.030 = \boxed{0.037 \ M}$$

(e)
$$pH = pK_a + \log \frac{\text{tris}^0}{\text{tris}^+}$$

$$pH = 8.1 + \log \frac{0.037}{0.163}$$

$$pH = 8.1 - \log \frac{0.163}{0.037}$$

$$pH = 8.1 - \log 4.4$$

$$pH = 8.1 - 0.644$$

$$\boxed{pH = 7.456}$$

Problem II-44

According to the Henderson-Hasselbalch equation, the pH of a buffer depends only on the *ratio* of conjugate base activity to conjugate acid activity. Explain then why the pH of a buffer changes when it is diluted.

Solution

The pH of a buffer changes with dilution for several reasons:
1. *Changes in Activity Coefficients*—The activity coefficients of different ions are not the same at any given concentration and do not change in an identical manner with a given change in concentration. For example, we can see from Table AI-5 that $\gamma_{\text{citrate}^{-3}}$ is 0.18 in a 0.1 M solution and

0.796 in a 0.001 M solution. The activity coefficient of its conjugate acid ($\gamma_{\text{Hcitrate}^{-2}}$) is 0.45 in a 0.1 M solution and 0.903 in a 0.001 M solution. In general, dilution results in an increase in γ; γ approaches unity at infinite dilution. The greater the charge on the ion, the greater is the change in its activity coefficient for a given change in concentration. Consider a "0.2 M citrate buffer" containing equal molar amounts of citrate^{-3} and Hcitrate^{-2} (0.1 M of each ionic species). The exact pH of the solution can be calculated taking into account the activity coefficients of the two ions. We can also calculate the pH of the solution after it is diluted 100-fold, taking into account the change in activity coefficients.

$$\text{pH} = \text{p}K_{a_3} + \log \frac{a_{\text{citrate}^{-2}}}{a_{\text{Hcitrate}^{-3}}}$$

$$\text{pH} = \text{p}K_{a_3} + \log \frac{[\gamma_{\text{citrate}^{-3}}][\text{citrate}^{-3}]}{[\gamma_{\text{Hcitrate}^{-2}}][\text{Hcitrate}^{-2}]}$$

0.2 M Buffer	**0.002 M Buffer**
$\text{pH} = 5.40 + \log \dfrac{(0.18)(0.1)}{(0.45)(0.1)}$	$\text{pH} = 5.40 + \log \dfrac{(0.796)(0.001)}{(0.903)(0.001)}$
$= 5.40 - \log \dfrac{(0.45)}{(0.18)}$	$= 5.40 - \log \dfrac{(0.903)}{(0.796)}$
$= 5.40 - \log 2.50$	$= 5.40 - \log 1.13$
$= 5.40 - 0.40$	$= 5.40 - 0.05$
$\boxed{\text{pH} = 5.00}$	$\boxed{\text{pH} = 5.35}$

In general, the log $a_{\text{A}^-}/a_{\text{HA}}$ term of "acidic" buffers increases upon dilution, resulting in an increase in pH. In "basic" buffers, the log $a_{\text{R-NH}_2}/a_{\text{R-NH}_3^+}$ term decreases upon dilution, resulting in a decrease in pH.

2.   *Changes in the Degree of Dissociation of HA*—As shown in Problems II-15 and II-16, the degree of dissociation of a weak acid increases as the solution is diluted. In a solution of a weak acid alone (no added conjugate base), the acid is 10% dissociated when $[\text{HA}]_{\text{orig}} = 100\ K_a$ and 50% dissociated when $[\text{HA}]_{\text{orig}} = 2\ K_a$. Thus, the log A$^-$/HA term increases as the solution is diluted. In a buffer solution (weak acid plus added conjugate base), the A$^-$ tends to suppress the dissociation of HA. Consequently, in a buffer solution containing $[\text{HA}] = 2K_a$, the HA is somewhat less than 50% dissociated.

For example, consider a 0.02 M succinate buffer, prepared by dissolving 0.01 mole of succinic acid and 0.01 mole of monosodium succinate in sufficient water to make 1 liter final volume. The monosodium succinate ionizes completely and the succinic acid ionizes partially. Let

$$y = M \text{ of succinic acid that dissociates}$$
$$\therefore \quad y = M \text{ of } H^+ \text{ produced upon dissociation}$$

and

$$y = M \text{ of } HA^- \text{ produced upon dissociation of the acid}$$
$$\therefore \quad [HA^-] = 0.01 + y$$
$$[H_2A] = 0.01 - y$$
$$[H^+] = y$$
$$K_a = \frac{[H^+][HA^-]}{[H_2A]} = \frac{(y)(0.01 + y)}{(0.01 - y)} = 6.4 \times 10^{-5}$$

Because the concentrations of H_2A and HA^- are more than 100 times the K_{a_1} value, y is small compared to 0.01 and may be neglected.

$$\therefore \quad [H^+] = K_{a_1} \qquad pH = pK_{a_1} = 4.19$$

Now let us dilute the above buffer 100 times.

$$[HA^-] = 10^{-4} + y$$
$$[H_2A] = 10^{-4} - y$$
$$[H^+] = y$$

Now the concentration of buffer components is of the same order of magnitude as the K_a value; that is, the H_2A is more than 10% dissociated and y is not small compared to 10^{-4}. Consequently, we must solve for y exactly.

$$K_a = \frac{(y)(10^{-4} + y)}{(10^{-4} - y)} = 6.4 \times 10^{-5}$$

$$6.4 \times 10^{-9} - 6.4 \times 10^{-5}y = 10^{-4}y + y^2$$
$$y^2 + 10 \times 10^{-5}y + 6.4 \times 10^{-5}y - 6.4 \times 10^{-9} = 0$$
$$y^2 + 16.4 \times 10^{-5}y - 6.4 \times 10^{-9} = 0$$

$$y = \frac{-b \pm \sqrt{b^2 - 4ac}}{2a}$$

where
$$a = 1$$
$$b = 16.4 \times 10^{-5}$$
$$c = 6.4 \times 10^{-9}$$

$$y = \frac{-16.4 \times 10^{-5} \pm \sqrt{(16.4 \times 10^{-5})^2 - 4(-6.4 \times 10^{-9})}}{2}$$

$$y = \frac{-16.4 \times 10^{-5} \pm \sqrt{269 \times 10^{-10} + 246 \times 10^{-10}}}{2}$$

$$y = \frac{-16.4 \times 10^{-5} \pm \sqrt{515 \times 10^{-10}}}{2}$$

$$y = \frac{-16.4 \times 10^{-5} \pm 22.7 \times 10^{-5}}{2} = \frac{6.3 \times 10^{-5}}{2}$$

$$y = 3.15 \times 10^{-5}$$

$$[H^+] = 3.15 \times 10^{-5}\,M \qquad \therefore \quad pH = 4.5$$
$$[H_2A] = (10 \times 10^{-5}) - (3.15 \times 10^{-5}) = 6.85 \times 10^{-5}\,M$$
$$[HA^-] = (10 \times 10^{-5}) + (3.15 \times 10^{-5}) = 13.15 \times 10^{-5}\,M$$

Note that the $[HA^-]/[H_2A]$ ratio that is essentially 1 in the $10^{-2}\,M$ buffer changes to about 2 when the buffer is diluted 100-fold.

3. Finally, as the buffer is diluted extensively, its contribution toward the H^+ or OH^- ion concentration of the solution approaches that of water and the pH approaches 7.

Problem II-45

Suppose that you prepare a buffer by dissolving 0.10 mole per liter of a weak acid, HA, and 0.10 mole per liter of its sodium salt, A^-. Assume that $pK_a = 3$. (a) What is the pH of the buffer? (b) How much does the buffer have to be diluted for the pH to increase by 1 unit? Neglect changes in activity coefficients.

Solution

(a) Let
$$y = M \text{ of HA that dissociates}$$
$$\therefore \quad [HA] = 0.10 - y$$
$$[A^-] = 0.10 + y$$
$$[H^+] = y$$

$$K_a = \frac{[H^+][A^-]}{[HA]} = \frac{(y)(0.10 + y)}{(0.10 - y)} = 10^{-3}$$

Because the concentrations of HA and A^- are very much larger than

K_a, y is small and may be neglected in the HA and A$^-$ terms.

$$K_a = \frac{(y)(0.10)}{(0.10)} = 10^{-3}$$

$$y = 10^{-3} = [H^+]$$

$$\boxed{\text{pH} = 3}$$

(b)   If the pH increases by 1 unit upon dilution (to pH 4.0), then $[H^+] = 10^{-4} M = y$. Calculate the new "original" concentrations of HA and A$^-$ that we must start with (or dilute the original buffer to) so that when the addition and subtraction of y are made (when the HA dissociates) the ratio of A$^-$/HA is 10/1.

Let C = new "original" M of HA and A$^-$.

$$K_a = \frac{[H^+][A^-]}{[HA]}$$

$$10^{-3} = \frac{(10^{-4})(C + 10^{-4})}{(C - 10^{-4})}$$

$$10^{-3} C - 10^{-7} = 10^{-4} C + 10^{-8}$$

$$0.9 \times 10^{-3} C = 1.1 \times 10^{-7}$$

$$C = \frac{1.1 \times 10^{-7}}{0.9 \times 10^{-3}} = 1.22 \times 10^{-4}$$

In other words, if we start by dissolving 1.22×10^{-4} moles per liter each of HA and A$^-$, $1 \times 10^{-4} M$ HA dissociates producing $2.22 \times 10^{-4} M$ A$^-$ and leaving $0.22 \times 10^{-4} M$ HA.

Check:

$$10^{-3} = \frac{(10^{-4})(2.22 \times 10^{-4})}{(0.22 \times 10^{-4})} = (10^{-4})(10)$$

We will obtain exactly the same result by diluting the 0.2 M buffer the above new concentrations.

$$\left. \begin{array}{c} 0.10\ M\ \text{HA} \\ + \\ 0.10\ M\ \text{A}^- \end{array} \right\} \xrightarrow{\text{dilution...}} \begin{array}{c} 1.22 \times 10^{-4}\ M\ \text{HA} \\ + \\ 1.22 \times 10^{-4}\ M\ \text{A}^- \end{array}$$

...results in further dissociation of HA

$$\begin{array}{c} 0.22 \times 10^{-4}\ M\ \text{HA} \\ 2.22 \times 10^{-4}\ M\ \text{A}^- \end{array}$$

$$\boxed{\begin{array}{l} \therefore \quad \textbf{The original buffer must be} \\ \textbf{diluted 820-fold.} \\ \frac{1}{820} \times 0.10\ M = 1.22 \times 10^{-4}\ M \end{array}}$$

Problem II-46

An enzyme-catalyzed reaction was carried out in a solution containing 0.15 M phosphate buffer, pH 7.2. As a result of the reaction, 0.02 mole/liter of H^+ was utilized. (a) What were the major ionic species of phosphate present in the buffer? (b) Write the reaction showing how the buffer maintained a near constant pH during the reaction. (c) What were the concentrations of conjugate base and conjugate acid at the start of the reaction? (d) What were the concentrations of conjugate acid and conjugate base at the end of the reaction? (e) What was the final pH? (f) What would the final pH have been if no buffer were present?

Solution

(a) The pH of the buffer equals pK_{a_2} for phosphoric acid. In the neighborhood of pK_{a_2}, the major species present are $H_2PO_4^-$ (conjugate acid) and $HPO_4^=$ (conjugate base).

(b) The utilization of H^+ ions upsets the equilibrium described by K_{a_2}.

$$H_2PO_4^- \rightleftharpoons HPO_4^= + H^+$$

As a result, more $H_2PO_4^-$ ionizes to reestablish the equilibrium. The net effect is that a large portion of the H^+ utilized is replaced; some $H_2PO_4^-$ is converted to $HPO_4^=$.

(c) At the start:

$$pH = pK_{a_2} + \log \frac{[HPO_4^=]}{[H_2PO_4^-]}$$

$$pH = pK_{a_2}$$

$$\therefore \quad \log \frac{[HPO_4^=]}{[H_2PO_4^-]} = 0$$

$$\therefore \quad \frac{[HPO_4^=]}{[H_2PO_4^-]} = 1$$

or

$$[HPO_4^=] = [H_2PO_4^-]$$

Thus, in the 0.15 M buffer,

$$[HPO_4^=] = 0.075\ M \quad \text{and} \quad [H_2PO_4^-] = 0.075\ M$$

(d) As a result of the reaction, the concentrations of $HPO_4^=$ and $H_2PO_4^-$ change.

$$[H_2PO_4^-] = 0.075 - 0.020 = 0.055\ M$$
$$[HPO_4^=] = 0.075 + 0.020 = 0.095\ M$$

(e)   At the end of the reaction:

$$pH = pK_{a_2} + \log \frac{[HPO_4^=]}{[H_2PO_4^-]}$$

$$pH = 7.2 + \log \frac{(0.095)}{(0.055)}$$

$$pH = 7.2 + \log 1.73$$

$$pH = 7.2 + 0.238$$

$$\boxed{pH = 7.438}$$

(f)   If no buffer were present, the utilization of 0.02 mole/liter of H^+ would cause more water to dissociate to reestablish the $H_2O \rightleftharpoons H^+ + OH^-$ equilibrium. The net effect is the same as if 0.02 mole/liter of OH^- were added.

$$pOH = \log \frac{1}{[OH^-]} = \log \frac{1}{0.02} = \log 50$$

$$pOH = 1.7$$

$$\therefore \quad pH = 14 - 1.7$$

$$\boxed{pH = 12.3}$$

Problem II-47

Describe the preparation of 3 liters of a 0.2 M acetate buffer, pH 5.00, starting from solid sodium acetate trihydrate (MW 136) and a 1 M solution of acetic acid.

Solution

Although the wording of this problem is different, the solution is essentially identical to that of Problem II-40.

First calculate the molarities of OAc^- and HOAc present. Any of the three methods shown in Problem II-40 may be used to obtain $[OAc^-] = 0.126$ M and $[HOAc] = 0.074$ M. We need 3 liters of the 0.2 M buffer.

3 liters $\times$ 0.2 M = 0.6 mole *total* (HOAc plus OAc^-)

The total of 0.6 mole is obtained from two sources:

3 liters $\times$ 0.126 M = 0.378 mole OAc^-

3 liters $\times$ 0.074 M = 0.222 mole HOAc

The 0.378 mole of OAc⁻ comes from solid NaOAc.

$$\text{number of moles} = \frac{\text{wt}_g}{\text{MW}}$$

$$0.378 = \frac{\text{wt}_g}{136}$$

$$\boxed{\text{wt}_g = 51.4 \text{ g}}$$

The 0.222 mole of HOAc comes from a 1 M stock solution.

$$\text{number of moles} = \text{liters} \times M$$

$$0.222 = \text{liters} \times 1$$

$$\text{liters} = 0.222 \quad \text{or} \quad \boxed{222 \text{ ml}}$$

Therefore, to prepare 3 liters of the buffer, dissolve 51.4 g of the sodium acetate in some water, add 222 ml of the 1 M acetic acid, and then dilute to a total final volume of 3.0 liters.

Problem II-48

Describe the preparation of 5 liters of a 0.3 M acetate buffer, pH 4.47, starting from a 2 M solution of acetic acid and a 2.5 M solution of KOH.

Solution

As in the previous problem, you must first calculate the proportions of the two acetate species present.

$$\text{pH} = \text{p}K_a + \log \frac{[\text{OAc}^-]}{[\text{HOAc}]}$$

$$4.47 = 4.77 + \log \frac{[\text{OAc}^-]}{[\text{HOAc}]}$$

$$-0.30 = \log \frac{[\text{OAc}^-]}{[\text{HOAc}]}$$

$$+0.30 = \log \frac{[\text{HOAc}]}{[\text{OAc}^-]}$$

$$\frac{[\text{HOAc}]}{[\text{OAc}^-]} = \text{antilog of } 0.3 = 2 = \frac{2}{1} \text{ ratio}$$

$\therefore$ $\frac{2}{3}$ of the total acetate is present as HOAc and $\frac{1}{3}$ of the total acetate is present as OAc⁻. The final solution contains:

$$\tfrac{2}{3} \times 0.3\ M = 0.2\ M\ \text{HOAc (1 mole in 5 liters)}$$

$$\tfrac{1}{3} \times 0.3\ M = 0.1\ M\ \text{OAc}^- \text{ (0.5 mole in 5 liters)}$$

In this buffer, *all* of the acetate must be provided by the HOAc. The buffer is prepared by converting the *proper proportion* of the HOAc to OAc⁻ by adding KOH. We need 5 liters $\times$ 0.3 M = 1.5 moles *total* acetate. Calculate how much stock 2 M HOAc is needed to obtain 1.5 moles.

$$\text{liters} \times M = \text{number of moles}$$

$$\text{liters} \times 2 = 1.5$$

$$\text{liters} = \frac{1.5}{2} = 0.75$$

$\therefore$ 750 ml of the 2 M HOAc is required.

Next, convert $\frac{1}{3}$ of the 1.5 moles to OAc⁻ by adding the proper amount of 2.5 M KOH.

$$\tfrac{1}{3} \times 1.5\ \text{moles} = 0.5\ \text{mole KOH needed}$$

$$\text{liters} \times M = \text{number of moles}$$

$$\text{liters} \times 2.5 = 0.5\ \text{mole}$$

$$\text{liters} = \frac{0.5}{2.5} = 0.2\ \text{liter}$$

$\therefore$ Add 200 ml of 2.5 M KOH.

The solution now contains 1 mole of HOAc and 0.5 mole of OAc⁻. Finally, add sufficient water to bring the volume up to 5 liters. The final solution contains 0.2 M HOAc and 0.1 M OAc⁻.

Problem II-49

Describe the preparation of 2.5 liters of a 0.3 M ammonia buffer, pH 9.00, starting from 2 M NH$_3$ and 1.5 M HCl (pK_a of NH$_4^+$ = 9.26).

Solution

An "ammonia" buffer contains NH$_3$ (conjugate base) and NH$_4^+$ (conjugate acid). First calculate the amounts of the two species in the final solution.

$$pH = pK_a + \log \frac{[NH_3]}{[NH_4^+]}$$

$$9.00 = 9.26 + \log \frac{[NH_3]}{[NH_4^+]}$$

$$-0.26 = \log \frac{[NH_3]}{[NH_4^+]}$$

$$+0.26 = \log \frac{[NH_4^+]}{[NH_3]}$$

$$\frac{[NH_4^+]}{[NH_3]} = \text{antilog of } 0.26 = 1.82 = \frac{1.82}{1} \text{ ratio}$$

$\therefore$ Of the total "ammonia," 1.82/2.82 is in the NH_4^+ form and 1/2.82 is in the NH_3 form.

The buffer contains 2.5 liters $\times$ 0.3 M = 0.75 mole of total "ammonia."

$$\therefore \quad \frac{1.82}{2.82} \times 0.75 = 0.484 \text{ mole of } NH_4^+$$

$$\frac{1}{2.82} \times 0.75 = 0.266 \text{ mole of } NH_3$$

(or 0.75 mole total $-$ 0.484 mole NH_4^+ = 0.266 mole NH_3).

Start by taking the proper amount of stock NH_3.

$$\text{liters} \times M = \text{number of moles}$$

$$\text{liters} \times 2 = 0.75 \text{ mole}$$

$$\text{liters} = \frac{0.75}{2} = 0.375 \text{ liter}$$

$\therefore$ Take 375 ml of the 2 M NH_3.

Next, convert the proper amount of the NH_3 to NH_4^+ by adding HCl.

$$NH_3 + H^+ \rightarrow NH_4^+$$

To form 0.484 mole of NH_4^+, 0.484 mole of H^+ must be added.

$$\text{liters} \times M = \text{number of moles}$$

$$\text{liters} \times 1.5 = 0.484 \text{ mole}$$

$$\text{liters} = \frac{0.484}{1.5} = 0.323 \text{ liter}$$

$\therefore$ Add 323 ml of the 1.5 M HCl.

The solution now contains the proper *proportion* of NH_3 and NH_4^+. Add sufficient water to bring the final volume to 2.5 liters to obtain the proper concentration.

The *concentrations* of the two species in the final solution can be calculated from the known ratio.

$$\frac{1.82}{2.82} \times 0.3 \ M = 0.1935 \ M \ NH_4^+$$

$$\frac{1}{2.82} \times 0.3 \ M = 0.1065 \ M \ NH_3$$

(or $0.300 \ M$ total $- 0.1935 \ M \ NH_4^+ = 0.1065 \ M \ NH_3$).

The concentrations of the two species can also be calculated from the known amount of each present:

$$\frac{0.484 \ \text{mole} \ NH_4^+}{2.5 \ \text{liters}} = 0.1935 \ M \ NH_4^+$$

$$\frac{0.266 \ \text{mole} \ NH_3}{2.5 \ \text{liters}} = 0.1065 \ M \ NH_3$$

Problem II-50

Describe the preparation of 1 liter of $0.45 \ M$ potassium phosphate buffer, pH 7.5.

Solution

The pH of this buffer is a little above the pK_{a_2} of H_3PO_4 as shown in the titration curve in Figure II-2. Consequently, the two major ionic species present are $H_2PO_4^-$ (conjugate acid) and $HPO_4^=$ (conjugate base) with the $HPO_4^=$ predominating.

The buffer can be prepared in any one of several ways: (1) by mixing KH_2PO_4 and K_2HPO_4 in the proper proportions, (2) by starting with H_3PO_4 and converting it to KH_2PO_4 plus K_2HPO_4 by adding the proper amount of KOH, (3) by starting with KH_2PO_4 and converting a portion of it to K_2HPO_4 by adding KOH, (4) by starting with K_2HPO_4 and converting a portion of it to KH_2PO_4 by adding a strong acid such as HCl, (5) by starting with K_3PO_4 and converting it to KH_2PO_4 plus K_2HPO_4 by adding HCl, and (6) by mixing K_3PO_4 and KH_2PO_4 in the proper proportions. Regardless of which method is used, the first step involves calculating the proportion and amounts of the two ionic species in the buffer.

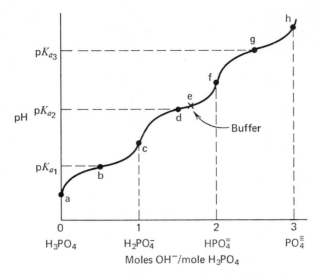

Figure II-2   Titration curve of phosphoric acid

The buffer contains a total of 1 liter $\times$ 0.45 M = 0.45 mole of phosphate.

$$pH = pK_{a_2} + \log \frac{[HPO_4^=]}{[H_2PO_4^-]}$$

$$7.5 = 7.2 + \log \frac{[HPO_4^=]}{[H_2PO_4^-]}$$

$$0.3 = \log \frac{[HPO_4^=]}{[H_2PO_4^-]}$$

$$\frac{[HPO_4^=]}{[H_2PO_4^-]} = \text{antilog of } 0.3 = 2 = \frac{2}{1} \text{ ratio.}$$

$\therefore$ Therefore $\frac{2}{3} \times 0.45$ mole = 0.30 mole of $HPO_4^=$ is needed and $\frac{1}{3} \times 0.45$ mole = 0.15 mole of $H_2PO_4^-$ is needed.

1. *From KH_2PO_4 and K_2HPO_4*—Weigh out 0.30 mole of K_2HPO_4 (52.8 g) and 0.15 mole of KH_2PO_4 (20.4 g) and dissolve in sufficient water to make 1 liter final volume. Or, if stock solutions of the two phosphates are available, measure out the appropriate volumes of each and dilute to 1 liter.

2.   *From H_3PO_4 and KOH*—Start with 0.45 mole of H_3PO_4 (position a in Figure II-2) and add sufficient KOH to titrate *completely* 1 hydrogen (to position c) and $\frac{2}{3}$ of the second hydrogen (to position e).

$$H_3PO_4 \xrightarrow{\text{OH}^-} H_2PO_4^- \xrightarrow{\text{OH}^-} HPO_4^=$$

For example, suppose we have available only stock concentrated (15 *M*) H_3PO_4 and a standard solution of 1.5 *M* KOH. We need 0.45 mole of H_3PO_4.

$$\text{liters} \times M = \text{number of moles}$$
$$\text{liters} \times 15 = 0.45 \text{ mole}$$
$$\text{liters} = \frac{0.45}{15} = 0.03 \text{ liter}$$

$\therefore$ Take 30 ml of H_3PO_4. Add 0.45 mole of KOH to convert *all* the H_3PO_4 to $H_2PO_4^-$; then add another $\frac{2}{3} \times 0.45 = 0.30$ mole of KOH to convert 0.3 mole of $H_2PO_4^-$ to $HPO_4^=$. In other words, a total of 0.75 mole of KOH is required.

Since we have available 1.5 *M* KOH, we can calculate how much of this solution to add.

$$\text{liters} \times M = \text{number of moles}$$
$$\text{liters} = \frac{\text{number of moles}}{M} = \frac{0.75}{1.5} = 0.500$$

$\therefore$ Add 500 ml of KOH to the 30 ml of concentrated H_3PO_4; then add sufficient water to bring the final volume to 1 liter.

3.   *From KH_2PO_4 and KOH*—We can start with KH_2PO_4 (position c) and add sufficient KOH to convert $\frac{2}{3}$ of the $H_2PO_4^-$ ion to $HPO_4^=$ (position e).

$$H_2PO_4^- \xrightarrow{\text{OH}^-} HPO_4^=$$

For example, suppose we have available only solid KH_2PO_4 and KOH. We need 0.45 mole of KH_2PO_4.

$$\frac{\text{wt}_g}{\text{MW}} = \text{number of moles}$$
$$\text{wt}_g = (0.45)(136) = 61.2 \text{ g}$$

Dissolve the KH_2PO_4 in some water; then add 0.30 mole of KOH (solid or dissolved in some water).

$$\text{wt}_g = (0.30)(56) = 16.8 \text{ g of KOH}$$

Next, add sufficient water to bring the volume to 1 liter.

4. *From K_2HPO_4 and HCl*—The $HPO_4^=$ (position f) may be converted to $H_2PO_4^-$ by adding HCl.

$$HPO_4^= \xrightarrow{\text{H}^+} H_2PO_4^-$$

Because we want to end up with an $HPO_4^=/H_2PO_4^-$ of $\frac{2}{1}$, we want to convert only $\frac{1}{3}$ of the $HPO_4^=$ to $H_2PO_4^-$ (to reach position e). Suppose we have available solid K_2HPO_4 and a 2 M solution of HCl. Weigh out 0.45 moles of K_2HPO_4.

$$\text{wt}_g = \text{number of moles} \times \text{MW}$$
$$\text{wt}_g = (0.45)(174) = 78.4 \text{ g}$$

Dissolve the KH_2PO_4 in some water; then add $\frac{1}{3} \times 0.45 = 0.15$ mole of HCl.

$$\text{liters} \times M = \text{number of moles}$$
$$\text{liters} = \frac{\text{number of moles}}{M} = \frac{0.15}{2} = 0.075 \text{ liter}$$

∴   Add 75 ml of 2 M HCl. Then add sufficient water to bring the volume to 1 liter.

5. *From K_3PO_4 and HCl*—Start with 0.45 mole of K_3PO_4 (position h) and add sufficient HCl to convert all the $PO_4^\equiv$ to $HPO_4^=$ (position f). Then add additional HCl to convert $\frac{1}{3}$ of the $HPO_4^=$ to $H_2PO_4^-$ (position e).

$$PO_4^\equiv \xrightarrow{\text{H}^+} HPO_4^= \xrightarrow{\text{H}^+} H_2PO_4^-$$

We need 0.45 mole of K_3PO_4.

$$\text{wt}_g = \text{number of moles} \times \text{MW}$$
$$\text{wt}_g = (0.45)(212) = 95.3 \text{ g}$$

Dissolve the K_3PO_4 in water. Add 0.45 mole of HCl to convert all the $PO_4^\equiv$ to $HPO_4^=$. Then add another $\frac{1}{3} \times 0.45 = 0.15$ mole of HCl to convert 0.15 mole of $HPO_4^=$ to $H_2PO_4^-$. The final solution then contains 0.15 mole of $H_2PO_4^-$ and 0.30 mole of $HPO_4^=$. Now add sufficient water to make 1 liter.

6. *From KH_2PO_4 and K_3PO_4*—The KH_2PO_4 and K_3PO_4 react to form $KHPO_4$. The $H_2PO_4^-$ acts as an acid and the $PO_4^\equiv$ acts as a base.

$$H_2PO_4^- + PO_4^\equiv \rightleftharpoons 2HPO_4^=$$

The reaction is the reverse of the disproportionation reaction. Note that each mole of $H_2PO_4^-$ and $PO_4^\equiv$ yields 2 moles of $HPO_4^=$. Thus, to produce 0.30 mole of $HPO_4^=$, 0.15 mole of $H_2PO_4^-$ and 0.15 mole of $PO_4^\equiv$ are required. But, in addition to the 0.30 mole of $HPO_4^=$, the final solution contains 0.15 mole of $H_2PO_4^-$. Therefore, dissolve 0.30 mole of KH_2PO_4 and 0.15 mole of K_3PO_4 in water. Of the original 0.30 mole of KH_2PO_4, 0.15 mole reacts with the $PO_4^\equiv$ to produce 0.30 mole of $HPO_4^=$, leaving 0.15 mole as $H_2PO_4^-$. Then add sufficient water to make 1 liter.

Problem II-51

What is the "buffer capacity" of a 0.25 M citrate buffer, pH 3.1? The pK_a values for citric acid are 3.1, 4.75, and 5.4.

Solution

The ability of a buffer to resist changes in pH is referred to as the "buffer capacity." "Buffer capacity" can be defined in two ways: (1) the number of moles per liter of H^+ or OH^- required to cause a given change in pH (e.g., 1 unit), or (2) the pH change that occurs upon addition of a given amount of H^+ or OH^- (e.g., 1 mole/liter). The first definition is better because it can be applied to buffers of any concentration. It should be obvious that a 0.25 M buffer, for example, has a greater buffer capacity than a 0.1 M buffer composed of the same conjugate base and conjugate acid at the same pH; it takes more acid or base to change the pH of the more concentrated buffer compared to the more dilute buffer.

A citrate buffer at pH 3.1 contains unionized citric acid [H_3citrate] as the conjugate acid and H_2citrate^{-1} as the conjugate base. Because pH $=$ pK_{a_1}, a 0.25 M "citrate" buffer contains 0.125 M H_3citrate and 0.125 M H_2citrate^{-1}. Let us calculate how much H^+ must be added to decrease the pH by 1 unit, to pH 2.1.

$$pH = pK_{a_1} + \log \frac{[H_2\text{citrate}^{-1}]}{[H_3\text{citrate}]}$$

$$2.1 = 3.1 + \log \frac{[H_2\text{citrate}^{-1}]}{[H_3\text{citrate}]}$$

$$-1 = \log \frac{[H_2\text{citrate}^{-1}]}{[H_3\text{citrate}]}$$

$$+1 = \log \frac{[H_3\text{citrate}]}{[H_2\text{citrate}^{-1}]}$$

$$\frac{[H_3\text{citrate}]}{[H_2\text{citrate}^{-1}]} = \text{antilog of } 1 = 10 = \tfrac{10}{1} \text{ ratio}$$

∴ When the pH decreases by 1 unit, $\tfrac{10}{11}$ of the total citrate will be H_3citrate and $\tfrac{1}{11}$ will be H_2citrate^{-1}.

$$\tfrac{10}{11} \times 0.25 \ M = 0.227 \ M \ H_3\text{citrate}$$

$$\tfrac{1}{11} \times 0.25 \ M = 0.0227 \ M \ H_2\text{citrate}^{-1}$$

We started with 0.125 M $H_2citrate^{-1}$ and 0.125 M $H_3citrate$. We must calculate how much H^+ had to be added to increase the $H_3citrate$ to 0.277 M. The reaction that occurred was:

$$H_2citrate^{-1} + H^+ \rightarrow H_3citrate$$

$$[\text{conjugate acid}]_{orig} + [H^+] = [\text{conjugate acid}]_{final}$$

$$0.125\ M + [H^+] = 0.277\ M$$

or

$$[\text{conjugate base}]_{orig} - [H^+] = [\text{conjugate base}]_{final}$$

$$0.125\ M - [H^+] = 0.0227\ M$$

$$[H^+] = 0.227\ M - 0.125\ M = 0.102\ M$$

or

$$[H^+] = 0.125\ M - 0.0227\ M = 0.102\ M$$

$$\boxed{[H^+] = \textbf{buffer capacity} = \textbf{0.102}\ \textbf{M}}$$

∴   It takes 0.102 mole per liter of H^+ to decrease the pH of a 0.25 M citrate buffer from pH 3.1 to pH 2.1.

We can also calculate the amount of OH^- that must be added to increase the pH from 3.1 to 4.1. The reaction involved is:

$$H_3citrate + OH^- \rightarrow H_2O + H_2citrate^{-1}$$

Using the Henderson-Hasselbalch equation, we find that at pH 4.1 the $H_2citrate^{-1}/H_3citrate$ ratio is 10:1. The final concentrations of conjugate base and acid are 0.227 M $H_2citrate^{-1}$ and 0.0227 M $H_3citrate$. By calculations similar to those shown above, we can see that the buffer capacity in terms of $[OH^-]$ required for a pH change of 1 unit comes out the same as the buffer capacity in terms of $[H^+]$ required. As a general rule, therefore, the buffer capacity for a buffer at its pK is identical in both the acidic and basic direction.

Problem II-52

Compare the buffer capacity of a (a) 0.1 M acetate buffer, pH 4.7, with that of a (b) 0.2 M acetate buffer, pH 4.7. Assume that $pK_a = 4.7$.

Solution

Calculate the amount of H^+ required by each buffer for a pH change of 1 unit to pH 3.7.

(a) In the 0.1 M buffer:

$$3.7 = 4.7 + \log \frac{[A^-]}{[HA]}$$

$$-1.0 = \log \frac{[A^-]}{[HA]}$$

$$+1.0 = \log \frac{[HA]}{[A^-]}$$

$$\frac{[HA]}{[A^-]} = \text{antilog of } 1.0 = 10$$

$$\therefore \quad \tfrac{10}{11} \times 0.1 \ M = 0.091 \ M \ [HA]_{final}$$

and

$$\tfrac{1}{11} \times 0.1 \ M = 0.0091 \ M \ [A^-]_{final}$$

At the start, $[HA] = [A^-]$ because $pH = pK_a$.

$$\therefore \quad [A^-]_{orig} = 0.05 \ M \qquad [HA]_{orig} = 0.05 \ M$$

To calculate the $[H^+]$ required:

$$[A^-]_{orig} - [H^+] = [A^-]_{final}$$
$$[HA]_{orig} + [H^+] = [HA]_{final}$$
$$0.05 + [H^+] = 0.091$$
$$[H^+] = \text{buffer capacity} = 0.091 - 0.05$$

$$\boxed{[H^+] = \mathbf{0.041 \ M}}$$

(b) In the 0.2 M buffer:

$$[HA]_{orig} = 0.1 \ M \qquad [A^-]_{orig} = 0.1 \ M$$

The final HA and A^- concentrations can be calculated as shown above.

$$[HA]_{final} = \tfrac{10}{11} \times 0.2 \ M = 0.182 \ M$$

$$[A^-]_{final} = \tfrac{1}{11} \times 0.2 \ M = 0.0182 \ M$$

$$[HA]_{orig} + [H^+] = [HA]_{final}$$

$$[H^+] = \text{buffer capacity} = [HA]_{final} - [HA]_{orig} = (0.182) - (0.1)$$

$$\boxed{[H^+] = \mathbf{0.082 \ M}}$$

We can see that the buffer capacity of the 0.2 M buffer is *twice* the buffer capacity of the 0.1 M buffer.

Problem II-53

Calculate the buffer capacity (a) in the acidic direction and (b) in the alkaline direction of a 0.3 M acetate buffer, pH 5.0. Assume that $pK_a = 4.7$.

Solution

The pH of this buffer is greater than the pK_a. We can see immediately from the titration curve in Figure II-3 that the buffer capacity is greater in the acidic direction than in the alkaline direction. We can easily calculate the buffer capacity in both directions. First calculate the original [HA] and [A⁻] concentrations.

$$pH = pK_a + \log \frac{[A^-]}{[HA]}$$

$$5.0 = 4.7 + \log \frac{[A^-]}{[HA]}$$

$$0.3 = \log \frac{[A^-]}{[HA]}$$

$$\frac{[A^-]}{[HA]} = \text{antilog of } 0.3 = 2 = \tfrac{2}{1}$$

$\therefore$ At pH 5.0, $\tfrac{2}{3} \times 0.3\ M = 0.2\ M$ [A⁻] and $\tfrac{1}{3} \times 0.3\ M = 0.1\ M$ [HA].

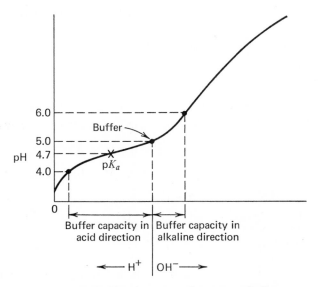

Figure II-3 Titration curve of an acetate buffer

(a) Calculate the [HA] and [A⁻] concentrations at 1 pH unit lower than the original pH.

$$\text{pH} = \text{p}K_a + \log \frac{[\text{A}^-]}{[\text{HA}]}$$

$$4.0 = 4.7 + \log \frac{[\text{A}^-]}{[\text{HA}]}$$

$$-0.7 = \log \frac{[\text{A}^-]}{[\text{HA}]}$$

$$+0.7 = \log \frac{[\text{HA}]}{[\text{A}^-]}$$

$$\frac{[\text{HA}]}{[\text{A}^-]} = \text{antilog of } 0.7 = 5 = \tfrac{5}{1}$$

∴   At pH 4.0, $\tfrac{5}{6} \times 0.3\,M = 0.25\,M$ [HA] and $\tfrac{1}{6} \times 0.3\,M = 0.05\,M$ [A⁻].

Next calculate the [H⁺] required to increase the [HA] from 0.1 M to 0.25 M, *or* the [H⁺] required to decrease the [A⁻] from 0.20 M to 0.05 M.

$$[\text{HA}]_{\text{orig}} + [\text{H}^+] = [\text{HA}]_{\text{final}}$$

$$0.1 + [\text{H}^+] = 0.25$$

$$\boxed{[\text{H}^+] = 0.15\,M}$$

or

$$[\text{A}^-]_{\text{orig}} - [\text{H}^+] = [\text{A}^-]_{\text{final}}$$

$$0.20 - [\text{H}^+] = 0.05$$

$$0.20 - 0.05 = [\text{H}^+]$$

$$\boxed{[\text{H}^+] = 0.15\,M}$$

(b) Calculate the [HA] and [A⁻] concentrations at 1 pH unit higher than the original pH.

$$\text{pH} = \text{p}K_a + \log \frac{[\text{A}^-]}{[\text{HA}]}$$

$$6.0 = 4.7 + \log \frac{[\text{A}^-]}{[\text{HA}]}$$

$$1.3 = \log \frac{[\text{A}^-]}{[\text{HA}]}$$

$$\frac{[\text{A}^-]}{[\text{HA}]} = \text{antilog of } 1.3 = 20 = \tfrac{20}{1}$$

$\therefore$ At   pH 6.0, $\frac{20}{21} \times 0.3\ M = 0.286\ M$ [A$^-$]   and $\frac{1}{21} \times 0.3\ M = 0.014\ M$ [HA].

Next calculate the [OH$^-$] required to increase the [A$^-$] from 0.20 M to 0.286 M, *or* the [OH$^-$] required to decrease the [HA] from 0.10 M to 0.014 M.

$$[A^-]_{orig} + [OH^-] = [A^-]_{final}$$
$$0.20 + [OH^-] = 0.286$$

$$\boxed{[OH^-] = 0.086\ M}$$

$$[HA]_{orig} - [OH^-] = [HA]_{final}$$
$$(0.100) - [OH^-] = 0.014$$
$$0.100 - 0.014 = [OH^-]$$

$$\boxed{[OH^-] = 0.086\ M}$$

The above calculations are based on the reactions shown below.

$$HA + OH^- \rightarrow A^- + H_2O$$
$$A^- + H^+ \rightarrow HA$$

Thus, the addition of a given amount of OH$^-$ to the buffer causes an equivalent amount of HA to disappear and an equivalent amount of A$^-$ to appear. Similarly, the addition of H$^+$ to the buffer causes an equivalent amount of A$^-$ to disappear and an equivalent amount of HA to appear. As shown earlier, these assumptions are quite valid.

Problem II-54

(a) How much H$^+$ is required to decrease the pH of 1 liter of 0.3 M acetate buffer from pH 5.0 to pH 4.5? (b) What will be the final pH after 0.075 mole/liter of H$^+$ is added to the buffer? Assume that the volume of the buffer remains constant.

Solution

In the preceding problem, we saw that 0.15 mole/liter of H$^+$ is required to decrease the pH of the buffer by **1 unit**. A titration curve is not a straight line and the pH scale is logarithmic, not linear. Hence, there is no direct proportionality between H$^+$ added and ΔpH. Thus, it does *not* take half as much H^+ to decrease the pH by 0.5 unit as it took to decrease the pH by 1.0 unit. The exact amount of H^+ required can easily be calculated as shown below.

(a) First calculate the concentrations of HA and A^- in the buffer at pH 4.5.

$$pH = pK_a + \log \frac{[A^-]}{[HA]}$$

$$4.5 = 4.7 + \log \frac{[A^-]}{[HA]}$$

$$-0.2 = \log \frac{[A^-]}{[HA]}$$

$$+0.2 = \log \frac{[HA]}{[A^-]}$$

$$\frac{[HA]}{[A^-]} = \text{antilog of } 0.2 = 1.584$$

$$\therefore \quad \frac{1.584}{2.584} \times 0.3 \ M = 0.1837 \ M \ [HA]$$

$$0.3000 - 0.1837 = 0.1163 \ M \ [A^-]$$

Next calculate the $[H^+]$ required to change the original [HA] to the final [HA], or the amount of $[H^+]$ required to change the original $[A^-]$ to the final $[A^-]$.

$$[HA]_{orig} + [H^+] = [HA]_{final}$$
$$(0.10) + [H^+] = (0.1837)$$
$$[H^+] = 0.1837 - 0.1$$

$$\boxed{[H^+] = 0.0837 \ M}$$

(b) The effect of adding 0.075 mole H^+/liter can also be calculated. First calculate the new [HA] and $[A^-]$ concentrations.

$$[HA]_{final} = [HA]_{orig} + [H^+]$$
$$[HA]_{final} = 0.10 + 0.075 = 0.175 \ M$$
$$[A^-]_{final} = [A^-]_{orig} - [H^+]$$
$$[A^-]_{final} = 0.200 - 0.075 = 0.125 \ M$$

Next calculate the pH by using the Henderson-Hasselbalch equation.

$$pH = pK_a + \log \frac{[A^-]}{[HA]}$$

$$pH = 4.7 + \log \frac{0.125}{0.175}$$

$$pH = 4.7 - \log \frac{0.175}{0.125}$$

$$pH = 4.7 - \log 1.4$$

$$pH = 4.700 - 0.146$$

$$\boxed{pH = 4.554}$$

We can see that more than half as much H^+ is required to change the pH by 0.5 unit as it took to decrease the pH by 1.0 unit; half as much H^+ changed the pH by much less than 0.5 unit.

Problem II-55

The pH of a sample of arterial blood is 7.42. Upon acidification of 10 ml of the blood, 5.91 ml of CO_2 (corrected for standard temperature and pressure S.T.P.) are produced. Calculate: (a) the total concentration of dissolved CO_2 in the blood $[CO_2 + HCO_3^-]$, (b) the concentrations of dissolved CO_2 and HCO_3^-, and (c) the partial pressure of the dissolved CO_2 in terms of mm Hg.

Solution

(a)   First calculate the number of moles of CO_2 represented by 5.91 ml at S.T.P. One mole of a "perfect" gas occupies 22.4 liters at S.T.P. The experimental value for CO_2 is 22.26 liters.

$$\therefore \quad \frac{5.91 \times 10^{-3} \text{ liters}}{22.26 \text{ liters/mole}} = 0.265 \times 10^{-3} \text{ moles}$$

This amount of CO_2 came from 10 ml of blood.

$$\therefore \quad \text{concentration of "total } CO_2\text{"} = \frac{26.5 \times 10^{-5} \text{ moles}}{10 \times 10^{-3} \text{ liters}}$$

$$= \boxed{\mathbf{2.65 \times 10^{-2} \ M}}$$

(b)
$$pH = pK_{a_1} + \log \frac{[HCO_3^-]}{[H_2CO_3]}$$

When dissolved in water, CO_2 reacts to a very slight extent to produce H_2CO_3.

$$CO_2 + H_2O \rightleftharpoons H_2CO_3$$

However, practically all of the non-HCO_3^--dissolved CO_2 is in the form of CO_2 rather than H_2CO_3. Consequently, we can express the Henderson-Hasselbalch equation in terms of HCO_3^- and CO_2.

$$pH = pK_{a_1} + \log \frac{[HCO_3^-]}{[CO_2]}$$

We can now determine the concentrations of each species present.

$$7.42 = 6.37 + \log \frac{[HCO_3^-]}{[CO_2]}$$

$$1.05 = \log \frac{[HCO_3^-]}{[CO_2]}$$

$$\frac{[HCO_3^-]}{[CO_2]} = \text{antilog of } 1.05 = 11.2$$

$$\therefore \quad \frac{[HCO_3^-]}{[CO_2]} = \frac{11.2}{1}$$

$$[HCO_3^-] = \frac{11.2}{12.2} \times 2.65 \times 10^{-2} \ M$$

$$\boxed{[HCO_3^-] = 2.43 \times 10^{-2} \ M}$$

$$[CO_2] = \frac{1}{12.2} \times 2.65 \times 10^{-2} \ M$$

$$\boxed{[CO_2] = 2.17 \times 10^{-3} \ M}$$

(c) An equilibrium exists between the concentration of a substance in solution and its concentration in the gaseous phase.

$$\frac{[CO_2]_{sol}}{[CO_2]_{gas}} = K$$

Another way of saying this is that the solubility of a gas is directly proportional to its partial pressure.

$$[CO_2]_{sol} = K[CO_2]_{gas} = K'P_{CO_2}$$

When the solubility is expressed in moles/liter (M) and the concentration in the gaseous phase (the partial pressure) is expressed in terms of mm Hg, the K' value of CO_2 at body temperature ($37°$ C) is 3.01×10^{-5}.

$$\therefore \quad [CO_2]_{sol} = (3.01 \times 10^{-5})P_{CO_2}$$

$$P_{CO_2} = \frac{[CO_2]_{sol}}{(3.01 \times 10^{-5})} = \frac{(2.17 \times 10^{-3})}{(3.01 \times 10^{-5})} \text{ mm Hg}$$

$$\boxed{P_{CO_2} = 72.2 \text{ mm Hg}}$$

That is, the concentration of dissolved CO_2 is that which is in equilibrium with an atmosphere containing 72.2 mm Hg of CO_2.

PRACTICE PROBLEM SET II

When solving the problems below, assume that $\gamma = 1$ for all substances unless otherwise indicated.

1. The weak acid, HA, is 2.4% dissociated in a $0.22\ M$ solution. Calculate: (a) the K_a, (b) the pH of the solution, (c) the amount of $0.1\ N$ KOH required to neutralize 550 ml of the weak acid solution, and (d) the number of H^+ ions in 550 ml of the weak acid solution.

2. The pH of a $0.27\ M$ solution of a weak acid, HA, is 4.3. (a) What is the H^+ ion concentration in the solution? (b) What is the degree of ionization of the acid? (c) What is the K_a?

3. The K_a of a weak acid, HA, is 3×10^{-4}. Calculate: (a) the OH^- ion concentration in the solution and (b) the degree of dissociation of the acid in a $0.15\ M$ solution.

4. At what concentration of weak acid, HA (in terms of its K_a), will the acid be: (a) 10% dissociated, (b) 25% dissociated, and (c) 90% dissociated?

5. (a) Calculate the pH of a $0.05\ M$ solution of ethanolamine, $K_b = 2.8 \times 10^{-5}$. (b) What is the degree of ionization of the amine?

6. A weak organic dicarboxylic acid, H_2A, ionizes in two steps. The K_{a_1} and K_{a_2} values are 2.6×10^{-5} and 7×10^{-7}, respectively. Calculate: (a) the pH of a $0.2\ M$ solution of H_2A, (b) the $[HA^-]$, (c) the $[A^=]$ concentrations in the $0.2\ M$ solution, and (d) the approximate degrees of primary and secondary ionizations.

7. Calculate the pK_a and pK_b of weak acids with K_a values of: (a) 6.23×10^{-4}, (b) 2.9×10^{-5}, (c) 3.4×10^{-5}, and (d) 7.2×10^{-6}.

8. Calculate the pK_b and pK_a of weak bases with K_b values of: (a) 2.1×10^{-5}, (b) 3.1×10^{-6}, (c) 7.8×10^{-5}, and (d) 9.2×10^{-4}.

9. Calculate the pH of a $0.2 M$ solution of an amine that has a pK_a of 9.5.

10. What is the pH of a $0.20 M$ solution of: (a) H_3PO_4, (b) KH_2PO_4, (c) K_2HPO_4, (d) K_3PO_4, (e) potassium acetate, (f) NH_4Br, (g) sodium phenolate, (h) trisodium citrate, (i) disodium citrate, (j) $NaHS$, (k) Na_2S, and (l) ethanolamine hydrochloride?

11. How many milliliters of $0.1 M$ KOH are required to titrate completely 270 ml of $0.4 M$ propionic acid?

12. How many milliliters of $0.2 M$ KOH are required to titrate completely 650 ml of $0.05 M$ citric acid?

13. What are the final hydrogen ion concentration and pH of a solution obtained by mixing 100 ml of $0.2 M$ KOH with 150 ml of $0.1 M$ HOAc?

14. What are the final hydrogen ion concentration and pH of a solution obtained by mixing 200 ml of $0.4 M$ aqueous NH_3 with 300 ml of $0.2 M$ HCl?

15. What are the final hydrogen ion concentration and pH of a solution obtained by mixing 250 ml of $0.1 M$ citric acid with 300 ml of $0.1 M$ KOH?

16. What are the final hydrogen ion concentration and pH of a solution obtained by mixing 400 ml of $0.2 M$ NaOH with 150 ml of $0.1 M$ H_3PO_4?

17. Show mathematically why a propanolamine buffer cannot maintain an absolutely constant pH upon addition of OH^- ions.

18. Calculate the appropriate values and draw the titration curve for the neutralization and overtitration of 200 ml of $0.1 M$ HOAc with $0.2 M$ KOH.

19. Calculate the appropriate values and draw the titration curve for the neutralization of 150 ml of $0.15 M$ succinic acid with $0.2 M$ NaOH.

20. What are the concentrations of NH_3 and NH_4Cl in a $0.15 M$ "ammonia" buffer, pH 9.6 ($K_b = 1.8 \times 10^{-5}$)?

21. (a) What is the pH of a solution containing $0.01 M$ $HPO_4^=$ and $0.01 M$ $PO_4^=$ (assume $\gamma = 1$)? (b) Calculate the actual pH by using the activity coefficients listed in Table AI-6, Appendix I.

22. What is the pH of a solution containing 0.3 M tris(hydroxymethyl)-aminomethane (free base) and 0.2 M tris hydrochloride?

23. What is the pH of a solution containing 0.2 g/liter Na_2CO_3 and 0.2 g/liter $NaHCO_3$?

24. What is the pH of a solution prepared by dissolving 5.35 g of NH_4Cl in a liter of 0.2 M NH_3?

25. Describe the preparation of 2 liters of 0.25 M formate buffer, pH 4.5, starting from 1 M formic acid and solid sodium formate (HCOONa).

26. Describe the preparation of 2 liters of 0.4 M phosphate buffer, pH 6.9, starting from: (a) a 2 M H_3PO_4 solution and a 1 M KOH solution, (b) a 0.8 M H_3PO_4 solution and solid NaOH, (c) a commercial concentrated H_3PO_4 solution and 1 M KOH, (d) 1 M solutions of KH_2PO_4 and Na_2HPO_4, (e) solid KH_2PO_4 and K_2HPO_4, (f) solid K_2HPO_4 and 1.5 M HCl, (g) 1.2 M K_2HPO_4 and 2 M H_2SO_4, (h) solid KH_2PO_4 and 2 M KOH, (i) 1.5 M KH_2PO_4 and 1 M NaOH, and (j) solid Na_3PO_4 and 1 M HCl.

27. What volume of glacial acetic acid and what weight of solid potassium acetate are required to prepare 5 liters of 0.2 M acetate buffer, pH 5.0?

28. An enzyme-catalyzed reaction was carried out in a solution buffered with 0.25 M phosphate, pH 7.2. As a result of the reaction, 0.05 mole/liter of acid was formed. (a) What was the pH at the end of the reaction? (b) What would the pH be if no buffer were present? (c) Write the chemical equation showing how the phosphate buffer resisted a large change in pH.

29. An enzyme-catalyzed reaction was carried out in a solution containing 0.2 M tris buffer. The pH of the reaction mixture at the start was 7.8. As a result of the reaction, 0.033 mole/liter of H^+ was consumed. (a) What was the ratio of tris^0 (free base) to tris$^+$ Cl$^-$ at the start of the reaction? (b) What was the tris^0/tris$^+$ ratio at the end of the reaction? (c) What was the final pH of the reaction mixture? (d) What would the final pH be if no buffer were present? (e) Write the chemical equations showing how the tris buffer maintained a near constant pH during the reaction.

30. Calculate the buffer capacity in the acid and alkaline directions of a 0.25 M phosphate buffer, pH 6.8.

AMINO ACIDS AND PEPTIDES

III

A. pH OF AMINO ACID SOLUTIONS

Problem III-1

"Glycine" can be obtained in three forms: (a) glycine hydrochloride, (b) isoelectric glycine (sometimes called glycine, free base), and (c) sodium glycinate. Draw the structures of these three forms.

Solution

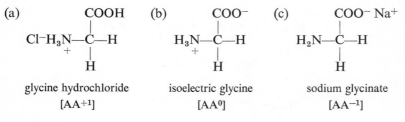

(a) glycine hydrochloride [AA^{+1}] (b) isoelectric glycine [AA^{0}] (c) sodium glycinate [AA^{-1}]

Problem III-2

Calculate the pH of a 0.1 M solution of: (a) glycine hydrochloride, (b) isoelectric glycine, and (c) sodium glycinate.

Solution

(a) Glycine hydrochloride is essentially a diprotic acid. Because the carboxyl group is so much stronger an acid ($pK_{a_1} = 2.34$) than the charged amino group ($pK_{a_2} = 9.6$), the pH of the solution is established almost exclusively by the extent to which the carboxyl ionizes.

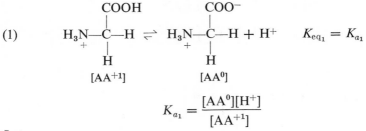

(1)

$$K_{a_1} = \frac{[AA^0][H^+]}{[AA^{+1}]}$$

Let

$$y = M \text{ of } AA^{+1} \text{ that ionizes}$$
$$\therefore \quad y = M \text{ of } H^+ \text{ produced}$$

and

$$y = M \text{ of } AA^0 \text{ produced}$$

and

$$0.1 - y = M \text{ of } AA^{+1} \text{ remaining at equilibrium}$$

Because the concentration of AA^{+1} is less than 100 K_{a_1}, we cannot ignore y in the denominator.

$$K_{a_1} = \frac{(y)(y)}{(0.1 - y)}$$

The pK_{a_1}, not K_{a_1}, was given. Therefore, first calculate K_{a_1}.

$$pK_{a_1} = -\log K_{a_1} = \log \frac{1}{K_{a_1}} = 2.34$$

$$\frac{1}{K_{a_1}} = \text{antilog of } 2.34 = 219$$

$$K_{a_1} = \frac{1}{219} = 4.57 \times 10^{-3}$$

$$4.57 \times 10^{-3} = \frac{(y)(y)}{(0.1 - y)}$$

$$4.57 \times 10^{-4} - 4.57 \times 10^{-3}y = y^2$$

$$y^2 + 4.57 \times 10^{-3}y - 4.57 \times 10^{-4} = 0$$

$$y = \frac{-b \pm \sqrt{b^2 - 4ac}}{2a}$$

where

$$a = 1$$

$$b = 4.57 \times 10^{-3}$$

$$c = -4.57 \times 10^{-4}$$

$$y = \frac{-4.57 \times 10^{-3} \pm \sqrt{(4.57 \times 10^{-3})^2 - 4(-4.57 \times 10^{-4})}}{2}$$

$$= \frac{-4.57 \times 10^{-3} \pm \sqrt{20.9 \times 10^{-6} + 18.3 \times 10^{-4}}}{2}$$

$$= \frac{-4.57 \times 10^{-3} \pm \sqrt{0.209 \times 10^{-4} + 18.3 \times 10^{-4}}}{2}$$

$$= \frac{-4.57 \times 10^{-3} \pm \sqrt{18.51 \times 10^{-4}}}{2}$$

$$= \frac{-4.57 \times 10^{-3} \pm 4.31 \times 10^{-2}}{2}$$

$$= \frac{-0.457 \times 10^{-2} \pm 4.31 \times 10^{-2}}{2}$$

$$y = \frac{3.85 \times 10^{-2}}{2} = 1.93 \times 10^{-2}$$

$$\therefore \quad [H^+] = 1.93 \times 10^{-2} \, M$$

The original glycine hydrochloride solution was 0.1 M. Therefore, the carboxyl group is 19.3 % ionized.

$$pH = \log \frac{1}{[H^+]}$$

$$pH = \log \frac{1}{1.93 \times 10^{-2}}$$

$$pH = \log 0.519 \times 10^2 = \log 51.9$$

$$\boxed{pH = 1.715}$$

(b) Isoelectric glycine is an intermediate ion of a polyprotic acid. It ionizes in solution as an acid (2) and as a base (3). The ionization as a base can be considered as the hydrolysis of the carboxylate ion.

$$(2) \quad \underset{\underset{+}{H_3N}}{\overset{COO^-}{\underset{|}{\underset{|}{C}}}}-H \rightleftharpoons \underset{H_2N}{\overset{COO^-}{\underset{|}{\underset{|}{C}}}}-H + H^+ \qquad K_{eq_2} = K_{a_2}$$

$$[AA^0] \qquad\qquad [AA^{-1}]$$

$$(3) \quad \underset{\underset{+}{H_3N}}{\overset{COO^-}{\underset{|}{\underset{H}{C}}}}-H + HOH \rightleftharpoons \underset{\underset{+}{H_3N}}{\overset{COOH}{\underset{|}{\underset{H}{C}}}}-H + OH^- \qquad K_{eq_3} = K_{h_1} = \frac{K_w}{K_{a_1}} = K_{b_2}$$

$$[AA^0] \qquad\qquad [AA^{+1}]$$

The pH of a solution of an intermediate ion is essentially independent of the concentration of the ion. The pH may be calculated from the pK_a values on either side of the ion; that is, from the pK_a of the next acid group to ionize and the pK_a of the previous acid group ionized.

$$pH = \frac{pK_{a_1} + pK_{a_2}}{2} = \frac{2.34 + 9.6}{2} = \frac{11.94}{2}$$

$$\boxed{pH = 5.97}$$

(c) Sodium glycinate is essentially a diprotic base. Both the unionized amino group and the carboxylate ion can accept a proton from water. However, because the amino group is a much stronger base than the carboxylate ion, the pH of the solution depends almost entirely on the extent to which the amino group ionizes. We can check the relative base strengths by calculating K_b for each group. For the amino group,

$$K_{b_1} = \frac{K_w}{K_{a_2}} = \frac{10^{-14}}{10^{-9.6}} = 10^{-4.4}$$

For the carboxylate ion,

$$K_{b_2} = \frac{K_w}{K_{a_1}} = \frac{10^{-14}}{10^{-2.34}} = 10^{-11.66}$$

$$(4) \quad \underset{H_2N}{\overset{COO^-}{\underset{|}{\underset{H}{C}}}}-H + HOH \rightleftharpoons \underset{\underset{+}{H_3N}}{\overset{COO^-}{\underset{|}{\underset{H}{C}}}}-H + OH^- \qquad K_{eq_4} = K_{b_1} = \frac{K_w}{K_{a_2}}$$

$$[AA^{-1}] \qquad\qquad [AA^0]$$

$$K_{b_1} = \frac{[AA^0][OH^-]}{[AA^{-1}]} = 10^{-4.4}$$

Let

$$y = M \text{ of } AA^{-1} \text{ that ionizes}$$

$$\therefore \quad y = M \text{ of } OH^- \text{ produced}$$

and

$$y = M \text{ of } AA^0 \text{ produced}$$

and

$$0.1 - y = M \text{ of } AA^{-1} \text{ remaining at equilibrium}$$

Because $10^{-4.4}$ is a little awkward for us to work with, we can recalculate K_{b_1}.

First calculate K_{a_2}

$$pK_{a_2} = -\log K_{a_2} = \log \frac{1}{K_{a_2}}$$

$$9.6 = \log \frac{1}{K_{a_2}}$$

$$\frac{1}{K_{a_2}} = \text{antilog of } 9.6$$

$$\frac{1}{K_{a_2}} = 3.98 \times 10^{-9}$$

$$K_{a_2} = \frac{1}{3.98 \times 10^9} = 0.251 \times 10^{-9}$$

$$\boxed{K_{a_2} = 2.51 \times 10^{-10}}$$

$$K_{b_1} = \frac{K_w}{K_{a_2}} = \frac{1 \times 10^{-14}}{2.51 \times 10^{-10}}$$

$$K_{b_1} = 0.398 \times 10^{-4}$$

$$\boxed{K_{b_1} = 3.98 \times 10^{-5}}$$

Or, calculate K_{b_1} directly

$$K_{b_1} = 10^{-4.4}$$

$$= 10^{-5} \times 10^{+0.6}$$

Look up antilog of 0.6.

$$= 10^{-5} \times 3.98$$

$$\boxed{K_{b_1} = 3.98 \times 10^{-5}}$$

$$K_{b_1} = \frac{(y)(y)}{(0.1 - y)} = 3.98 \times 10^{-5}$$

Because the concentration of sodium glycinate is large compared to

K_{b_1}, we can neglect the y term in the denominator.

$$3.98 \times 10^{-5} = \frac{y^2}{0.1}$$

$$y^2 = 3.98 \times 10^{-6}$$

$$y = \sqrt{3.98 \times 10^{-6}} = 1.995 \times 10^{-3}$$

$$[OH^-] = 1.995 \times 10^{-3}\ M \simeq 2 \times 10^{-3}\ M$$

$$[H^+] = \frac{K_w}{[OH^-]} = \frac{1 \times 10^{-14}}{2 \times 10^{-3}} = 0.5 \times 10^{-11}$$

$$[H^+] = 5 \times 10^{-12}\ M$$

$$pH = \log \frac{1}{[H^+]} = \log \frac{1}{5 \times 10^{-12}} = \log 0.2 \times 10^{12}$$

$$pH = \log 2 \times 10^{11}$$

$$pH = \log 2 + \log 10^{11}$$

$$pH = 0.3 + 11$$

$$\boxed{pH = 11.3}$$

Problem III-3

(a)   Draw the structures of the various forms of "aspartic acid" that may be obtained. (b)–(e) Show how each form ionizes in water.

Solution

(a)   "Aspartic acid" may be obtained in four forms: aspartic hydrochloride [AA^{+1}], isoelectric aspartic acid [AA^0], monosodium aspartate [AA^{-1}], and disodium aspartate [AA^{-2}]. The structure are shown below.

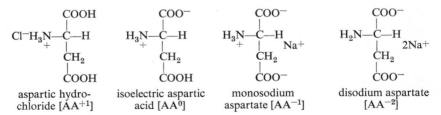

(b)   Aspartic hydrochloride ionizes as a typical polyprotic acid. The pK_a values for the three acidic groups are 2.1 (α-COOH), 3.86 (β-COOH), and 9.82 (α-NH$_3^+$). Because the α-COOH is so much stronger an acid than the other two groups, the pH of an aspartic hydrochloride solution depends almost exclusively on the concentration of aspartic hydrochloride and the extent to which the α-COOH ionizes.

(1)

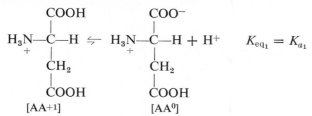

$$[AA^{+1}] \qquad\qquad [AA^0]$$

The pH calculations may be made exactly as described in the preceding problem for glycine hydrochloride.

$$K_{a_1} = \frac{[AA^0][H^+]}{[AA^{+1}]}$$

(c)   Disodium aspartate ionizes as a typical polyprotic base. The pK_b values for the three basic groups can be calculated from their respective pK_a values as shown in Table III-1.

$$pK_b = 14 - pK_a$$

TABLE III-1

Conjugate Acid	pK_a	Conjugate Base	pK_b
α-COOH	2.1 (pK_{a_1})	α-COO$^-$	11.9 (pK_{b_3})
β-COOH	3.86 (pK_{a_2})	β-COO$^-$	10.14 (pK_{b_2})
α-NH$_3^+$	9.82 (pK_{a_3})	α-NH$_2$	4.18 (pK_{b_1})

Note that the pK_a values of the conjugate acids are numbered in decreasing order of acid strength. The pK_b values are numbered in decreasing order of base strength. Therefore, the α-NH$_3^+$ group is the weakest acid and its pK_a value is designated pK_{a_3}. The conjugate base of the α-NH$_3^+$ group is the α-NH$_2$ group which is the strongest of the basic groups. Hence, its pK_b value is designated pK_{b_1}. Because pK_{b_2} is almost 2 pH units less than pK_{b_1} (K_{b_2} is almost 100 times less than K_{b_1}), the pH of a disodium aspartate solution is established almost exclusively by the concentration of the salt and the extent of ionization of the α-NH$_2$ group.

(2)

$$\underset{[AA^{-2}]}{\begin{array}{c} COO^- \\ | \\ H_2N-C-H \\ | \\ CH_2 \\ | \\ COO^- \end{array}} + HOH \rightleftharpoons \underset{[AA^{-1}]}{\begin{array}{c} COO^- \\ | \\ H_3N-C-H \\ + \; | \\ CH_2 \\ | \\ COO^- \end{array}} + OH^- \qquad K_{eq_2} = K_{b_1}$$

Calculations of pH are done exactly as described in the previous problem for sodium glycinate.

$$K_{b_1} = \frac{[AA^{-1}][OH^-]}{[AA^{-2}]}$$

(d)   The two remaining forms of aspartic acid are intermediate ions of polyprotic acids. For example, consider isoelectric aspartic acid. The compound ionizes both as an acid and a base.

(3)

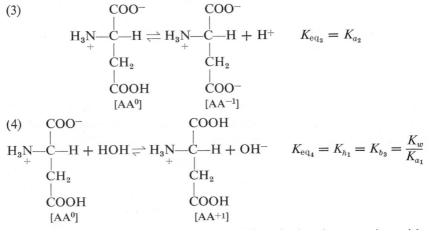

(4)

Remember that the ionization as an acid really involves reaction with H_2O as a conjugate base and should properly be written as shown in equation 5. Note that the group shown ionizing as an acid is the strongest remaining acidic group. The charged amino group is a much weaker acid than the β-COOH group and remains essentially in the NH_3^+ form. Its ionization contributes negligibly to the pH of the solution.

(5)

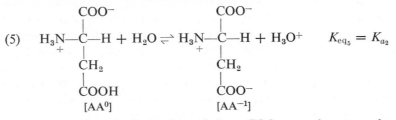

Also note that the ionization of the α-COO$^-$ as a base may be considered an "hydrolysis" reaction. The pH of solutions of isoelectric aspartic acid may be calculated from the pK_{a_1} and pK_{a_2} values (the pK_a values on either side of isoelectric aspartic acid in its titration curve).

$$pH = \frac{pK_{a_1} + pK_{a_2}}{2} = \frac{2.1 + 3.86}{2} = \frac{5.96}{2} = 2.98$$

(e)   The remaining form, monosodium aspartate, is also an intermediate ion of a polyprotic acid. Its ionizations as an acid and as a base are shown below.

(6)

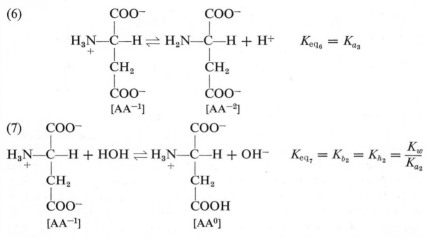

$$H_3\overset{+}{N}-\underset{\underset{\displaystyle COO^-}{\overset{\displaystyle |}{\underset{|}{CH_2}}}}{\overset{|}{C}}-H \rightleftharpoons H_2N-\underset{\underset{\displaystyle COO^-}{\overset{\displaystyle |}{\underset{|}{CH_2}}}}{\overset{|}{C}}-H + H^+ \qquad K_{eq_6} = K_{a_3}$$

$$[AA^{-1}] \qquad\qquad [AA^{-2}]$$

(7)

$$H_3\overset{+}{N}-\underset{\underset{\displaystyle COO^-}{\overset{\displaystyle |}{\underset{|}{CH_2}}}}{\overset{|}{C}}-H + HOH \rightleftharpoons H_3\overset{+}{N}-\underset{\underset{\displaystyle COOH}{\overset{\displaystyle |}{\underset{|}{CH_2}}}}{\overset{|}{C}}-H + OH^- \qquad K_{eq_7} = K_{b_2} = K_{h_2} = \frac{K_w}{K_{a_2}}$$

$$[AA^{-1}] \qquad\qquad\qquad [AA^0]$$

The group shown ionizing as an acid (the α-NH_3^+) is the only remaining acidic group. The group shown ionizing as a base (the β-COO^- group) is the stronger of the two basic groups present. The ionization of the α-COO^- group as a base contributes little toward establishing the pH of the solution because it is so much weaker a base than is the β-COO^- group. The pH of solutions of monosodium aspartate may be calculated from the pK_{a_2} and pK_{a_3} values.

$$pH = \frac{pK_{a_2} + pK_{a_3}}{2} = \frac{3.86 + 9.82}{2} = \frac{13.68}{2} = 6.84$$

From this problem and the one preceding, it is obvious that pH problems involving amino acids may be treated in exactly the same manner as those involving any polyprotic acid or polyprotic base.

Problem III-4

(a)   Draw the structures of the various ionic forms of lysine. (b)–(e) Show how each form ionizes in water. The pK_a values of lysine are 2.18 (α-$COOH$), 8.95 (α-NH_3^+), and 10.53 (ϵ-NH_3^+).

Solution

(a)   "Lysine" may be obtained in four different forms: lysine dihydro-chloride $[AA^{+2}]$, lysine monohydrochloride $[AA^{+1}]$, isoelectric lysine $[AA^0]$, and sodium lysinate $[AA^{-1}]$. The structures are shown at top of p. 132.

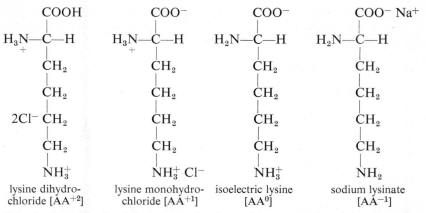

lysine dihydro- lysine monohydro- isoelectric lysine sodium lysinate
chloride $[AA^{+2}]$ chloride $[AA^{+1}]$ $[AA^0]$ $[AA^{-1}]$

(b)   Lysine dihydrochloride can ionize only as an acid. The strongest acidic group (α-COOH) ionizes first. The charged amino groups are weak as acids compared to the α-COOH. Consequently, the pH of a solution of lysine dihydrochloride is established almost exclusively by the concentration of the salt and the extent to which the α-COOH ionizes. The pH calculations are done as shown in the previous problems for glycine and aspartic hydrochlorides.

(1)

$$
\begin{array}{ccc}
\text{COOH} & & \text{COO}^- \\
| & & | \\
\underset{+}{\text{H}_3\text{N}}\!-\!\text{C}\!-\!\text{H} & \rightleftharpoons & \underset{+}{\text{H}_3\text{N}}\!-\!\text{C}\!-\!\text{H} + \text{H}^+ \qquad K_{eq_1} = K_{a_1}\\
| & & | \\
[\text{CH}_2]_4 & & [\text{CH}_2]_4 \\
| & & | \\
\text{NH}_3^+ & & \text{NH}_3^+ \\
[AA^{+2}] & & [AA^{+1}]
\end{array}
$$

$$K_{a_1} = \frac{[AA^{+1}][H^+]}{[AA^{+2}]}$$

(c)   Sodium lysinate can ionize only as a base; three basic groups are present. However, the pH of the solution is established almost exclusively by the extent to which the *strongest* base (the ϵ-NH$_2$ group) ionizes and the concentration of the salt.

(2)

$$
\begin{array}{ccc}
\text{COO}^- & & \text{COO}^- \\
| & & | \\
\text{H}_2\text{N}\!-\!\text{C}\!-\!\text{H} + \text{HOH} & \rightleftharpoons & \text{H}_2\text{N}\!-\!\text{C}\!-\!\text{H} + \text{OH}^- \qquad K_{eq_2} = K_{b_1} = \dfrac{K_w}{K_{a_3}}\\
| & & | \\
[\text{CH}_2]_4 & & [\text{CH}_2]_4 \\
| & & | \\
\text{NH}_2 & & \text{NH}_3^+ \\
[AA^{-1}] & & [AA^0]
\end{array}
$$

$$K_{b_1} = \frac{[AA^0][OH^-]}{[AA^{-1}]}$$

(d) Lysine monohydrochloride is an intermediate ion. Its ionizations as an acid and as a base are shown below.

(3)

$$
\underset{\underset{+}{\text{H}_3\text{N}}}{\overset{\text{COO}^-}{|}}\!\!\!-\overset{|}{\underset{|}{\text{C}}}\!-\text{H} \;\rightleftharpoons\; \text{H}_2\text{N}\!-\overset{\text{COO}^-}{\underset{|}{\text{C}}}\!-\text{H} + \text{H}^+ \qquad K_{\text{eq}_3} = K_{a_2}
$$

$$
\begin{array}{cc}
[\text{CH}_2]_4 & [\text{CH}_2]_4 \\
\text{NH}_3^+ & \text{NH}_3^+ \\
[\text{AA}^{+1}] & [\text{AA}^0]
\end{array}
$$

Again, we consider only the ionization of the stronger of the two remaining acidic groups, the α-NH_3^+.

(4)

$$
\underset{\underset{+}{\text{H}_3\text{N}}}{\overset{\text{COO}^-}{|}}\!\!\!-\overset{|}{\underset{|}{\text{C}}}\!-\text{H} + \text{HOH} \;\rightleftharpoons\; \underset{+}{\text{H}_3\text{N}}\!-\overset{\text{COOH}}{\underset{|}{\text{C}}}\!-\text{H} + \text{OH}^- \qquad K_{\text{eq}_4} = K_{h_1} = K_{b_3} = \frac{K_w}{K_{a_1}}
$$

$$
\begin{array}{cc}
[\text{CH}_2]_4 & [\text{CH}_2]_4 \\
\text{NH}_3^+ & \text{NH}_3^+ \\
[\text{AA}^{+1}] & [\text{AA}^{+2}]
\end{array}
$$

$$
\text{pH} = \frac{\text{p}K_{a_1} + \text{p}K_{a_2}}{2} = \frac{2.18 + 8.95}{2} = \frac{11.13}{2} = 5.57
$$

(e) Isoelectric lysine is also an intermediate ion. Its ionizations are shown below.

(5)

$$
\text{H}_2\text{N}\!-\overset{\text{COO}^-}{\underset{|}{\text{C}}}\!-\text{H} \;\rightleftharpoons\; \text{H}_2\text{N}\!-\overset{\text{COO}^-}{\underset{|}{\text{C}}}\!-\text{H} + \text{H}^+ \qquad K_{\text{eq}_5} = K_{a_3}
$$

$$
\begin{array}{cc}
[\text{CH}_2]_4 & [\text{CH}_2]_4 \\
\text{NH}_3^+ & \text{NH}_2 \\
[\text{AA}^0] & [\text{AA}^{-1}]
\end{array}
$$

(6)

$$
\text{H}_2\text{N}\!-\overset{\text{COO}^-}{\underset{|}{\text{C}}}\!-\text{H} + \text{HOH} \;\rightleftharpoons\; \text{H}_3\text{N}^+\!\!-\overset{\text{COO}^-}{\underset{|}{\text{C}}}\!-\text{H} + \text{OH}^- \qquad K_{\text{eq}_6} = K_{b_2} = \frac{K_w}{K_{a_2}}
$$

$$
\begin{array}{cc}
[\text{CH}_2]_4 & [\text{CH}_2]_4 \\
\text{NH}_3^+ & \text{NH}_3^+ \\
[\text{AA}^0] & [\text{AA}^{+1}]
\end{array}
$$

Again, of the two basic groups present (α-NH$_2$ and α-COO$^-$), only the ionization of the stronger (α-NH$_2$) plays a significant role in establishing the pH of the solution.

$$\text{pH} = \frac{pK_{a_2} + pK_{a_3}}{2} = \frac{8.95 + 10.53}{2} = \frac{19.48}{2} = 9.74$$

When dealing with intermediate ions of amino acids, you also can calculate the pH of their solutions indirectly after first calculating the concentrations of the various ionic species present, employing the disproportionation reaction equations.

B. TITRATIONS OF AMINO ACIDS

Problem III-5

Calculate the volume of 0.2 M KOH required to titrate completely 200 ml of 0.1 M glycine hydrochloride.

Solution

Glycine hydrochloride is essentially a diprotic acid. The reactions involved in its titration are shown below.

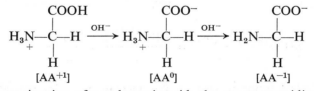

As in any titration of a polyprotic acid, the strongest acidic group is neutralized first and then the weaker acid groups in order. The calculations are exactly the same as, for example, those for succinic or sulfuric acid. As shown above, 1 mole of OH$^-$ is required per mole of AA^{+1} to convert all of the glycine hydrochloride to AA^0; another mole of OH$^-$ is required to convert the AA^0 to AA^{-1}.

$$\text{liters} \times M = \text{number of moles AA}^{+1}$$
$$0.20 \times 0.1 = 0.02 \text{ mole AA}^{+1}$$
$$\therefore \quad 2 \times 0.02 = 0.04 \text{ mole OH}^- \text{ required}$$
$$\text{liters} \times M = \text{number of moles of OH}^-$$
$$\text{liters} \times 0.2 = 0.04 \text{ mole}$$
$$\text{liters} = \frac{0.04}{0.2} = 0.20$$

$$\boxed{\therefore \quad \textbf{200 ml of 0.2 } \textit{\textbf{M}} \textbf{ KOH required}}$$

Problem III-6

Calculate the volume of: (a) 0.2 M KOH and (b) 0.1 M HCl required to titrate completely 200 ml of 0.15 M isoelectric aspartic acid.

Solution

Isoelectric aspartic acid is an intermediate ion. It may be titrated in both directions as shown below.

$$
\begin{array}{cccc}
\text{COOH} & \text{COO}^- & \text{COO}^- & \text{COO}^- \\
| & | & | & | \\
\overset{+}{\text{H}_3\text{N}}\text{—C—H} \xleftarrow{\text{H}^+} & \overset{+}{\text{H}_3\text{N}}\text{—C—H} \xrightarrow{\text{OH}^-} & \overset{+}{\text{H}_3\text{N}}\text{—C—H} \xrightarrow{\text{OH}^-} & \text{H}_2\text{N—C—H} \\
| & | & | & | \\
\text{CH}_2 & \text{CH}_2 & \text{CH}_2 & \text{CH}_2 \\
| & | & | & | \\
\text{COOH} & \text{COOH} & \text{COO}^- & \text{COO}^- \\
[\text{AA}^{+1}] & [\text{AA}^0] & [\text{AA}^{-1}] & [\text{AA}^{-2}]
\end{array}
$$

We can see that 2 moles of OH$^-$ are required per mole of AA^0 to titrate completely the amino acid with the base; only 1 mole of H$^+$ per mole of AA^0 is required for the titration in the acid direction.

(a)   liters $\times$ M = number of moles AA^0

$$0.20 \times 0.15 = 0.03 \text{ mole of AA}^0$$

$$\therefore \quad 2 \times 0.03 = 0.06 \text{ mole OH}^- \text{ required}$$

liters $\times$ M = number of moles OH$^-$

liters $\times$ 0.20 = 0.06 mole

$$\text{liters} = \frac{0.06}{0.2} = 0.30 \text{ liter}$$

$$\boxed{\therefore \quad \textbf{300 ml of 0.2 } M \textbf{ KOH required}}$$

(b)   To titrate 0.03 mole of AA^0 to AA^{+1}, 0.03 mole of H$^+$ is required.

liters $\times$ M = number of moles H$^+$

$$\text{liters} = \frac{0.03}{0.10} = 0.3 \text{ liter}$$

$$\boxed{\therefore \quad \textbf{300 ml of 0.1 } M \textbf{ HCl required}}$$

Problem III-7

Calculate the appropriate values and plot the pH curve for the titration of 200 ml of 0.1 M glycine hydrochloride with 0.1 M KOH.

Solution   (See Figures AI-5 and AI-6.)

Titration curve calculations for amino acids are done in exactly the same way as for other polyprotic acids.

1.   First calculate the starting pH as shown in Problem III-2; pH = 1.715.

2.   Next calculate the pH at the first equivalence point. At the first equivalence point, all the glycine hydrochloride $[AA^{+1}]$ is converted to isoelectric glycine $[AA^0]$. The pH at the first equivalence point is 5.97, as shown in Problem III-2.

3.   Next calculate the pH after various additions of KOH to the glycine hydrochloride solution.   Use the Henderson-Hasselbalch equation. Because we are in the region between AA^{+1} and AA^0, the pK_a value to be used is pK_{a_1} (i.e., K_{a_1} describes the equilibrium between AA^{+1} and AA^0). Assume that the number of moles of AA^{+1} remaining = moles AA^{+1}_{orig} − moles $AA^{+1}_{titrated}$, and that the number of moles of AA^0 produced = moles $AA^{+1}_{titrated}$.

For example, we started with 0.2 liter of 0.1 M AA^{+1} (0.02 mole). After adding 50 ml of 0.1 M KOH (0.005 mole), 0.005 mole of AA^0 is formed, and $0.020 − 0.005 = 0.015$ mole of AA^{+1} remain. The volume of the solution at this point is $0.2 + 0.05$ liter = 0.25 liter. The concentrations of each ionic species can be calculated.

$$[AA^{+1}] = \frac{0.015 \text{ mole}}{0.25 \text{ liter}} = 0.06 \; M$$

$$[AA^0] = \frac{0.005 \text{ mole}}{0.25 \text{ liter}} = 0.02 \; M$$

The total concentration at this point has decreased to 0.08 M because of dilution by the KOH solution.

$$pH = pK_{a_1} + \log \frac{[\text{conjugate base}]}{[\text{conjugate acid}]}$$

The AA^{+1} contains more ionizable hydrogens than AA^0. ∴ AA^{+1} = conjugate acid and AA^0 = conjugate base.

$$pH = pK_{a_1} + \log \frac{[AA^0]}{[AA^{+1}]}$$

$$pH = 2.34 + \log \frac{0.02}{0.06}$$

$$= 2.34 - \log \frac{0.06}{0.02}$$

$$= 2.34 - \log 3$$

$$= 2.34 - 0.466$$

$$\boxed{pH = 1.874}$$

Check: The pH at the start was 1.715. The addition of base should increase the pH. When $[AA^{+1}] = [AA^0]$, pH $= pK_{a_1}$. When $[AA^{+1}] >$ $[AA^0]$, the pH should be less than pK_{a_1}.

4. Next calculate the pH at the second equivalence point. At the second equivalence point, all the glycine is in the sodium glycinate $[AA^{-1}]$ form. The pH at the second equivalence point is established by the ionization of the strongest basic group (the α-NH_2 group). Although we started with $0.1\ M$ glycine hydrochloride, the concentration is less at the end of the titration because of dilution by the KOH. To titrate completely 0.02 mole of AA^{+1} to AA^{-1}, 0.04 mole of OH^- must be added.

$$\text{liters} \times M = \text{number of moles } OH^-$$
$$\text{liters} \times 0.1 = 0.04 \text{ mole}$$
$$\text{liters} = \frac{0.04}{0.1} = 0.4 \text{ liter} = 400 \text{ ml}$$

The final volume is 200 ml + 400 ml = 600 ml. The concentration of

$$AA^{-1} = \frac{0.02 \text{ mole}}{0.6 \text{ liter}} = 0.033\ M$$

$$K_{b_1} = \frac{[AA^0][OH^-]}{[AA^{-1}]}$$

The $[OH^-]$ and subsequently $[H^+]$ and pH may be calculated as shown in Problem III-2. Let:

$$y = M \text{ of } AA^0 \text{ formed}$$

and $$\therefore\ y = M \text{ of } OH^- \text{ formed}$$

$$0.033 - y = M \text{ of } AA^{-1} \text{ remaining at equilibrium}$$

5. Next calculate the pH during the titration between the first and second equivalence points.

$$\text{pH} = pK_{a_2} + \log \frac{[AA^{-1}]}{[AA^0]}$$

Again you may assume that the number of moles of $AA^0 =$ moles AA^0_{orig} − moles $AA^0_{titrated}$, and the number of moles of $AA^{-1} =$ moles $AA^0_{titrated}$. Calculate the concentrations of the ionic species, taking into account the increased volume caused by addition of KOH.

6. The pH beyond the last equivalence point depends on the concentration of excess OH^- present.

$$\text{pOH} = -\log [OH^-] \qquad \text{pH} = 14 - \text{pOH}$$

Problem III-8

Sketch the pH curve for the titration of 100 ml of $0.1\ M$ alanine hydrochloride with KOH solution (a) in the absence and (b) in the presence of excess formaldehyde.

Solution

(a)   The titration curve resembles that of a typical diprotic acid with two buffering plateaus in the regions of the pK_a values. It takes 1 mole of OH^- per mole of amino acid to go from the starting point to the first end (equivalence) point and 0.5 mole of OH^- per mole of amino acid to get to the midway (the pK_{a_1}) position. To go from the first end point to the second end (equivalence) point, another mole of KOH per mole of amino acid is required; 1.5 moles of KOH per mole of amino acid hydrochloride bring the pH to the point midway between the first and second equivalence points (to the pK_{a_2} value).

(b)   Formaldehyde reacts with amino groups to form methylol derivatives.

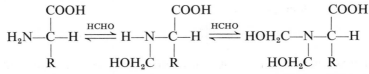

The methylol derivatives are stronger acids (weaker bases) than are the original unsubstituted amino groups. In other words, the pK_{a_2} value for the substituted amino acid is lower than the pK_{a_2} value for the original amino acid. The titration curves are sketched in Figure III-1. Note that formaldehyde has no effect on the amounts of KOH required to titrate the amino acid to pK_{a_1}, pK_{a_2} (or pK'_{a_2}), and the equivalence points. Also note that only the pK_{a_2} value is shifted; formaldehyde has no effect on

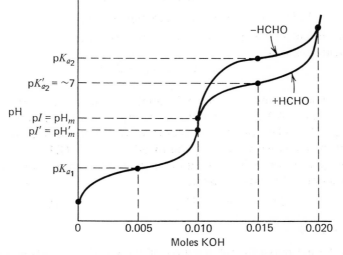

Figure III-1   Titration curve of alanine hydrochloride in the presence and in the absence of formaldehyde

the α-COOH group. Because the value of pK_{a_2} is lower in the presence of formaldehyde, the isoelectric point, pI, also is lower. (See Problem III-10 for calculations of pI.)

Problem III-9

Sketch the pH curve for the titration of 1 mole of the hydrochlorides of: (a) alanylglutamylserine, (b) dilysylglycine, (c) glutamylserylglutamylvaline, (d) γ-glutamylglycine, and (e) histidyltyrosine.

Solution

You can sketch the titration curves for peptides by employing the same reasoning as for amino acids.† Bear in mind, however, that the amino and carboxyl groups used in forming the peptide bonds are no longer present and, consequently, cannot be titrated. Furthermore, if two or more identical groups are present, it takes an equivalent amount of titrant to neutralize them. For clarity, the vertical axes of the curves below are not drawn to scale.

(a) Alanylglutamylserine Hydrochloride—The titratable groups are shown with an asterisk.

$$(pK_{a_3} = \sim 9.5)\ \overset{*}{\underset{+}{H_3N}}\text{—}C\text{—}C\text{—}N\text{—}C\text{—}C\text{—}N\text{—}\overset{*}{C}\text{—}COOH\ (pK_{a_1} = \sim 2.5)$$

$$Cl^- \qquad O \qquad CH_2 \qquad H \quad CH_2$$

$$CH_3 \qquad CH_2 \qquad OH$$

$$*COOH$$

$$(pK_{a_2} = \sim 4)$$

The curve in Figure III-2 shows the titration of one α-COOH group, one γ-COOH group, and one α-NH$_3^+$ group.

(b) Dilysylglycine Hydrochloride

$$(pK_{a_2} = \sim 9.5)\ \overset{*}{\underset{+}{H_3N}}\text{—}C\text{—}C\text{—}N\text{—}C\text{—}C\text{—}N\text{—}\overset{*}{C}\text{—}COOH\ (pK_{a_1} = \sim 2.5)$$

$$Cl^- \qquad H \qquad O \qquad H$$

$$[CH_2]_4 \qquad [CH_2]_4$$

$$*NH_3^+Cl^- \quad *NH_3^+Cl^-$$

$$(pK_{a_3} = \sim 10.5) \quad (pK_{a_3} = \sim 10.5)$$

† For convenience we shall assume that the pK_a values of the ionizable groups in peptides are the same as they are in the free amino acids. The actual values depend on the size of the peptide and the proximity of other ionizable groups.

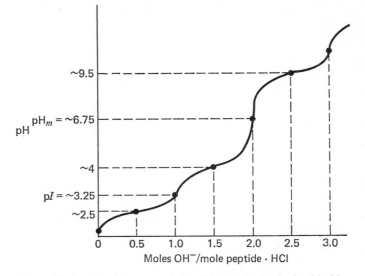

Figure III-2   Titration curve of alanylglutamylserine hydrochloride

In sketching the curve in Figure III-3, we assumed that the two
ϵ-NH$_3^+$ groups of lysine retain a pK_a of ~10.5 in the peptide. The curve
shows the titration of one α-COOH group, one α-NH$_3^+$ group, and two
ϵ-NH$_3^+$ groups.

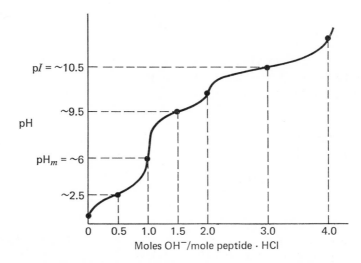

Figure III-3   Titration curve of dilysylglycine hydrochloride

(c)   Glutamylserylglutamylvaline Hydrochloride

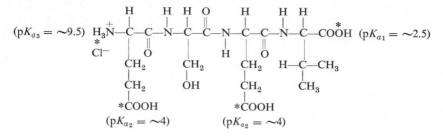

The curve in Figure III-4 shows the titration of one α-COOH group, two γ-COOH groups, and one α-NH$_3^+$ group. We assumed that the two γ-COOH groups of glutamic acid retain pK_a values of $\sim$4 in the peptide.

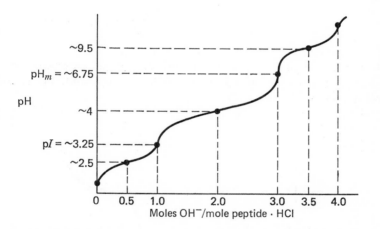

Figure III-4   Titration curve of glutamylserylglutamylvaline hydrochloride

(d) γ-Glutamylglycine Hydrochloride

In sketching the curve in Figure III-5, we assumed that the two α-COOH groups retain a pK_a of ~2.5 in the peptide. Actually, the α-COOH groups of glutamic acid and glycine have slightly different pK_a values to begin with and their values may change slightly when the amino acids form peptides.

(e) Histidyltyrosine Hydrochloride

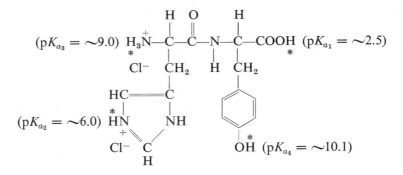

The curve in Figure III-6 shows the titration of one α-carboxyl group, one histidine imidazole group, one α-amino group, and one tyrosine phenol group.

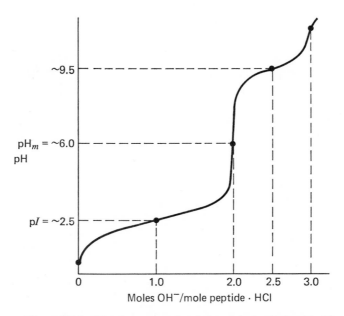

Figure III-5 Titration curve of γ-glutamylglycine hydrochloride

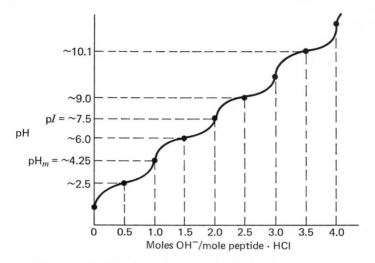

Figure III-6 Titration curve of histidyltyrosine hydrochloride

Problem III-10

Calculate the "isoelectric point," pI, and the pH at which the compound carries the maximum number of charges, pH_m, for: (a) glycine, (b) aspartic acid, (c) lysine, (d) lysylalanine, and (e) lysyllysine.

Solution

The isoelectric point, pI, is the pH at which the amino acid or peptide carries no net charge; that is, the predominant ionic form is the isoelectric species, AA^0, and (because the isoelectric form ionizes both as an acid and as a base) there are equal amounts of the ionic forms AA^{+1} and AA^{-1}. (The ionization of AA^0 to form AA^{+1} and AA^{-1} is a disproportionation reaction, as described earlier.) The pI may also be thought of as the pH of a solution of the isoelectric form of the amino acid.

The pI of amino acids is the pH at one equivalence point along the titration curve, specifically the equivalence point at which all the AA^{+1} is converted to AA^0. The pH at this point is, as usual, the average of the pK_a value to follow and the pK_a value just passed. Similarly, pH_m is the pH at one equivalence point and may be similarly calculated. To determine pI and pH_m simply sketch the titration curve and indicate the predominant ionic species present at each key point. Or, prepare a table showing the ionic form of each titratable group at key points. For simplicity, assume that you are starting with the maximally protonated amino acid or peptide.

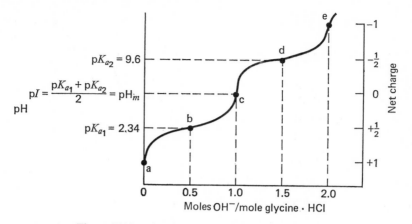

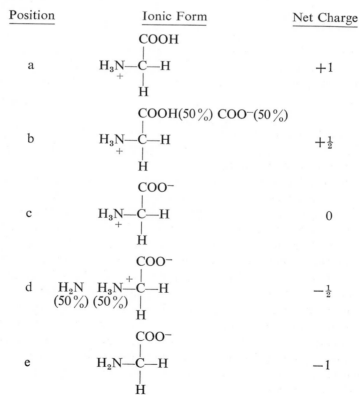

Figure III-7 Titration curve of glycine hydrochloride

(a)   Glycine—The following positions on the titration curve are shown in Figure III-7.

Position	Ionic Form	Net Charge
a	$\begin{array}{c} \text{COOH} \\ | \\ \overset{+}{H_3N}\text{—C—H} \\ | \\ \text{H} \end{array}$	$+1$
b	$\begin{array}{c} \text{COOH(50\%) COO}^-\text{(50\%)} \\ | \\ \overset{+}{H_3N}\text{—C—H} \\ | \\ \text{H} \end{array}$	$+\frac{1}{2}$
c	$\begin{array}{c} \text{COO}^- \\ | \\ \overset{+}{H_3N}\text{—C—H} \\ | \\ \text{H} \end{array}$	0
d	$\begin{array}{cc} & \text{COO}^- \\ H_2N\ \ \overset{+}{H_3N}\text{—C—H} \\ (50\%)\ (50\%)\ \ | \\ & \text{H} \end{array}$	$-\frac{1}{2}$
e	$\begin{array}{c} \text{COO}^- \\ | \\ H_2N\text{—C—H} \\ | \\ \text{H} \end{array}$	-1

We can see that glycine has a net zero charge at position c. At position c, glycine also bears the maximum total number of charges. Therefore, pI is also the pH_m for glycine.

$$pI = pH_m = \frac{pK_{a_1} + pK_{a_2}}{2} = \frac{2.34 + 9.6}{2} = \frac{11.94}{2} = 5.97$$

A table containing the same information could be arranged as shown in Table III-2.

TABLE III-2

Ionizable Group	Start (a)	pK_{a_1}(b)	First Equivalence (c)	pK_{a_2}(d)	Second Equivalence (e)
		Predominant Ionic Form at Different Positions along the Titration Curve			
α-carboxyl	COOH	COOH($\frac{1}{2}$) COO$^-$($\frac{1}{2}$)	COO$^-$	COO$^-$	COO$^-$
α-amino	NH$_3^+$	NH$_3^+$	NH$_3^+$	NH$_3^+$($\frac{1}{2}$) NH$_2$($\frac{1}{2}$)	NH$_2$
Net charge	+1	+$\frac{1}{2}$	0	−$\frac{1}{2}$	−1

(b) Aspartic acid—We can see from Table III-3 and Figure III-8 that aspartic acid carries no net charge at the first equivalence point.

TABLE III-3

Ionizable Group	Start	pK_{a_1}	First Equivalence	pK_{a_2}	Second Equivalence	pK_{a_3}	Third Equivalence
			Predominant Ionic Form at Different Positions along the Titration Curve				
α-carboxyl	COOH	COOH($\frac{1}{2}$) COO$^-$($\frac{1}{2}$)	COO$^-$	COO$^-$	COO$^-$	COO$^-$	COO$^-$
β-carboxyl	COOH	COOH	COOH	COOH($\frac{1}{2}$) COO$^-$($\frac{1}{2}$)	COO$^-$	COO$^-$	COO$^-$
α-amino	NH$_3$+	NH$_3$+	NH$_3$+	NH$_3$+	NH$_3$+	NH$_3$+($\frac{1}{2}$) NH$_2$($\frac{1}{2}$)	NH$_2$
Net charge	+1	+$\frac{1}{2}$	0	−$\frac{1}{2}$	−1	−1$\frac{1}{2}$	−2

$$pI = \frac{pK_{a_1} + pK_{a_2}}{2} = \frac{2.1 + 3.86}{2} = \frac{5.96}{2} = 2.98$$

We can also see that aspartic acid carries the maximum total number of charges at the second equivalence point.

$$pH_m = \frac{pK_{a_2} + pK_{a_3}}{2} = \frac{3.86 + 9.82}{2} = \frac{13.68}{2} = 6.84$$

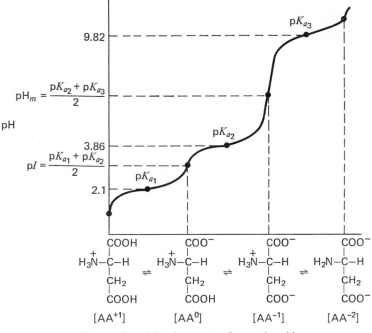

Figure III-8 Titration curve of aspartic acid

When constructing Table III-3, we assumed that at the first equivalence point the α-carboxyl is completely ionized and that the β-carboxyl is completely unionized. These assumptions, of course, are not entirely true; the actual degree to which the α- and β-carboxyls are ionized can be calculated using the Henderson-Hasselbalch equation. If we carry out the calculation, we find that the proportion of α-carboxyl that is still in the COOH form exactly equals the proportion of β-carboxyl in the COO^- form. (At pH 2.98, we are just as far above the pK_{a_1} for the α-carboxyl as we are below the pK_{a_2} for the β-carboxyl.) Thus, to determine the net charge on the molecule, we are justified in tallying only the predominant ionic forms at each key point along the titration curve.

(c) Lysine—We see from Table III-4 and Figure III-9 that lysine carries no net charge at the second equivalence point.

$$pI = \frac{pK_{a_2} + pK_{a_3}}{2} = \frac{8.95 + 10.53}{2} = \frac{19.48}{2} = 9.74$$

We also see that lysine carries the maximum total number of charges at

TABLE III-4

Ionizable Group	Predominant Ionic Form at Different Positions along the Titration Curve						
	Start	pK_{a_1}	First Equivalence	pK_{a_2}	Second Equivalence	pK_{a_3}	Third Equivalence
α-carboxyl	COOH	COOH(½) COO⁻(½)	COO⁻	COO⁻	COO⁻	COO⁻	COO⁻
α-amino	NH₃⁺	NH₃⁺	NH₃⁺	NH₃⁺(½) NH₂(½)	NH₂	NH₂	NH₂
ε-amino	NH₃⁺	NH₃⁺	NH₃⁺	NH₃⁺	NH₃⁺	NH₃⁺(½) NH₂(½)	NH₂
Net charge	+2	+1½	+1	+½	0	-½	-1

the first equivalence point (all ionizable groups charged).

$$pH_m = \frac{pK_{a_1} + pK_{a_2}}{2} = \frac{2.18 + 8.95}{2} = \frac{11.13}{2} = 5.57$$

We assumed that at the second equivalence point the α-amino group is completely uncharged and the ε-amino group is completely ionized. These assumptions are valid for calculating the net charge on lysine for the same reasons described earlier concerning aspartic acid.

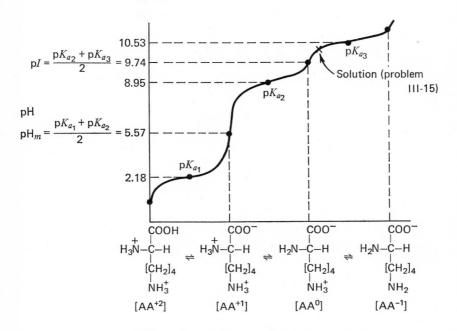

Figure III-9 Titration curve of lysine

(d) Lysylalanine

$$(\text{p}K_{a_2} = 8.95)\ \underset{+}{H_3N}\!\!-\!\!\overset{\displaystyle \overset{H}{|}}{\underset{\displaystyle \underset{[CH_2]_4}{|}}{C}}\!\!-\!\!\overset{\displaystyle \overset{O}{\|}}{C}\!\!-\!\!\overset{H}{N}\!\!-\!\!\overset{\displaystyle \overset{H}{|}}{\underset{\displaystyle \underset{CH_3}{|}}{C}}\!\!-\!\!COOH\ (\text{p}K_{a_1} = 2.34)$$

$$\underset{+}{NH_3}$$

$$(\text{p}K_{a_3} = 10.53)$$

TABLE III-5

Ionizable Group	Start	$\text{p}K_{a_1}$	First Equivalence	$\text{p}K_{a_2}$	Second Equivalence	$\text{p}K_{a_3}$	Third Equivalence
			Predominant Ionic Form at Different Positions along the Titration Curve				
α-carboxyl	COOH	COOH($\frac{1}{2}$) COO$^-$($\frac{1}{2}$)	COO$^-$	COO$^-$	COO$^-$	COO$^-$	COO$^-$
α-amino	NH$_3{}^+$	NH$_3{}^+$	NH$_3{}^+$	NH$_3{}^+$($\frac{1}{2}$) NH$_2$($\frac{1}{2}$)	NH$_2$	NH$_2$	NH$_2$
ε-amino	NH$_3{}^+$	NH$_3{}^+$	NH$_3{}^+$	NH$_3{}^+$	NH$_3{}^+$	NH$_3{}^+$($\frac{1}{2}$) NH$_2$($\frac{1}{2}$)	NH$_2$
Net charge	+2	+1$\frac{1}{2}$	+1	+$\frac{1}{2}$	0	−$\frac{1}{2}$	−1

$\text{p}I = \text{pH at second equivalence point}$

$$\text{p}I = \frac{\text{p}K_{a_3} + \text{p}K_{a_2}}{2} = \frac{10.53 + 8.95}{2} = \frac{19.48}{2} = 9.74$$

$\text{pH}_m = \text{pH at first equivalence point}$

$$\text{pH}_m = \frac{\text{p}K_{a_1} + \text{p}K_{a_2}}{2} = \frac{2.34 + 8.95}{2} = \frac{11.29}{2} = 5.65$$

(e) Lysyllysine

$$(\text{p}K_{a_2} = 8.95)\ \underset{+}{H_3N}\!\!-\!\!\overset{\displaystyle \overset{H}{|}}{\underset{\displaystyle \underset{[CH_2]_4}{|}}{C}}\!\!-\!\!\overset{\displaystyle \overset{O}{\|}}{C}\!\!-\!\!\overset{H}{N}\!\!-\!\!\overset{\displaystyle \overset{H}{|}}{\underset{\displaystyle \underset{[CH_2]_4}{|}}{C}}\!\!-\!\!COOH\ (\text{p}K_{a_1} = 2.2)$$

$$\underset{+}{NH_3}\qquad\qquad \underset{+}{NH_3}$$

$$(\text{p}K_{a_3} = 10.53)\ (\text{p}K_{a_3} = 10.53)$$

Note that for this peptide pI is not at an equivalence point but rather at a pK value.

$$\text{p}I = \text{p}K_{a_3} = 10.53$$

$\text{pH}_m = \text{pH at first equivalence}$

$$= \frac{\text{pH}_{a_2} + \text{p}K_{a_1}}{2} = \frac{8.95 + 2.2}{2} = \frac{11.15}{2} = 5.58$$

TABLE III-6

Ionizable Group	Start	pK_{a_1}	First Equivalence	pK_{a_2}	Second Equivalence	pK_{a_3}	Third Equivalence
			Predominant Ionic Form at Different Positions along the Titration Curve				
α-carboxyl	COOH	COOH($\frac{1}{2}$) COO⁻($\frac{1}{2}$)	COO⁻	COO⁻	COO⁻	COO⁻	COO⁻
α-amino	NH₃⁺	NH₃⁺	NH₃⁺	NH₃⁺($\frac{1}{2}$) NH₂($\frac{1}{2}$)	NH₂	NH₂	NH₂
ε-amino (1)	NH₃⁺	NH₃⁺	NH₃⁺	NH₃⁺	NH₃⁺	NH₃⁺($\frac{1}{2}$) NH₂($\frac{1}{2}$)	NH₂
(2)	NH₃⁺	NH₃⁺	NH₃⁺	NH₃⁺	NH₃⁺	NH₃⁺($\frac{1}{2}$) NH₂($\frac{1}{2}$)	NH₂
Net charge	+3	+2$\frac{1}{2}$	+2	+1$\frac{1}{2}$	+1	0	−1

Note also that it takes 4 moles of OH⁻ per mole of peptide hydrochloride to get to the third equivalence point because of the two ε-amino groups.

Problem III-11

In which direction in an electric field will the peptide diglycylaspartyllysine move at the following pH values: (a) 1.0, (b) 3.0, (c) 4.0, (d) 6.75, (e) 8.0, (f) 9.5, and (g) 10.5?

Solution

An amino acid or peptide migrates toward the anode (the positive electrode) when it carries a net negative charge and toward the cathode (the negative electrode) when it carries a net positive charge. To determine the direction in which diglycylaspartyllysine migrates at the above pH values, simply determine the p*I* of the peptide. At pH values above p*I*, the molecule carries a net negative charge; at pH values below p*I*, the molecule carries a net positive charge. Diglycylaspartyllysine has a p*I* of about 6.75 (calculated as shown in the previous problems).

C. DISPROPORTIONATION REACTIONS OF AMINO ACIDS

For a general discussion of disproportionation reactions, see Appendix I-K. Also review Problems II-23, II-24, and II-25. Disproportionation equilibria relationships are useful for calculating the concentrations of the different ionic species present in a solution of an intermediate ion of an amino acid.

Problem III-12

Write equations showing the ionization and disproportionation reactions of isoelectric glycine in water.

Solution

As shown earlier, isoelectric glycine ionizes both as an acid and as a base in water.

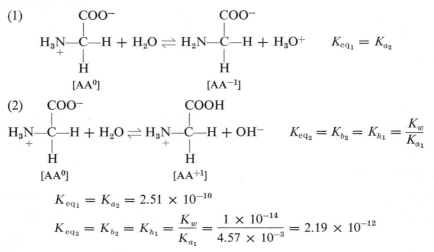

(1)

$$\underset{[AA^0]}{\overset{COO^-}{H_3\overset{+}{N}-\underset{H}{\overset{|}{C}}-H}} + H_2O \rightleftharpoons \underset{[AA^{-1}]}{\overset{COO^-}{H_2N-\underset{H}{\overset{|}{C}}-H}} + H_3O^+ \qquad K_{eq_1} = K_{a_2}$$

(2)

$$\underset{[AA^0]}{\overset{COO^-}{H_3\overset{+}{N}-\underset{H}{\overset{|}{C}}-H}} + H_2O \rightleftharpoons \underset{[AA^{+1}]}{\overset{COOH}{H_3\overset{+}{N}-\underset{H}{\overset{|}{C}}-H}} + OH^- \qquad K_{eq_2} = K_{b_2} = K_{h_1} = \frac{K_w}{K_{a_1}}$$

$$K_{eq_1} = K_{a_2} = 2.51 \times 10^{-10}$$

$$K_{eq_2} = K_{b_2} = K_{h_1} = \frac{K_w}{K_{a_1}} = \frac{1 \times 10^{-14}}{4.57 \times 10^{-3}} = 2.19 \times 10^{-12}$$

The larger value of K_{a_2} indicates that the solution will be acidic. However, the solution will not be as acidic as we might expect because of a compensating reaction that takes place. A large portion of the excess H_3O^+ produced in reaction 1 reacts with unreacted isoelectric glycine according to equation 3.

(3)

$$\underset{[AA^0]}{\overset{COO^-}{H_3\overset{+}{N}-\underset{H}{\overset{|}{C}}-H}} + H_3O^+ \rightleftharpoons \underset{[AA^{+1}]}{\overset{COOH}{H_3\overset{+}{N}-\underset{H}{\overset{|}{C}}-H}} + H_2O \qquad K_{eq_3} = \frac{1}{K_{a_1}}$$

$$K_{eq_3} = \frac{1}{4.57 \times 10^{-3}} = 2.19 \times 10^2$$

Thus the major reactions taking place in solution (i.e., those with the greatest K_{eq} values) are reactions 1 and 3. The overall effect is disproportionation reaction 4.

(1)

$$\underset{[AA^0]}{\overset{COO^-}{H_3\overset{+}{N}-\underset{H}{\overset{|}{C}}-H}} + H_2O \rightleftharpoons \underset{[AA^{-1}]}{\overset{COO^-}{H_2N-\underset{H}{\overset{|}{C}}-H}} + H_3O^+ \qquad K_{eq_1} = K_{a_2}$$

(3)

$$\underset{\underset{+}{}}{H_3N}-\underset{\underset{H}{\overset{\overset{\displaystyle COO^-}{|}}{\underset{|}{C}}}}{}-H + H_3O^+ \rightleftharpoons \underset{\underset{+}{}}{H_3N}-\underset{\underset{H}{\overset{\overset{\displaystyle COOH}{|}}{\underset{|}{C}}}}{}-H + H_2O \qquad K_{eq_3} = \frac{1}{K_{a_1}}$$

$$[AA^0] \qquad\qquad\qquad [AA^{+1}]$$

Sum (4) $\qquad\qquad 2AA^0 \rightleftharpoons AA^{-1} + AA^{+1} \qquad K_{eq_4} = K_{eq_1} \times K_{eq_3}$

$$= \frac{K_{a_2}}{K_{a_1}} = K_{dis}$$

Reaction 4 is the major reaction establishing the concentrations of the various ionic species in a solution of isoelectric glycine. The pH of the solution can be calculated as usual from the pK_{a_1} and pK_{a_2} values.

$$pH = \frac{pK_{a_1} + pK_{a_2}}{2} = \frac{2.34 + 9.6}{2} = \frac{11.94}{2} = 5.97$$

Problem III-13

Write equations showing the ionization and disproportionation reactions of isoelectric lysine in water.

Solution

As shown earlier, isoelectric lysine ionizes as an acid (reaction 1) and as a base (reaction 2).

(1)

$$H_2N-\underset{\underset{\displaystyle NH_3^+}{\overset{\overset{\displaystyle [CH_2]_4}{|}}{\underset{|}{\underset{|}{C}}}}}{}-H + H_2O \rightleftharpoons H_2N-\underset{\underset{\displaystyle NH_2}{\overset{\overset{\displaystyle [CH_2]_4}{|}}{\underset{|}{\underset{|}{C}}}}}{}-H + H_3O^+ \qquad K_{eq_1} = K_{a_3}$$

$$[AA^0] \qquad\qquad\qquad\qquad [AA^{-1}]$$

(2)

$$H_2N-\underset{\underset{\displaystyle NH_3^+}{\overset{\overset{\displaystyle [CH_2]_4}{|}}{\underset{|}{\underset{|}{C}}}}}{}-H + H_2O \rightleftharpoons \underset{+}{H_3N}-\underset{\underset{\displaystyle NH_3^+}{\overset{\overset{\displaystyle [CH_2]_4}{|}}{\underset{|}{\underset{|}{C}}}}}{}-H + OH^- \qquad K_{eq_2} = K_{b_2}$$

$$[AA^0] \qquad\qquad\qquad\qquad [AA^{+1}]$$

$$K_{eq_1} = K_{a_3} = 2.95 \times 10^{-11}$$
$$K_{eq_2} = K_{b_2} = 8.91 \times 10^{-6}$$

We can see from the relative values of K_{a_3} and K_{b_2} that a solution of isoelectric lysine will be basic but not quite as basic as we might expect. The compensating reaction in this case involves the excess OH^- produced in reaction 2 with another molecule of unreacted isoelectric lysine according to reaction 3.

(3)

$$
\begin{array}{ccc}
\text{COO}^- & & \text{COO}^- \\
| & & | \\
\text{H}_2\text{N}-\text{C}-\text{H} + \text{OH}^- \rightleftharpoons \text{H}_2\text{N}-\text{C}-\text{H} + \text{H}_2\text{O} \\
| & & | \\
[\text{CH}_2]_4 & & [\text{CH}_2]_4 \\
| & & | \\
\text{NH}_3^+ & & \text{NH}_2 \\
[\text{AA}^0] & & [\text{AA}^{-1}]
\end{array}
\qquad K_{eq_3} = \frac{1}{K_{b_1}}
$$

$$
K_{eq_3} = \frac{1}{K_{b_1}} = \frac{1}{3.39 \times 10^{-4}} = 2.95 \times 10^3
$$

We can see that the two major reactions taking place in solution are reactions 2 and 3.

(2) $AA^0 + H_2O \rightleftharpoons AA^{+1} + OH^-$ $K_{eq_2} = K_{b_2}$

(3) $AA^0 + OH^- \rightleftharpoons AA^{-1} + H_2O$ $K_{eq_3} = \dfrac{1}{K_{b_1}}$

Sum: (4) $2AA^0 \rightleftharpoons AA^{+1} + AA^{-1}$ $K_{eq_4} = \dfrac{K_{b_2}}{K_{b_1}} = K_{dis}$

$$
K_{b_2} = \frac{K_w}{K_{a_2}} \qquad K_{b_1} = \frac{K_w}{K_{a_3}}
$$

$$
K_{dis} = \frac{K_{b_2}}{K_{b_1}} = \frac{K_w}{K_{a_2}} \times \frac{K_{a_3}}{K_w} = \frac{K_{a_3}}{K_{a_2}}
$$

As usual, the K_{dis} is equal to the ratio of the K_a for the *next* acid ionization to the K_a for the *previous* acid ionization. The disproportionation reaction is the major reaction taking place that establishes the concentrations of AA^0, AA^{+1}, and AA^{-1} in solution. The pH of the solution can be calculated from the pK_{a_3} and pK_{a_2} values.

$$
\text{pH} = \frac{pK_{a_3} + pK_{a_2}}{2} = \frac{10.53 + 8.95}{2} = \frac{19.48}{2} = 9.74
$$

Problem III-14

What are the concentrations of the major ionic species present in a 0.1 M solution of monosodium aspartate?

Solution

Monosodium aspartate (AA^{-1}) undergoes disproportionation to yield equimolar amounts of the ionic species on either side of it on the aspartic acid titration curve (see Figure III-8). The major ionic species present then are AA^{-1}, AA^0, and AA^{-2}. Only an extremely small amount of A^+A^1 forms.

In method 1:

$$2AA^{-1} \rightleftharpoons AA^0 + AA^{-2} \qquad K_{dis} = \frac{K_{a3}}{K_{a2}}$$

Let:

$$y = M \text{ of } AA^{-1} \text{ that undergoes disproportionation}$$

$$\therefore \quad \frac{y}{2} = M \text{ of } AA^0 \text{ that forms}$$

and

$$\frac{y}{2} = M \text{ of } AA^{-2} \text{ that forms}$$

and

$$(0.1 - y) = M \text{ of } AA^{-1} \text{ remaining at equilibrium}$$

$$K_{dis} = \frac{K_{a3}}{K_{a2}} = \frac{1.51 \times 10^{-10}}{1.38 \times 10^{-4}} = 1.095 \times 10^{-6}$$

$$K_{dis} = \frac{(y/2)(y/2)}{(0.1 - y)^2} = 1.095 \times 10^{-6}$$

$$\frac{y^2}{4(0.1 - y)^2} = 1.095 \times 10^{-6} \qquad \frac{y^2}{(0.1 - y)^2} = 4.38 \times 10^{-6}$$

$$\sqrt{\frac{y^2}{(0.1 - y)^2}} = \sqrt{4.38 \times 10^{-6}} \qquad \frac{y}{0.1 - y} = 2.09 \times 10^{-3}$$

$$2.09 \times 10^{-4} - 2.09 \times 10^{-3}y = y$$

$$1.00209y = 2.09 \times 10^{-4} \qquad y = \frac{2.09 \times 10^{-4}}{1.0021}$$

$$y = 2.09 \times 10^{-4}$$

$$[AA^0] = \frac{y}{2} = \boxed{\textbf{1.045} \times \textbf{10}^{-4} \textbf{\textit{M}}}$$

$$[AA^{-2}] = \frac{y}{2} = \boxed{\textbf{1.045} \times \textbf{10}^{-4} \textbf{\textit{M}}}$$

$$[AA^{-1}] = 0.1 - 2.09 \times 10^{-4} = \boxed{\textbf{0.0998 \textit{M}}}$$

The pH of the solution may be calculated from any K_{eq} expression involving H^+ or OH^- and AA^0, AA^{-1}, or AA^{-2}. Alternately, we know that the pH equals the average of pK_{a_2} and pK_{a_3}. Knowing the pH, we could calculate the concentrations of AA^{-1}, AA^0, and AA^{-2} using the Henderson-Hasselbalch equation as shown below.

In method 2:

$$pH = \frac{pK_{a_2} + pK_{a_3}}{2} = \frac{3.86 + 9.82}{2} = \frac{13.68}{2} = 6.84$$

$$pH = pK_{a_2} + \log \frac{[AA^{-1}]}{[AA^0]} \qquad\qquad pH = pK_{a_3} + \log \frac{[AA^{-2}]}{[AA^{-1}]}$$

$$6.84 = 3.86 + \log \frac{[AA^{-1}]}{[AA^0]} \qquad\qquad 6.84 = 9.82 + \log \frac{[AA^{-2}]}{[AA^{-1}]}$$

$$2.98 = \log \frac{[AA^{-1}]}{[AA^0]} \qquad\qquad -2.98 = \log \frac{[AA^{-2}]}{[AA^{-1}]}$$

$$\frac{[AA^{-1}]}{[AA^0]} = \text{antilog of } 2.98 \qquad\qquad +2.98 = \log \frac{[AA^{-1}]}{[AA^{-2}]}$$

$$\qquad\qquad\qquad\qquad\qquad \frac{[AA^{-1}]}{[AA^{-2}]} = \text{antilog of } 2.98$$

$$\frac{[AA^{-1}]}{[AA^0]} = 955 = \frac{955}{1} \qquad\qquad \frac{[AA^{-1}]}{[AA^{-2}]} = 955 = \frac{955}{1}$$

$\therefore$ The ratios of AA^0, AA^{-1}, and AA^{-2} are $1:955:1$; that is, $955/957$ of the total ($0.1\ M$) are AA^{-1}, $1/957$ of the total is AA^0, and $1/957$ of the total is AA^{-2}.

$$\tfrac{955}{957} \times 0.1\ M = \boxed{\mathbf{0.0998\ M\ AA^{-1}}}$$

$$\tfrac{1}{957} \times 0.1\ M = \boxed{\mathbf{1.045 \times 10^{-4}\ M\ AA^{-2}}}$$

and

$$\tfrac{1}{957} \times 0.1\ M = \boxed{\mathbf{1.045 \times 10^{-4}\ M\ AA^{0}}}$$

Problem III-15

What are the concentrations of the major ionic species present in $0.2\ M$ solution of lysine at pH 10.0?

Solution

The titration curve for lysine is shown in Figure III-9. We can see that at pH 10.0, the major ionic form present is isoelectric lysine, AA^0. However, because of the proximity of the pK_{a_3} value, an appreciable amount of AA^{-1} is also present. The amount of AA^{+1} present is much less, and the amount of AA^{+2} is extremely small. Nevertheless, it is possible to calculate the concentrations of each species present using the Henderson-Hasselbalch equation and the appropriate pK_a values. First calculate the ratio of AA^0/AA^{-1} using pK_{a_3}.

$$pH = pK_{a_3} + \log \frac{[AA^{-1}]}{[AA^0]}$$

$$10.0 = 10.53 + \log \frac{[AA^{-1}]}{[AA^0]}$$

$$-0.53 = \log \frac{[AA^{-1}]}{[AA^0]}$$

$$+0.53 = \log \frac{[AA^0]}{[AA^{-1}]}$$

$$\frac{[AA^0]}{[AA^{-1}]} = \text{antilog of } 0.53 = 3.38 = \frac{3.38}{1}$$

Next, calculate the ratio of AA^0/AA^{+1} using pK_{a_2}.

$$pH = pK_{a_2} + \log \frac{[AA^0]}{[AA^{+1}]}$$

$$10.0 = 8.95 + \log \frac{[AA^0]}{[AA^{+1}]}$$

$$1.05 = \log \frac{[AA^0]}{[AA^{+1}]}$$

$$\frac{[AA^0]}{[AA^{+1}]} = \text{antilog of } 1.05 = 11.2 = \frac{11.2}{1}$$

Thus, the ratios of the major ionic species are:

AA^{+1}	:	AA^0	:	AA^{-1}
		3.38	:	1
1	:	11.2		

The ratio can be expressed in terms of a common component, AA^0. For example, if there is 1 part of AA^{+1} for every 11.2 parts of AA^0, calculate how many parts of AA^{+1} there are for every 3.38 parts of AA^0.

$$\frac{1}{11.2} = \frac{X}{3.38} \qquad X = \frac{3.38}{11.2} = 0.3$$

AA^{+1}	:	AA^0	:	AA^{-1}	Total
0.3	:	3.38	:	1	4.68 parts

Check: The pH 10 is near the second equivalence point and closer to the third equivalence point than to the first. Therefore, the relative proportions of ionic forms should be $AA^0 > AA^{-1} > AA^{+1}$. The concentrations of the ionic forms can be calculated from their ratios as usual.

$$[AA^0] = \frac{3.38}{4.68} \text{ of the total } (0.2 \ M)$$

$$= \frac{3.38}{4.68} \times 0.2 \ M = \boxed{\mathbf{0.1442 \ M}}$$

$$[AA^{-1}] = \frac{1}{4.68} \times 0.2 \ M = \boxed{\mathbf{0.0428 \ M}}$$

$$[AA^{+1}] = \frac{0.3}{4.68} \times 0.2 \ M = \boxed{\mathbf{0.0129 \ M}}$$

The concentration of the minor form AA^{+2} can be calculated by using the Henderson-Hasselbalch equation and pK_{a_1}.

$$pH = pK_{a_1} + \log \frac{[AA^{+1}]}{[AA^{+2}]}$$

$$10.0 = 2.18 + \log \frac{[AA^{+1}]}{[AA^{+2}]}$$

$$7.82 = \log \frac{[AA^{+1}]}{[AA^{+2}]}$$

$$\frac{[AA^{+1}]}{[AA^{+2}]} = \text{antilog of } 7.82 = 10^{7.82}$$

Thus, at pH 10 the concentration of AA^{+2} is less than one ten-millionth of AA^{+1}. Because $[AA^{+1}]$ itself is very small, the concentration of AA^{+2} is quite negligible.

D. AMINO ACID BUFFERS

Amino acids and peptides can be used to prepare buffers. Calculations are performed in exactly the same manner as for other carboxylic acids and amines.

Problem III-16

Describe the preparation of 2 liters of a 0.1 M lysine buffer, pH 9.25, starting from solid lysine dihydrochloride [AA^{+2}] and 1 M KOH.

Solution

The desired pH is in the region of pK_{a_2} as shown on the titration curve in Figure III-10. The major ionic species present are isoelectric lysine [AA^0] and lysine monohydrochloride [AA^{+1}], with the former predominating. To obtain the desired pH, sufficient base must first be added to convert all of the original AA^{+2} to AA^{+1}. Then additional base must be added to obtain the proper ratio of [AA^0]/[AA^{+1}]. We start by dissolving 0.2 mole of lysine dihydrochloride in some water. Next, add 0.2 mole of KOH (200 ml of 1 M solution). Next calculate the ratio of [AA^0]/[AA^{+1}].

$$pH = pK_{a_2} + \log \frac{[AA^0]}{[AA^{+1}]}$$

$$9.25 = 8.95 + \log \frac{[AA^0]}{[AA^{+1}]}$$

$$0.30 = \log \frac{[AA^0]}{[AA^{+1}]}$$

$$\frac{[AA^0]}{[AA^{+1}]} = \text{antilog of } 0.3 = 2.0 = \tfrac{2}{1}$$

Thus $\tfrac{2}{3}$ of the total lysine are in the AA^0 form and $\tfrac{1}{3}$ of the total lysine remains in the AA^{+1} form. To convert $\tfrac{2}{3}$ of the total to AA^0, add $\tfrac{2}{3} \times 0.2$ mole $= 0.133$ mole more of KOH (133 ml of 1 M solution). Finally, add sufficient water to bring the total volume to 2 liters.

Problem III-17

Describe the preparation of 3 liters of 0.1 M glutamate buffer, pH 2.5, starting with solid monosodium glutamate [AA^{-1}] and 0.5 M HCl.

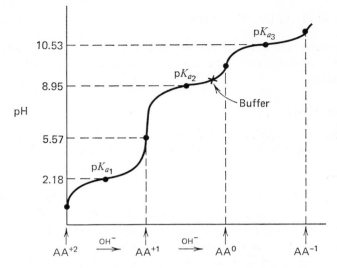

Figure III-10   Titration curve of lysine dihydrochloride

Solution

The titration curve for glutamic acid is shown in Figure III-11. The desired pH is a little above pK_{a_1}. The major ionic species present are glutamic hydrochloride $[AA^{+1}]$ and isoelectric glutamic acid $[AA^0]$, with the latter predominating. To obtain the desired pH, sufficient HCl must first be added to convert all of the original AA^{-1} to AA^0; then additional HCl must be added to obtain the proper ratio of $[AA^0]/[AA^{+1}]$.

We start by dissolving 3 liters × 0.1 M = 0.3 mole of monosodium glutamate in some water.

$$wt_g = \text{number of moles} \times \text{MW}$$

$$= (0.3)(169)$$

$$\boxed{wt_g = 50.7 \text{ g}}$$

Next, add 0.3 mole of HCl.

$$\text{liters} \times M = \text{number of moles}$$

$$\text{liters} = \frac{0.3}{0.5} = 0.6 = 600 \text{ ml}$$

Now all the original AA^{-1} is present as AA^0.

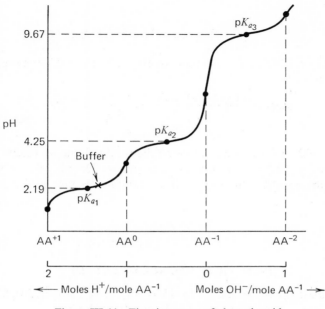

Figure III-11 Titration curve of glutamic acid

Next, calculate the amount of AA^0 that is to be converted to AA^{+1}. Note that the desired pH is greater than pK_{a_1}. Thus, less than half of the AA^0 will be converted.

$$pH = pK_{a_1} + \log \frac{[AA^0]}{[AA^{+1}]}$$

$$2.5 = 2.19 + \log \frac{[AA^0]}{[AA^{+1}]}$$

$$0.31 = \log \frac{[AA^0]}{[AA^{+1}]}$$

$$\frac{[AA^0]}{[AA^{+1}]} = \text{antilog of } 0.31 = 2.04$$

The buffer contains:

$$\frac{2.04}{3.04} \times 0.3 \text{ mole} = 0.201 \text{ mole } AA^0$$

and

$$\frac{1.00}{3.04} \times 0.3 = 0.099 \text{ mole } AA^{+1}$$

∴   Add an additional 0.099 mole of HCl to convert 0.099 mole of the AA^0 to AA^{+1}.

$$\text{liters} = \frac{\text{number of moles}}{M} = \frac{0.099}{0.5} = 0.198$$

$$= 198 \text{ ml}$$

Finally, add sufficient water to bring the total volume to 3 liters.

In summary, 50.7 g monosodium glutamate + 798 ml of 0.5 M HCl + H_2O to 3 liters yields 3 liters of a 0.1 M glutamate buffer, pH 2.5.

PRACTICE PROBLEM SET III

1. Calculate the pH of a 0.2 M solution of: (a) alanine hydrochloride, (b) isoelectric alanine, and (c) the sodium salt of alanine.

2. What is the degree of ionization of: (a) the α-COOH group in a 0.15 M solution of serine hydrochloride, and (b) the α-NH_2 group in a 0.1 M solution of disodium aspartate?

3. What is the degree of ionization of the NH_2 group in a 0.1 M solution of sodium glycinate?

4. Calculate the pH of a 0.05 M solution of: (a) monosodium glutamate, (b) glutamic hydrochloride, and (c) lysine dihydrochloride.

5. Calculate the volume of 0.1 M KOH required to titrate completely: (a) 450 ml of 0.25 M alanine hydrochloride, (b) 200 ml of 0.10 M isoelectric serine, (c) 400 ml of 0.15 M monosodium glutamate, and (d) 400 ml of 0.15 M isoelectric glutamic acid.

6. Calculate the volume of 0.2 M HCl required to titrate completely: (a) 200 ml of 0.25 M isoelectric leucine, (b) 375 ml of 0.25 M isoelectric glutamic acid, (c) 490 ml of 0.25 M isoelectric lysine, and (d) 125 ml of 0.25 M disodium salt of lysine.

7. Calculate the appropriate values and plot the pH curve for the titration of: (a) 250 ml of 0.25 M serine hydrochloride with 0.10 M NaOH, (b) 300 ml of 0.4 M isoelectric alanine with 0.4 M KOH and 0.2 M HCl, and (c) 200 ml of 0.1 M disodium glutamate with 0.25 M HCl.

8. Calculate the pH of a solution obtained by adding 20 ml of 0.2 M HCl to 30 ml of 0.1 M sodium glycinate.

9. Sketch the pH curve for the titration of: (a) 100 ml of 0.1 M sodium salt of alanylaspartylserine with 0.1 M HCl, (b) 250 ml of 0.2 M tyrosyl-glutamyllysylcysteine hodrochloride with 0.1 M KOH, and (c) 200 ml of 0.1 M β-aspartylalanine hydrochloride with 0.4 N NaOH.

10. Sketch the pH curve for the titration of 250 ml of 0.1 M disodium glutamate with 0.1 M HCl (a) in the absence and (b) in the presence of excess formaldehyde.

11. Calculate the pI and pH$_m$ of: (a) glutamic acid, (b) serine, (c) ornithine, (d) γ-glutamylalanine, (e) lysyltyrosylglycine, (f) lysyllysyl-valine, and (g) cysteinylglycine.

12. Sketch the titration curves for the compounds listed in Problem III-10, showing along the right-hand axis the net charge on the molecule at the various pH values.

13. In which direction in an electric field will the following compounds migrate at pH 1.5, 4.0, 7.0, 10.5, and 13: (a) aspartic acid, (b) aspartyl-glycine, (c) lysylalanine, (d) valyltyrosylglutamic acid, and (e) cysteinyl-tyrosine?

14. Write equations showing the ionization reactions of: (a) iso-electric glutamic acid, (b) lysine monohydrochloride, (c) monosodium glutamate, and (d) isoelectric valine in water.

15. Calculate the concentrations of the major ionic species present in 0.2 M solutions of each compound listed in Practice Problem 14.

16. Calculate the concentrations of the major ionic species present at pH 7.5 in 0.25 M solutions of: (a) leucine, (b) aspartic acid, and (c) lysine.

17. Calculate the concentrations of the major ionic species present at pH 5 in 0.3 M solutions of: (a) glutamic acid, (b) valine, and (c) lysine.

18. Calculate the concentrations of the major ionic species present at pH 10.5 in 0.2 M solutions of: (a) glycine, (b) glutamic acid, and (c) lysine.

19. Describe the preparation of the following amino acid buffers starting from the solid isoelectric amino acids and 1 M HCl or 1 M KOH: (a) 0.2 M histidine at pH 6.5, (b) 0.3 M glycine at pH 2.7, (c) 0.15 M lysine at pH 9.0, (d) 0.25 M alanine at pH 9.2, and (e) 0.15 M glutamate at pH 3.5.

BIOCHEMICAL ENERGETICS

IV

A. EQUILIBRIUM CONSTANTS AND FREE ENERGY CHANGES

Problem IV-1

Glucose-6-phosphate was hydrolyzed enzymatically (at pH 7 and 25° C) to glucose and inorganic phosphate. The concentration of glucose-6-phosphate was 0.1 M at the start. At equilibrium, only 0.05% of the original glucose-6-phosphate remained. Calculate: (a) K_{eq} for the hydrolysis of glucose-6-phosphate, (b) $\Delta F'$ for the hydrolysis reaction, (c) K_{eq} for he reaction by which glucose-6-phosphate is synthesized from inorganic phosphate and glucose, and (d) $\Delta F'$ for the synthesis reaction.

Solution

(a) The equation for the hydrolysis of glucose-6-phosphate is shown below.

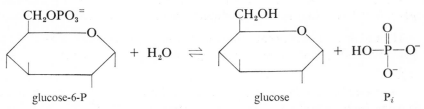

glucose-6-P   glucose P_i

The K_{eq} for the reaction is given by the expression shown below.

$$K_{eq} = \frac{[glucose][P_i]}{[glucose\text{-}6\text{-}P][H_2O]}$$

where $[P_i]$ represents the total inorganic phosphate concentration (a mixture of $HPO_4^=$ and $H_2PO_4^-$ at pH 7).

The concentration of water in dilute aqueous solutions is 55 M. The concentrations of the other components can easily be calculated since we know that the reaction went 99.95% to completion.

$$[\text{glucose-6-P}] = 0.10 \ M - 99.95\% \text{ of } 0.10 \ M$$

$$= 0.10 \ M - 0.09995 \ M = 0.00005 \ M = 0.5 \times 10^{-4} \ M$$

$$\therefore \quad [\text{glucose}] = 0.09995 \ M$$

and

$$[P_i] = 0.09995 \ M$$

$$K_{\text{eq}_{\text{actual}}} = \frac{(9.995 \times 10^{-2})(9.995 \times 10^{-2})}{(0.5 \times 10^{-4})(55)} = \frac{99.9 \times 10^{-4}}{27.5 \times 10^{-4}}$$

$$\boxed{K_{\text{eq}_{\text{actual}}} = 3.63}$$

Because the concentration of water is essentially constant in dilute aqueous solutions dealt with in biochemical reactions, we can combine this constant (55 M) and the actual K_{eq} and then define a new equilibrium constant, $K_{\text{eq}_{\text{new}}}$.

$$K_{\text{eq}_{\text{actual}}} \times M_{H_2O} = K_{\text{eq}_{\text{new}}}$$

$$K_{\text{eq}_{\text{new}}} = \frac{[\text{glucose}][P_i]}{[\text{glucose-6-P}]}$$

$$K_{\text{eq}_{\text{new}}} = \frac{(9.995 \times 10^{-2})(9.995 \times 10^{-2})}{(0.5 \times 10^{-4})} = \frac{99.9 \times 10^{-4}}{0.5 \times 10^{-4}}$$

$$\boxed{K_{\text{eq}_{\text{new}}} = 199.8}$$

This K_{eq} value is the same that we would obtain if we called the activity of water unity in the original expression.

Equilibrium constants discussed in all future problems will be calculated assuming a unit activity for water and will be designated simply as K_{eq} or K'_{eq}. Even though H^+ ions are not involved in this reaction, the symbols $\Delta F'$ and K'_{eq} are used to indicate that the reaction was carried out at a physiological pH. If H^+ ions are involved, the K'_{eq} value is generally corrected for the H^+ ion concentration as well as for water, as described in Problems IV-3 and IV-5.

(b) The relationship between K'_{eq} and $\Delta F'$ is given by the expression shown below.

$$\Delta F' = -RT \ln K'_{\text{eq}} = -2.3 \ RT \log K'_{\text{eq}}$$

where

$$2.3\ RT = (2.3)(1.98)(298) = 1363 \text{ at } 25°\text{ C}$$
$$\Delta F' = -1363 \log K'_{eq}$$
$$\Delta F' = -1363 \log 199.8$$
$$\log 199.8 = 2.3$$
$$\Delta F' = -1363\ (2.3)$$

$$\boxed{\Delta F' = -3140 \text{ cal/mole}}$$

In other words, under standard-state conditions in which the concentrations of glucose-6-phosphate, glucose, and P_i are all maintained at a steady-state level of unit activity, the conversion of 1 mole of glucose-6-phosphate to 1 mole of glucose and 1 mole of P_i liberates 3140 cal.

The equation for the hydrolysis of glucose-6-phosphate can be expressed in several ways as shown below. All three equations say the same thing:

$$\text{glucose-6-P} + H_2O \rightleftharpoons \text{glucose} + P_i + 3140 \text{ cal}$$
$$\text{glucose-6-P} + H_2O - 3140 \text{ cal} \rightleftharpoons \text{glucose} + P_i$$

or

$$\text{glucose-6-P} + H_2O \rightleftharpoons \text{glucose} + P_i \qquad \Delta F' = -3140 \text{ cal/mole}$$

(c)   The equilibrium constant for a reaction $A \rightarrow B$ is the reciprocal of the equilibrium constant for the reaction $B \rightarrow A$.

(1) $$A \longrightarrow B \qquad K'_{eq_1} = \frac{[B]}{[A]}$$

(2) $$B \longrightarrow A \qquad K'_{eq_2} = \frac{[A]}{[B]} = \frac{1}{[B]/[A]} = \frac{1}{K'_{eq_1}}$$

Therefore, for the reaction

$$\text{glucose} + P_i \rightleftharpoons \text{glucose-6-P}$$

$$K'_{eq} = \frac{[\text{glucose-6-P}]}{[\text{glucose}][P_i]} = \frac{1}{199.8} = 5 \times 10^{-3}$$

(d)   If the *hydrolysis* of glucose-6-phosphate *yields* 3140 cal/mole, the *synthesis* of glucose-6-phosphate *requires* 3140 cal/mole. The equation for the synthesis reaction can be expressed in three ways:

$$\text{glucose} + P_i + 3140 \text{ cal} \rightleftharpoons \text{glucose-6-P}$$
$$\text{glucose} + P_i \rightleftharpoons \text{glucose-6-P} - 3140 \text{ cal}$$

or

$$\text{glucose} + P_i \rightleftharpoons \text{glucose-6-P} \qquad \Delta F' = +3140 \text{ cal/mole}$$

$\Delta F'$ can also be calculated from K'_{eq}.

$$\Delta F' = -1363 \log K'_{eq} = -1363 \log 5 \times 10^{-3}$$
$$\Delta F' = -1363 (\log 5 + \log 10^{-3})$$
$$\Delta F' = -1363 (0.7 - 3)$$
$$\Delta F' = -1363 (-2.3)$$

$$\boxed{\Delta F' = +3140 \text{ cal/mole}}$$

Problem IV-2

Calculate the ΔF for the hydrolysis of glucose-6-phosphate at pH 7 and 37° C under steady-state conditions (such as might exist in a living cell) in which the concentrations of glucose-6-phosphate, glucose, and P_i are maintained at 10^{-3} M, 10^{-5} M, and 10^{-2} M, respectively.

Solution

The equation for the ΔF of the hydrolysis under nonstandard-state conditions is shown below.

$$\Delta F = -2.3RT \log K'_{eq} + 2.3RT \log \frac{[\text{glucose}][P_i]}{[\text{glucose-6-P}]}$$

or

$$\Delta F = \Delta F' + 2.3RT \log \frac{[\text{glucose}][P_i]}{[\text{glucose-6-P}]}$$

First evaluate $\Delta F'$ at 37° C.

$$T = 37° \text{ C} = 310° \text{ K}$$
$$\therefore \quad 2.3 RT = (2.3)(1.98)(310) = 1412$$
$$\Delta F' = -1412 \log K'_{eq} = -1412 \log 199.8 = (-1412)(2.3)$$

$$\boxed{\Delta F' = -3250 \text{ cal/mole}}$$

Next, calculate ΔF.

$$\Delta F = -3250 + 1412 \log \frac{(10^{-5})(10^{-2})}{(10^{-3})}$$
$$= -3250 + 1412 \log 10^{-4}$$
$$= -3250 + 1412(-4)$$
$$= -3250 - 5650$$

$$\boxed{\Delta F = -8900 \text{ cal/mole}}$$

Problem IV-3

Write equations and expressions for K_{eq} and $\Delta F°$ (or $\Delta F'$) for the hydrolysis of ethyl acetate: (a) at pH 0, (b) in a solution buffered at pH 7, and (c) in a solution buffered at pH 5.

Solution

(a)   The hydrolysis equation is shown below.   The pK_a of acetic acid is about 5.   Thus, at pH 0 the products are ethanol and unionized acetic acid.

(1) $\overset{\displaystyle O}{\overset{\displaystyle \|}{CH_3-C}}-O-CH_2-CH_3 + H_2O \rightleftharpoons CH_3COOH + CH_3CH_2OH$

$$K_{eq_1} = \frac{[CH_3COOH][CH_3CH_2OH]}{[CH_3COOCH_2CH_3]}$$

$$\Delta F_1° = -2.3\ RT \log K_{eq_1}$$

As in the previous problems, the activity of water is taken as unity. (b)   At pH 7, acetic acid is about 99% ionized.   Consequently, the hydrolysis products may be considered to be the acetate anion, a H^+ ion, and ethanol.

(2) $\overset{\displaystyle O}{\overset{\displaystyle \|}{CH_3-C}}-O-CH_2CH_3 + H_2O \rightleftharpoons CH_3COO^- + H^+ + CH_3CH_2OH$

$$K_{eq_2} = \frac{[CH_3COO^-][H^+][CH_3CH_2OH]}{[CH_3COOCH_2CH_3]}$$

We can see that K_{eq} formulated as above varies with pH.   For simplicity, we can define a K_{eq} at any given pH and combine the two constants (K_{eq_2} and H^+ ion concentration).

$$\frac{K_{eq_2}}{[H^+]} = K'_{eq} = \frac{[CH_3COO^-][CH_3CH_2OH]}{[CH_3COOCH_2CH_3]}$$

Similarly, a new standard-state ΔF may be formulated:

$$\Delta F' = -2.3RT \log K'_{eq}$$

The $\Delta F'$ represents the free energy change for the reaction under steady-state conditions where all products and reactants *except* H^+ are maintained at unit activity. The $[H^+]$ must then be specified. For most reactions of biochemical interest, the tabulated K'_{eq} and $\Delta F'$ values are those for some physiological pH, generally pH 7.

(c) At pH 5, approximately half of the acetic acid produced is ionized and half exists as unionized acetic acid. The reaction may be written as shown below.

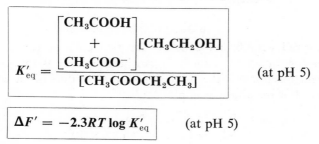

(3)

Under these conditions, the K_{eq} is generally defined in terms of *total* acetate (ionized plus unionized) and the pH is specified.

$$K'_{eq} = \frac{\begin{bmatrix} CH_3COOH \\ + \\ CH_3COO^- \end{bmatrix}[CH_3CH_2OH]}{[CH_3COOCH_2CH_3]} \qquad \text{(at pH 5)}$$

$$\boxed{\Delta F' = -2.3RT \log K'_{eq}} \qquad \text{(at pH 5)}$$

General Rules

1. In formulating the K_{eq} expression for reactions that involve water, the activity of water is taken as unity.

2. In formulating the K_{eq} expression for reactions that involve H^+ ions, the H^+ ion concentration generally is not included but rather is specified. The K_{eq} is then designated K'_{eq} or $K_{eq_{app}}$ (apparent K_{eq}).

3. In formulating the K_{eq} expression for reactions containing components that are present (at the specified pH) in more than one ionic form, *total* (analytical) concentrations are employed.

Problem IV-4

Calculate: (a) K'_{eq} and (b) $\Delta F'$ for the reaction between glucose and ATP catalyzed by hexokinase.

Solution

(a) Hexokinase catalyzes the phosphotransferase reaction between ATP and glucose.

$$\text{(1)} \qquad \text{glucose} + \text{ATP} \underset{}{\overset{\text{Mg}^{++}}{\rightleftharpoons}} \text{glucose-6-P} + \text{ADP}$$

The reaction may be considered the sum of two reactions.

(2) glucose + $P_i \rightleftharpoons$ glucose-6-P + H_2O $\qquad K'_{eq_2} = 5 \times 10^{-3}$

$\qquad\qquad\qquad\qquad\qquad\qquad\qquad\qquad\qquad \Delta F'_2 = +3140$ cal/mole

(3) ATP + $H_2O \rightleftharpoons$ ADP + P_i $\qquad\qquad K'_{eq_3} = 4.37 \times 10^5$

$\qquad\qquad\qquad\qquad\qquad\qquad\qquad\qquad\qquad \Delta F'_3 = -7700$ cal/mole

---

Sum: (1) $\qquad$ glucose + ATP $\underset{}{\overset{\text{Mg}^{++}}{\rightleftharpoons}}$ glucose-6-P + ADP

The K'_{eq_3} was calculated from the corresponding $\Delta F'$ value (Appendix II).

$$\Delta F'_3 = -1363 \log K'_{eq_3}$$
$$\log K'_{eq_3} = \frac{-7700}{-1363} = 5.64$$
$$K'_{eq_3} = \text{antilog of } 5.64 = 4.37 \times 10^5$$

The overall equilibrium constant for two coupled (or consecutive) reactions (or for any reaction that can be expressed as the sum of two component reactions) is the product of the K_{eq} values of the component reactions. For the above example:

$$K'_{eq_1} = K'_{eq_2} \times K'_{eq_3}$$
$$K'_{eq_1} = (5 \times 10^{-3})(4.37 \times 10^5)$$
$$K'_{eq_1} = 21.8 \times 10^2$$

$$\boxed{K'_{eq_1} = 2180}$$

We can satisfy ourselves that

$$K'_{eq_1} = K'_{eq_2} \times K'_{eq_3}$$

in the following way:

$$K'_{eq_2} = \frac{[\text{glucose-6-P}][H_2O]}{[\text{glucose}][P_i]}$$

$$K'_{eq_3} = \frac{[\text{ADP}][P_i]}{[\text{ATP}][H_2O]}$$

$$K'_{eq_1} = \frac{[\text{glucose-6-P}][\text{ADP}]}{[\text{glucose}][\text{ATP}]}$$

$$K'_{eq_2} \times K'_{eq_3} = \frac{[\text{glucose-6-P}][H_2O]}{[\text{glucose}][P_i]} \times \frac{[\text{ADP}][P_i]}{[\text{ATP}][H_2O]}$$

$$= \frac{[\text{glucose-6-P}][\text{ADP}]}{[\text{glucose}][\text{ATP}]} = K'_{eq_1}$$

(b) The overall $\Delta F'$ value for two coupled reactions (or for any reaction that can be expressed as the sum of two component reactions) is the sum of the $\Delta F'$ values for each component reaction. In the above example:

$$\Delta F_1' = \Delta F_2' + \Delta F_3'$$

$$\Delta F_1' = (+3140 \text{ cal/mole}) + (-7700 \text{ cal/mole}) = -4560 \text{ cal/mole}$$

We could also calculate $\Delta F_1'$ from the K_{eq_1}' value:

$$\Delta F_1' = -1363 \log K_{eq_1}' = -1363 \log 2180$$

$$\Delta F_1' = -1363(3.339)$$

$$\boxed{\Delta F_1' = -4560 \text{ cal/mole}}$$

Problem IV-5

The actual K_{eq} for the hydrolysis of ethyl acetate at 25° C in the presence of 1 M HCl is 0.33. Calculate the: (a) standard-state ΔF value ($\Delta F°$) for the hydrolysis, and (b) K_{eq}' and standard-state ΔF value ($\Delta F'$) for the esterase-catalyzed reaction at pH 7 and 25° C.

Solution

(a) $$K_{eq} = \frac{[CH_3COOH][CH_3CH_2OH]}{[CH_3-\underset{\underset{O}{\|}}{C}-O-CH_2-CH_3][H_2O]} = 0.33$$

$$\Delta F° = -1363 \log K_{eq}$$
$$\Delta F° = -1363 \log 0.33$$

$$\Delta F° = +1363 \log \frac{1}{0.33} = +1363 \log 3.0 = 1363(0.477)$$

$$\boxed{\Delta F° = 650 \text{ cal/mole}}$$

(b) At pH 7 the products of the hydrolysis are ethanol and acetate ion.

$$CH_3-\underset{\underset{O}{\|}}{C}-O-CH_2CH_3 + H_2O \underset{}{\overset{\text{esterase}}{\rightleftharpoons}} CH_3COO^- + H^+ + CH_3CH_2OH$$

We can consider the overall reaction as the sum of two consecutive reactions: (1) the hydrolysis to yield ethanol and unionized acetic acid, and (2) the spontaneous ionization of the acetic acid. The K_{eq} for the ionization of acetic acid (K_a) is 1.75×10^{-5}.

(1) $CH_3-C-OCH_2CH_3 + H_2O \rightleftharpoons CH_3COOH + CH_3CH_2OH$
         $\parallel$ $K_{eq_1} = 0.33$
         O

(2) $CH_3COOH \leftrightharpoons CH_3COO^- + H^+$ $K_{eq_2} = 1.75 \times 10^{-5}$

Sum: (3)

         $CH_3-C-OCH_2CH_3 + H_2O \rightleftharpoons CH_3COO^- + H^+ + CH_3CH_2OH$
         $\parallel$
         O

$$K_{eq_3} = \frac{[CH_3COO^-][H^+][CH_3CH_2OH]}{[CH_3COOCH_2CH_3][H_2O]}$$

$$= K_{eq_1} \times K_{eq_2} = 0.33(1.75 \times 10^{-5}) = 5.77 \times 10^{-6}$$

By convention, we shall incorporate the concentrations of water and H^+ ion into the equilibrium constant and define a new constant, K'_{eq}.

$$K'_{eq} = \frac{K_{eq_3}[H_2O]}{[H^+]} = \frac{(5.77 \times 10^{-6})(55)}{(10^{-7})}$$

$$= \frac{[CH_3COO^-][CH_3CH_2OH]}{[CH_3COOCH_2CH_3]} = 3.17 \times 10^3 \quad \text{(at pH 7)}$$

$$\Delta F' = -1363 \log K'_{eq} = -1363 \log 3.17 \times 10^3$$

$$\Delta F' = -1363(\log 3.17 + \log 10^3) = -1363(0.5 + 3) = -1363(3.5)$$

$$\boxed{\Delta F' = -4760 \text{ cal/mole}}$$

Problem IV-6

Compare the K'_{eq} and $\Delta F'$ values at pH 7 and 25° C for the synthesis of glucose-1-phosphate: (a) by direct reaction of glucose with inorganic phosphate, and (b) in the presence of ATP, Mg^{++}, and the enzymes hexokinase and phosphoglucomutase.

Solution

(a)   The $\Delta F'$ for the *hydrolysis* of glucose-1-phosphate is -5000 cal/mole (Appendix II).

$$\text{glucose-1-P} + H_2O \rightleftharpoons \text{glucose} + P_i \quad \Delta F' = -5000 \text{ cal/mole}$$

The K'_{eq} for this reaction may be calculated as usual.

$$\Delta F' = -1363 \log K'_{eq}$$

$$\log K'_{eq} = \frac{\Delta F'}{-1363}$$

$$K'_{eq} = \text{antilog of } 3.67$$

$$K'_{eq_{hyd}} = 4.68 \times 10^{+3}$$

The K'_{eq} for the *synthesis* of glucose-1-phosphate from glucose and P_i is $1/K'_{eq_{hyd}}$.

$$K'_{eq_{syn}} = \frac{1}{K'_{eq_{hyd}}} = \frac{1}{4.68 \times 10^{+3}} = 0.214 \times 10^{-3}$$

$$\boxed{K'_{eq_{syn}} = 2.14 \times 10^{-4}}$$

Because $\Delta F'$ for the hydrolysis is -5000 cal/mole, $\Delta F'$ for the synthesis reaction is $+5000$ cal/mole. The low value for $K'_{eq_{syn}}$ (or the high positive value for the $\Delta F'_{syn}$) indicates that the above reaction does not proceed to any great extent.

(b) Glucose may be converted to glucose-1-phosphate by two consecutive enzyme-catalyzed reactions.

(1) glucose $+$ ATP $\xrightarrow[\text{Mg}^{++}]{\text{hexokinase}}$ glucose-6-P $+$ ADP

$$K'_{eq_1} = 2180$$

$$\Delta F'_1 = -4560 \text{ cal/mole}$$

(2) glucose-6-P $\xrightarrow[\text{Mg}^{++}]{\text{phosphoglucomutase}}$ glucose-1-P $K'_{eq_2} = 0.0526$

$$\Delta F'_2 = 1710 \text{ cal/mole}$$

Sum: (3)   glucose $+$ ATP $\rightleftharpoons$ glucose-1-P $+$ ADP

$$K'_{eq_3} = \frac{[\text{glucose-1-P}][\text{ADP}]}{[\text{glucose}][\text{ATP}]}$$

The phosphoglucomutase reaction (reaction 2) attains equilibrium when the ratio of glucose-6-P/glucose-1-P is 19. Therefore, for the reaction as written,

$$K'_{eq_2} = \tfrac{1}{19} = 0.0526$$

The $\Delta F_2'$ may be calculated from K_{eq_2}'.

$$\Delta F_2' = -1363 \log 0.0526$$

$\Delta F_2' = -1363 \log 5.26 \times 10^{-2}$ or $\Delta F_2' = +1363 \log \dfrac{1}{0.0526}$

$$= -1363(\log 5.26 + \log 10^{-2})$$
$$= -1363(0.721 - 2)$$
$$= -1363(-1.279)$$

$$\boxed{\Delta F_2' = +1710 \text{ cal/mole}}$$

$$= +1363 \log 19$$
$$= +1363(1.279)$$

$$\boxed{\Delta F_2' = +1710 \text{ cal/mole}}$$

In the previous problem we dealt with a *single reaction* that *could be expressed* as the sum of two coupled or consecutive reactions. In this problem we are dealing with two actual consecutive reactions. The rules for calculating K_{eq}' and $\Delta F'$ are the same.

$$K_{eq_3}' = K_{eq_1}' \times K_{eq_2}'$$
$$K_{eq_3}' = (2180)(0.0526)$$

$$\boxed{K_{eq_3}' = 115}$$

$$\Delta F_3' = \Delta F_1' + \Delta F_2'$$
$$\Delta F_3' = (-4560 \text{ cal/mole}) + (+1710 \text{ cal/mole})$$

$$\boxed{\Delta F_3' = -2850 \text{ cal/mole}}$$

The high value for K_{eq}' (or the high negative value for $\Delta F'$) indicates that the overall (two-step) synthesis of glucose-1-phosphate from glucose and ATP is quite favorable and proceeds as written. The driving force of the overall reaction is reaction 1 in which an energy-rich ATP is utilized.

Problem IV-7

Calculate the overall K_{eq}' and $\Delta F'$ at pH 7 and 25° C for the conversion of fumaric acid to citric acid in the presence of the appropriate enzymes, cosubstrates, and cofactors.

Solution

The enzyme-catalyzed reactions by which fumaric acid is converted to citric acid are shown below.

(1)   fumarate + H_2O $\xrightleftharpoons{\text{fumarase}}$ malate $K'_{eq_1} = 4.5$

(2)   malate + DPN^+ $\xrightleftharpoons{\text{malic dehydrogenase}}$

oxalacetate + DPNH + H^+ $K'_{eq_2} = 1.3 \times 10^{-5}$

(3)   oxalacetate + acetyl CoA + H_2O $\xrightleftharpoons{\text{condensing enzyme}}$

citrate + CoASH $K'_{eq_3} = 3.2 \times 10^5$

Sum: (4) fumarate + $2H_2O$ + acetyl CoA + DPN^+ $\rightleftharpoons$

citrate + DPNH + H^+ + CoASH

$$K'_{eq_4} = \frac{[\text{citrate}][\text{DPNH}][\text{CoASH}]}{[\text{fumarate}][\text{acetyl CoA}][\text{DPN}^+]}$$

$$K'_{eq_4} = K'_{eq_1} \times K'_{eq_2} \times K'_{eq_3}$$

$$= (4.5)(1.3 \times 10^{-5})(3.2 \times 10^5)$$

$$\boxed{K'_{eq_4} = 18.7}$$

$$\Delta F'_4 = -1363 \log K'_{eq_4} = -1363 \log 18.7$$

$$\Delta F'_4 = -1363(1.272)$$

$$\boxed{\mathbf{\Delta F'_4 = -1735 \ cal/mole}}$$

We can see that the *overall* conversion of fumarate to citrate is favorable in spite of reaction 2 with its low K'_{eq}. This problem and those preceding it illustrate some general rules and principles summarized below.

General Rules

1. The overall K'_{eq} for any number of consecutive reactions, 1, 2, 3, 4, . . . , etc., is $K'_{eq_1} \times K'_{eq_2} \times K'_{eq_3} \times K'_{eq_4} \ldots$, etc.
2. The overall $\Delta F'$ for any number of consecutive reactions, 1, 2, 3, 4, . . . , etc., is $\Delta F'_1 + \Delta F'_2 + \Delta F'_3 + \Delta F'_4 \cdots$, etc. The $\Delta F'_{\text{overall}}$ can also be calculated from $K'_{eq_{\text{overall}}}$

$$\Delta F'_{\text{overall}} = -2.3RT \log K'_{eq_{\text{overall}}}$$

3. The K'_{eq} for a single reaction that can be expressed as the sum of two or more consecutive reactions, 1, 2, 3, . . . , etc., is $K'_{eq_1} \times K'_{eq_2} \times K'_{eq_3} \cdots$, etc. Similarly, the $\Delta F'$ for a single reaction that can be expressed

as the sum of two or more consecutive reactions is $\Delta F'_1 + \Delta F'_2 + \Delta F'_3 \cdots$, etc.

$$\Delta F'_{overall} = -2.3RT \log K'_{eq_{overall}}$$

Problem IV-8

The cleavage of citrate to acetate and oxalacetate has a $\Delta F'$ of -680 cal/mole. The K'_{eq} of the condensing enzyme reaction is 3.2×10^5. From this information, calculate the standard free energy of hydrolysis of acetyl CoA and the K'_{eq} for the hydrolysis.

Solution

The two reactions given are shown below.

(1) citrate $\xrightleftharpoons{\text{citrate lyase}}$ acetate + oxalacetate $\Delta F'_1 = -680$ cal/mole

(2) acetyl CoA + oxalacetate + H_2O $\xrightleftharpoons{\text{condensing enzyme}}$ citrate + CoASH

$$K'_{eq_2} = 3.2 \times 10^5$$

We can work the problem in terms of $\Delta F'$ or K'_{eq} values. First calculate the missing values.

$$\Delta F'_1 = -1363 \log K'_{eq_1} = -680 \text{ cal/mole}$$

$$\log K'_{eq_1} = \frac{-680}{-1363} = 0.498$$

$$K'_{eq_1} = \text{antilog of } 0.498$$

$$\boxed{K'_{eq_1} = 3.15}$$

$$\Delta F'_2 = -1363 \log K'_{eq_2}$$

$$= -1363 \log 3.2 \times 10^5 = -1363(\log 3.2 + \log 10^5)$$

$$= -1363(0.52 + 5) = -1363(5.52)$$

$$\boxed{\Delta F'_2 = -7520 \text{ cal/mole}}$$

The reaction we are interested in is the hydrolysis of acetyl CoA:

(3)   acetyl CoA + $H_2O \rightleftharpoons$ acetate + CoASH

Can this reaction be expressed in terms of the ones given? We can see that reaction 3 is the sum of reactions 1 and 2.

(1)   citrate $\rightleftharpoons$ acetate + oxalacetate $\Delta F_1' = -680$ cal/mole

$$K_{eq_1}' = 3.15$$

(2)   acetyl CoA + oxalacetate + $H_2O \rightleftharpoons$ citrate + CoASH,

$$\Delta F_2' = -7520 \text{ cal/mole}$$

$$K_{eq_2}' = 3.2 \times 10^5$$

Sum: (3)   acetyl CoA + $H_2O \rightleftharpoons$ acetate + CoASH

$$\Delta F_3' = \Delta F_1' + \Delta F_2'$$

$$= (-680) + (-7520)$$

$$\boxed{\Delta F_3' = -8200 \text{ cal/mole}}$$

$$\Delta F_3' = -1363 \log K_{eq_3}'$$

$$-8200 = -1363 \log K_{eq_3}'$$

$$\frac{-8200}{-1363} = \log K_{eq_3}' = 6$$

$$K_{eq_3}' = \text{antilog of } 6$$

$$\boxed{K_{eq_3}' = 10^6}$$

$$K_{eq_3}' = K_{eq_1}' \times K_{eq_2}'$$

$$= (3.15)(3.2 \times 10^5)$$

$$K_{eq_3}' = 10 \times 10^5$$

$$\boxed{K_{eq_3}' = 10^6}$$

$$\Delta F_3' = -1363 \log K_{eq_3}'$$

$$= -1363 \log 10^6$$

$$\Delta F_3' = -1363(6)$$

$$\boxed{\Delta F_3' = -8200 \text{ cal/mole}}$$

We could also determine the $\Delta F'$ of hydrolysis of acetyl CoA by a different line of reasoning. If the *cleavage* of citrate to acetate and oxal-acetate *liberates* 680 cal/mole, then the *synthesis* of citrate from acetate and oxalacetate *requires* 680 cal/mole. When the synthesis is carried out from acetyl CoA (an "activated" form of acetate), 7520 cal/mole are released. The acetyl CoA then must have contained sufficient energy to form the new carbon-carbon bond (680 cal/mole) *plus* have 7520 cal/mole left over; that is, the acetyl CoA is worth 680 *plus* 7520 = 8200 cal/mole in terms of "group transfer potential" (or "free energy of hydrolysis"). The condensing enzyme reaction may be thought of as the sum of two intimately coupled reactions:

acetate + oxalacetate + 680 cal $\rightleftharpoons$ citrate

acetyl CoA + $H_2O \rightleftharpoons$ acetate + CoASH + 8200 cal

Sum:

acetyl CoA + H_2O + oxalacetate $\rightleftharpoons$ citrate + CoASH + 7520 cal

The situation is analagous to the hexokinase reaction. An energy-rich compound provides both the energy to drive an endergonic condensation as well as the particular group that is transferred to an acceptor.

General Principle

Endergonic reactions may be driven toward completion by coupling them to highly exergonic reactions. The coupling may be intimate so that the overall coupled reaction appears as a single step (for example, the hexokinase reaction or the condensing enzyme reaction), or the coupling may take place in two or more consecutive steps (for example, the fumarate → citrate sequence). In sequential reactions, we can think of a subsequent exergonic reaction as removing the product of a preceding endergonic reaction as it is formed, thereby driving the overall sequence to the right.

Problem IV-9

Estimate the $\Delta F'$ values for the following reactions: (a) ATP + GDP $\rightleftharpoons$ GTP + ADP, (b) glycerol + ATP $\rightleftharpoons$ α-glycerophosphate + ADP, and (c) 3-phosphoglycerate + ATP $\rightleftharpoons$ 1,3,-diphosphoglycerate + ADP.

Solution

(a) In this reaction the energy (group-transfer potential) of ATP is utilized to transfer its terminal phosphate to GDP. The product (GTP) is itself "energy rich"; in other words, the group-transfer potential of GTP is as high as that of ATP. Thus, $\Delta F' = 0$ and $K'_{eq} = 1$. We can verify these results mathematically by considering the reaction as the sum of two component reactions.

(1) $\qquad\qquad\qquad$ ATP + $H_2O \rightleftharpoons$ ADP + P_i + 7700 cal

$$\Delta F'_1 = -7700 \text{ cal/mole}$$

(2) $\qquad$ GDP + P_i + 7700 cal $\rightleftharpoons$ GTP + H_2O

$$\Delta F'_2 = +7700 \text{ cal/mole}$$

Sum: (3) $\qquad$ ATP + GDP $\rightleftharpoons$ GTP + ADP $\quad \boxed{\Delta F'_3 = 0}$

(b) In this reaction the energy of ATP is used to transfer its terminal phosphate to glycerol to form an "energy-poor" phosphate ester. The $\Delta F'$ of hydrolysis of α-glycerophosphate is about -2000 cal/mole while that of ATP (terminal phosphate) is about -7700 cal/mole. Thus, of the original 7700 cal, only 2000 cal are conserved. The $\Delta F'$ of the reaction is the difference between -7700 cal and -2000 cal, or -5700 cal/mole. This energy (most of which appears as heat) drives the overall reaction. We can verify this mathematically by considering the reaction as the sum of two component reactions.

(1) $ATP + H_2O \rightleftharpoons ADP + P_i + 7700$ cal

$$\Delta F_1' = -7700 \text{ cal/mole}$$

(2)   glycerol $+ P_i + 2000$ cal $\rightleftharpoons$ α-glycerophosphate

$$\Delta F_2' = +2000 \text{ cal/mole}$$

Sum: (3) $ATP + $ glycerol $\rightleftharpoons ADP + $ α-glycerophosphate $+ 5700$ cal

$$\boxed{\Delta F_3' = -5700 \text{ cal/mole}}$$

(c) The product of this reaction, 1,3-diphosphoglycerate, is more "energy rich" than ATP ($\Delta F_{hyd}' = -12,000$ cal/mole). Thus, the reaction as written is endergonic. We can consider the reaction as the sum of two component reactions.

(1) $ATP + H_2O \rightleftharpoons ADP + P_i + 7700$ cal

$$\Delta F_1' = -7700 \text{ cal/mole}$$

(2) $3\text{-PGA} + P_i + 12,000$ cal $\rightleftharpoons 1,3\text{-DiPGA} + H_2O$

$$\Delta F_2' = +12,000 \text{ cal/mole}$$

Sum: (3) $ATP + 3\text{-PGA} + 4300$ cal $\rightleftharpoons ADP + 1,3\text{-DiPGA}$

$$\boxed{\Delta F_3' = +4300 \text{ cal/mole}}$$

Problem IV-10

The ATP/ADP ratio in an actively respiring yeast cell is about 10. What would the intracellular 3-phosphoglycerate/1,3-diphosphoglycerate ratio have to be to make the phosphoglycerate kinase reaction thermodynamically favorable in the direction of 1,3-diphosphoglycerate synthesis? Assume a temperature of $25°$ C.

Solution

The reaction catalyzed by phosphoglycerate kinase is shown below.

$$ATP + 3\text{-PGA} \xrightleftharpoons{\text{phosphoglycerate kinase}} ADP + 1,3\text{-DiPGA}$$

$$\Delta F' = +4300 \text{ cal/mole}$$

We have seen from the previous problem that under *standard-state* conditions the reaction is endergonic in the direction of 1,3-DiPGA synthesis ($\Delta F' = 4300$ cal/mole). In other words, if the concentrations of the four components of the reaction are maintained at steady-state levels of 1 M, the reaction goes spontaneously in the direction of ATP and 3-PGA formation with the liberation of 4300 cal/mole. In a living cell, however, the concentrations of ATP, ADP, 3-PGA, and 1,3-DiPGA are not maintained at 1 M. The reaction can be made to proceed spontaneously in the direction of 1,3-DiPGA and ADP formation if the concentrations of the components are maintained at suitable levels. Qualitatively, we know that the reaction can be made to go spontaneously from left to right if the concentrations of ATP and 3-PGA are increased sufficiently, or if the concentrations of ADP and 1,3-DiPGA are decreased sufficiently, or a combination of both. A quantitative expression relating the ΔF of the reaction (an indication of the spontaneous direction) and the concentrations of the reaction components is shown below.

$$\Delta F = \Delta F' + 1363 \log \frac{[\text{ADP}][\text{1,3-DiPGA}]}{[\text{ATP}][\text{3-PGA}]}$$

The reaction proceeds spontaneously as written if ΔF is negative. First calculate the ratio of 1,3-DiPGA/3-PGA that would yield a $\Delta F = 0$. At this ratio the reaction would be at equilibrium.

$$0 = \Delta F' + 1363 \log \frac{[\text{ADP}][\text{1,3-DiPGA}]}{[\text{ATP}][\text{3-PGA}]}$$

$$-\Delta F' = 1363 \log \frac{[\text{ADP}][\text{1,3-DiPGA}]}{[\text{ATP}][\text{3-PGA}]}$$

$$= -4300 \text{ cal/mole}$$

$$\log \frac{(1)[\text{1,3-DiPGA}]}{(10)[\text{3-PGA}]} = \frac{-4300}{1363} = -3.15$$

$$\log \frac{(10)[\text{3-PGA}]}{[\text{1,3-DiPGA}]} = +3.15$$

$$\log 10 + \log \frac{[\text{3-PGA}]}{[\text{1,3-DiPGA}]} = 3.15$$

$$1 + \log \frac{[\text{3-PGA}]}{[\text{1,3-DiPGA}]} = 3.15$$

$$\log \frac{[\text{3-PGA}]}{[\text{1,3-DiPGA}]} = 3.15 - 1 = 2.15$$

$$\frac{[\text{3-PGA}]}{[\text{1,3-DiPGA}]} = \text{antilog of } 2.15 = \boxed{\textbf{141}}$$

Thus, when the ratio of ATP/ADP is 10 and the ratio of 3-PGA/1,3-DiPGA is 141, the phosphoglycerate kinase reaction would be at equilibrium. Any slight increase in either ratio would force the reaction in the direction of ADP and 1,3-DiPGA formation—it would make ΔF a negative value. This problem illustrates another general principle.

General Principle

The $\Delta F'$ (or K'_{eq}) values provide a convenient way to classify and tabulate various kinds of reactions but they do not indicate the direction in which a reaction proceeds in a living cell. The spontaneous direction *in vivo* (the nonstandard-state ΔF value) depends on the intracellular concentrations (activities) of the reaction components.

Problem IV-11

The conversion of glucose to lactic acid has an overall $\Delta F'$ of $-52,000$ cal/mole. In an anaerobic cell, this conversion is coupled to the synthesis of 2 moles of ATP per mole of glucose. (a) Calculate the $\Delta F'$ of the overall coupled reaction. (b) Calculate the efficiency of energy conservation in the anaerobic cell. (c) At the same efficiency, how many moles of ATP per mole of glucose could be obtained in an aerobic organism in which glucose is completely oxidized to CO_2 and H_2O ($\Delta F' = -686,000$ cal/mole)? (d) Calculate the $\Delta F'$ for the overall oxidation coupled to ATP synthesis.

Solution

(a)

(1) $C_6H_{12}O_6 \rightarrow 2CH_3$—CHOH—COOH $\Delta F'_1 = -52,000$ cal/mole
      glucose   lactic acid

(2) $2ADP + 2P_i \rightarrow 2ATP$ $\Delta F'_2 = +8,000$ cal/mole $\times$ 2

                                                $= +16,000$ cal/mole

---

Sum: (3) glucose $+ 2ADP + 2P_i \rightarrow$ 2 lactic acid $+ 2ATP$

$\Delta F'_3 = (-52,000) + (16,000)$

$$\boxed{\Delta F'_3 = -36,000 \text{ cal/mole}}$$

(b)
$$\text{efficiency} = \frac{\text{energy conserved}}{\text{energy made available}} \times 100\%$$

$$= \frac{16,000 \text{ cal}}{52,000 \text{ cal}} \times 100\%$$

$$\boxed{\text{efficiency} = 30.8\%}$$

(c)

(4) $C_6H_{12}O_6 + 6O_2 \rightarrow 6CO_2 + 6H_2O \qquad \Delta F_4' = -686,000 \text{ cal/mole}$

(5) $nADP + nP_i \rightarrow nATP \qquad \Delta F_5' = n(+8000 \text{ cal/mole})$

Sum: (6) $C_6H_{12}O_6 + 6O_2 + nADP + nP_i \rightarrow 6CO_2 + 6H_2O + nATP$

At 30.8% efficiency the total energy that can be conserved is $0.308 \times 686,000 \text{ cal/mole} = 211,000 \text{ cal/mole}$. If each mole of ATP requires about 8000 cal for its synthesis:

$$\frac{211,000}{8000} = 26.4 \text{ moles of ATP formed}$$

$$= \boxed{26 \text{ moles of ATP}} \qquad \text{(nearest whole number)}$$

(d)   For the overall coupled reaction:

$$\Delta F_6' = \Delta F_4' + \Delta F_5'$$

$$= (-686,000) + 26(+8000) = (-686,000) + (208,000)$$

$$\boxed{\Delta F_6' = -478,000 \text{ cal/mole}}$$

Aerobic cells are actually more efficient than 30.8%. One mole of glucose is capable of yielding 38 moles of ATP when completely oxidized.

38 moles $\times$ 8000 cal/mole = 304,000 cal conserved/mole glucose oxidized

$$\frac{304,000}{686,000} \times 100 = 44.3\% \text{ efficiency}$$

$$\Delta F_6' = (-686,000) + (304,000)$$

$$\boxed{\Delta F_6' = -382,000 \text{ cal/mole}}$$

Problem IV-12

(a)   Calculate the $\Delta F'$ for the complete oxidation of lactic acid to CO_2 and H_2O given the information below. (b) How many moles of ATP could be synthesized in the process?

(1) $\qquad$ glucose $\rightarrow$ 2 lactic acid $\qquad \Delta F_1' = -52,000$ cal/mole

(2) $\quad$ glucose $+ 6O_2 \rightarrow 6CO_2 + 6H_2O$ $\quad \Delta F_2' = -686,000$ cal/mole

Solution

(a) The reaction we are interested in is shown below.

(3) $\qquad\qquad$ lactic acid $+ 3O_2 \rightarrow 3CO_2 + 3H_2O$

The overall oxidation of glucose to CO_2 and H_2O can be written in two steps.

(1) $\qquad\qquad$ glucose $\rightarrow$ 2 lactic acid $\quad \Delta F_1' = -52,000$ cal/mole

(4) $\quad$ 2 lactic acid $+ 6O_2 \rightarrow 6CO_2 + 6H_2O$ $\quad \Delta F_4' = ?$

Sum: (2) $\quad$ glucose $+ 6O_2 \rightarrow 6CO_2 + 6H_2O$ $\quad \Delta F_2' = -686,000$ cal/mole

The $\Delta F_4'$ can be calculated easily.

$$\Delta F_2' = \Delta F_1' + \Delta F_4'$$
$$-686,000 = -52,000 + \Delta F_4'$$
$$\Delta F_4' = -686,000 + 52,000 = -634,000 \text{ cal}$$

Reaction 3 is half of reaction 4.

$$\therefore \quad \Delta F_3' = \frac{-634,000}{2} = \boxed{-317,000 \text{ cal/mole of lactic acid}}$$

The oxidation of 2 moles of lactic acid yields 634,000 cal. Therefore, the oxidation of 1 mole would yield 317,000 cal.

(b) At 40% efficiency:

$$0.40 \times 317,000 = 127,000 \text{ cal conserved}$$

Each mole of ATP requires 8000 cal.

$$\frac{127,000}{8000} = \boxed{\sim 16 \text{ moles of ATP}}$$

Problem IV-13

Calculate the total number of moles of ATP that can be produced from the complete oxidation of tributyrin to CO_2 and H_2O in an aerobic organism.

Solution

The overall reaction sequence by which tributyrin is oxidized is shown in Figure IV-1.

B. CALCULATIONS OF EQUILIBRIUM CONCENTRATIONS

Problem IV-14

The K'_{eq} for the fructose-1,6-diphosphate aldolase reaction at 25° C and pH 7, written in the direction of triose phosphate formation, is about 10^{-4}. Calculate: (a) $\Delta F'$ for the reaction as written, (b) the equilibrium concentrations of fructose-1,6-diphosphate (FDP), glyceraldehyde-3-phosphate (GAP), and dihydroxyacetone-phosphate (DHAP) when the initial concentration of FDP was 1 M, (c) the equilibrium concentrations of all components of the reaction when the initial concentration of FDP was 0.01 M, (d) the equilibrium concentrations of all components when the initial concentration of FDP was 2 × 10^{-4} M, and (e) the equilibrium concentrations of all components when the initial concentration of FDP was 10^{-5} M.

Solution

The reaction catalyzed by FDP-aldolase is shown below.

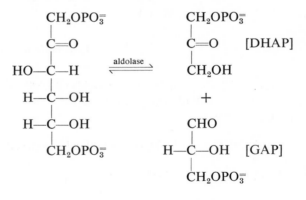

(a) $\Delta F' = -1363 \log K'_{eq} = -1363 \log 10^{-4} = -1363(-4)$

$$\boxed{\Delta F' = +5460 \text{ cal/mole}}$$

(b) $$K'_{eq} = \frac{[\text{DHAP}][\text{GAP}]}{[\text{FDP}]}$$

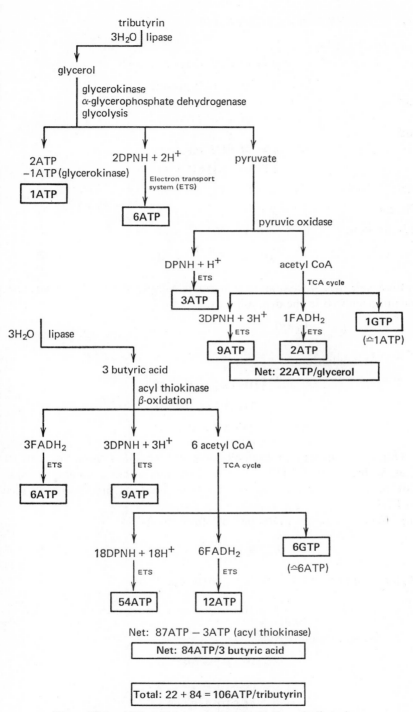

Figure IV-1 ATP yield from complete oxidation of tributyrin

Let

$$y = M \text{ FDP that disappears}$$

$$\therefore \quad y = M \text{ DHAP produced}$$

and

$$y = M \text{ GAP produced}$$

and

$$1 - y = M \text{ FDP remaining}$$

$$FDP \rightleftharpoons DHAP + GAP$$

Start:	1	0	0
Change:	$-y$	$+y$	$+y$
Equilibrium:	$1 - y$	y	y

$$K'_{eq} = \frac{(y)(y)}{(1-y)} = 10^{-4}$$

First calculate y, assuming y is small compared to 1 M. Therefore, y can be ignored in the denominator.

$$K'_{eq} = \frac{(y)(y)}{(1)} = 10^{-4} \qquad y^2 = 10^{-4} \qquad y = 10^{-2}$$

$\therefore$ At equilibrium:

$$\boxed{\begin{array}{l} \textbf{[DHAP]} = \mathbf{10^{-2}} \textbf{\textit{M}} \\ \textbf{[GAP]} = \mathbf{10^{-2}} \textbf{\textit{M}} \\ \textbf{[FDP]} = \mathbf{0.99} \textbf{\textit{M}} \end{array}}$$

Our assumption that y is small compared to 1 M is valid. We can see that at high initial FDP concentrations the reaction comes to equilibrium when the concentrations of DHAP and GAP are small compared to FDP.

(c)   When the initial FDP concentration was 0.01 M:

$$K'_{eq} = \frac{(y)(y)}{(0.01 - y)} = 10^{-4}$$

Again assume that y is small compared to 0.01 M.

$$K'_{eq} = \frac{(y)(y)}{(0.01)} = 10^{-4} \qquad y^2 = 10^{-6} \qquad y = 10^{-3}$$

$\therefore$ At equilibrium:

$$\boxed{\begin{array}{l} \textbf{[DHAP]} = \mathbf{10^{-3}} \textbf{\textit{M}} \\ \textbf{[GAP]} = \mathbf{10^{-3}} \textbf{\textit{M}} \\ \textbf{[FDP]} = \mathbf{0.009} \textbf{\textit{M}} \end{array}}$$

Under these conditions y is about 10% of [FDP]. Thus, the elimination of y from the denominator still gives a reasonably correct answer.

(d)   When the initial FDP concentration was $2 \times 10^{-4}\ M$:

$$K'_{eq} = \frac{(y)(y)}{(2 \times 10^{-4} - y)} = 10^{-4}$$

We have seen from the trend established in parts b and c that y becomes relatively larger compared to [FDP] as the initial concentration of FDP decreases. For an accurate solution in part d, we can no longer neglect y in the denominator.

Cross-multiplying:

$$y^2 = (10^{-4})(2 \times 10^{-4} - y) = 2 \times 10^{-8} - 10^{-4}y$$

$$y^2 + 10^{-4}y - 2 \times 10^{-8} = 0$$

$$y = \frac{-b \pm \sqrt{b^2 - 4ac}}{2a}$$

where

$$a = 1$$
$$b = 10^{-4}$$
$$c = -2 \times 10^{-8}$$

$$y = \frac{-10^{-4} \pm \sqrt{(10^{-4})^2 - 4(1)(-2 \times 10^{-8})}}{2}$$

$$y = \frac{-10^{-4} \pm \sqrt{10^{-8} + 8 \times 10^{-8}}}{2} = \frac{-10^{-4} \pm \sqrt{9 \times 10^{-8}}}{2}$$

$$y = \frac{-10^{-4} \pm 3 \times 10^{-4}}{2} = \frac{-4 \times 10^{-4}}{2} \quad \text{and} \quad \frac{+2 \times 10^{-4}}{2}$$

$$= -2 \times 10^{-4} \quad \text{and} \quad 1 \times 10^{-4}$$

The negative value is obviously incorrect.

$$\therefore \quad y = 10^{-4}$$

At equilibrium:

> **[DHAP]** $= 10^{-4}\ M$
>
> **[GAP]** $= 10^{-4}\ M$
>
> **[FDP]** $= 2 \times 10^{-4}\ M - 10^{-4}\ M = 10^{-4}\ M$

Thus, when the initial concentration of FDP is $2 \times 10^{-4}\ M$, the reaction comes to equilibrium when the concentrations of all three components are equal at $10^{-4}\ M$ each.

Check:

$$K'_{eq} = \frac{[DHAP][GAP]}{[FDP]} = 10^{-4} = \frac{(10^{-4})(10^{-4})}{(10^{-4})}$$

(e)   When the initial FDP concentration was 10^{-5} M:

$$K'_{eq} = \frac{(y)(y)}{(10^{-5} - y)} = 10^{-4}$$

$$y^2 = 10^{-9} - 10^{-4}y$$

$$y^2 + 10^{-4}y - 10^{-9} = 0$$

$$y = \frac{-b \pm \sqrt{b^2 - 4ac}}{2a}$$

where

$$a = 1$$
$$b = 10^{-4}$$
$$c = -10^{-9}$$

$$y = \frac{-10^{-4} \pm \sqrt{(10^{-4})^2 - 4(1)(-10^{-9})}}{2} = \frac{-10^{-4} \pm \sqrt{10^{-8} + 4 \times 10^{-9}}}{2}$$

$$y = \frac{-10^{-4} \pm \sqrt{10^{-8} + 0.4 \times 10^{-8}}}{2} = \frac{-10^{-4} \pm \sqrt{1.4 \times 10^{-8}}}{2}$$

$$y = \frac{-10^{-4} \pm 1.183 \times 10^{-4}}{2} = \frac{-2.183 \times 10^{-4}}{2} \quad \text{and} \quad \frac{+0.183 \times 10^{-4}}{2}$$

$$y = 0.0915 \times 10^{-4} = 0.915 \times 10^{-5}, \quad \text{neglecting the negative value.}$$

$\therefore$ At equilibrium:

$$\boxed{\begin{array}{l} \textbf{[DHAP]} = \textbf{0.915} \times \textbf{10}^{-5} \ \textbf{\textit{M}} = \textbf{9.15} \times \textbf{10}^{-6} \ \textbf{\textit{M}} \\ \textbf{[GAP]} = \textbf{0.915} \times \textbf{10}^{-5} \ \textbf{\textit{M}} = \textbf{9.15} \times \textbf{10}^{-6} \ \textbf{\textit{M}} \\ \textbf{[FDP]} = \textbf{1} \times \textbf{10}^{-5} \ \textbf{\textit{M}} - \textbf{0.915} \times \textbf{10}^{-5} \ \textbf{\textit{M}} \\ \qquad = \textbf{0.085} \times \textbf{10}^{-5} \ \textbf{\textit{M}} = \textbf{8.5} \times \textbf{10}^{-7} \ \textbf{\textit{M}} \end{array}}$$

We have seen from this problem that although K'_{eq} is small (10^{-4}) and $\Delta F'$ is positive ($+5460$ cal/mole), we cannot automatically assume that at equilibrium there will always be more FDP than DHAP or GAP. The relative proportions of the reaction components depend on the initial concentration of the starting compound(s). This phenomenon will be observed whenever there is an unequal number of components on both sides of the equation. For a reaction $A \rightleftharpoons B$ with a $K_{eq} > 1$, there will always be more B than A at equilibrium regardless of the initial concentration of A or B. Similarly, for a reaction $A + B \rightleftharpoons C + D$ with a $K_{eq} > 1$, there will always be more C and D than A and B, regardless of the initial concentrations (assuming that we start with equal concentrations of A and B or C and D).

Problem IV-15

Fructose-6-phosphate (F-6-P) was allowed to react at 25° C and pH 7 in the presence of Mg^{++} and the enzymes phosphohexoisomerase (PHI) and phosphoglucomutase (PGM). At equilibrium the solution contained unreacted F-6-P, glucose-6-phosphate (G-6-P), and glucose-1-phosphate (G-1-P). Calculate the concentrations of each reaction component if the initial F-6-P concentration was 0.1 M.

Solution

The reactions catalyzed by the two enzymes are shown below.

(1) $\qquad$ F-6-P $\overset{\text{PHI}}{\rightleftharpoons}$ G-6-P $\qquad K'_{eq_1} = 2$

(2) $\qquad$ G-6-P $\underset{Mg^{++}}{\overset{\text{PGM}}{\rightleftharpoons}}$ G-1-P $\qquad K'_{eq_2} = 0.0526$

$$K'_{eq_1} = \frac{[\text{G-6-P}]}{[\text{F-6-P}]} \qquad K'_{eq_2} = \frac{[\text{G-1-P}]}{[\text{G-6-P}]}$$

Let

$$y = M \text{ G-6-P present at equilibrium}$$
$$z = M \text{ G-1-P present at equilibrium}$$
$$\therefore \quad (y + z) = M \text{ F-6-P used up}$$

and

$$0.1 - (y + z) = M \text{ F-6-P present at equilibrium}$$

$$K'_{eq_1} = \frac{y}{0.1 - (y + z)} = 2 \qquad K'_{eq_2} = \frac{z}{y} = 0.0526$$

The problem requires solving for two unknowns, y and z. Two simultaneous equations with the same two unknowns can be solved. First solve for y in terms of z, or vice versa. For example,

$$z = 0.0526y \qquad \text{or} \qquad y = \frac{z}{0.0526} = 19z$$

Next, substitute $0.0526y$ for z or $19z$ for y in the K'_{eq_1} expression.

$$K'_{eq_1} = \frac{y}{0.1 - (y + z)} = \frac{19z}{0.1 - (19z + z)} = \frac{19z}{0.1 - 20z} = 2$$

Cross-multiplying:

$$2(0.1 - 20z) = 19z$$

$$0.2 - 40z = 19z$$

$$0.2 = 59z$$

$$z = \frac{0.2}{59}$$

$$z = 0.00339$$

$$\therefore \quad [\text{G-1-P}] = \boxed{\mathbf{3.39 \times 10^{-3} \ M}}$$

$$[\text{G-6-P}] = y = 19z = (19)(3.39 \times 10^{-3}) \ M = \boxed{\mathbf{6.44 \times 10^{-2} \ M}}$$

$$[\text{F-6-P}] = 0.1 - (y + z) = 0.1 - (67.79 \times 10^{-3}) = \boxed{\mathbf{0.0322 \ M}}$$

Problem IV-16

Glucose and ATP were allowed to react in the presence of Mg^{++} and the enzymes hexokinase (HK) and phosphoglucomutase (PGM). Calculate the concentrations of glucose, ATP, ADP, glucose-6-P (G-6-P), and glucose-1-phosphate (G-1-P) at equilibrium if the initial concentrations of ATP and glucose were each 0.1 M.

Solution

The reactions involved are shown below.

(1) $\text{glucose} + \text{ATP} \underset{\text{Mg}^{++}}{\overset{\text{HK}}{\rightleftharpoons}} \text{G-6-P} + \text{ADP} \qquad K'_{eq_1} = 2180$

(2) $\text{G-6-P} \overset{\text{PGM}}{\rightleftharpoons} \text{G-1-P} \qquad K'_{eq_2} = 0.0526$

An exact solution can be obtained by setting up and solving two simultaneous equations. Let:

$$y = M \text{ G-6-P present at equilibrium}$$

$$z = M \text{ G-1-P present at equilibrium}$$

$$\therefore \quad (y + z) = M \text{ glucose used up}$$

and

$$0.1 - (y + z) = M \text{ glucose remaining at equilibrium}$$

Each mole of glucose used up required 1 mole of ATP and produced 1 mole of ADP.

$$\therefore \quad (y + z) = M \text{ ATP used up}$$
$$0.1 - (y + z) = M \text{ ATP remaining at equilibrium}$$
$$(y + z) = M \text{ ADP present at equilibrium}$$

Substituting these values into the appropriate K'_{eq} expressions:

$$K'_{eq_1} = \frac{[\text{ADP}][\text{G-6-P}]}{[\text{ATP}][\text{glucose}]} = 2180 \qquad K'_{eq_2} = \frac{[\text{G-1-P}]}{[\text{G-6-P}]} = 0.0526$$

$$K'_{eq_1} = \frac{(y + z)(y)}{[0.1 - (y + z)][0.1 - (y + z)]} = 2180 \qquad K'_{eq_2} = \frac{z}{y} = 0.0526$$

First solve for y in terms of z using the K'_{eq_2} expression.

$$\frac{z}{y} = 0.0526 \qquad y = \frac{z}{0.0526} \qquad y = 19z$$

Next, substitute $19z$ for y in the K'_{eq_1} expression.

$$K'_{eq_1} = \frac{(y + z)(y)}{[0.1 - (y + z)]^2} = \frac{(19z + z)(19z)}{[0.1 - (19z + z)]^2} = \frac{(20z)(19z)}{[0.1 - (20z)]^2} = 2180$$

$$K'_{eq_1} = \frac{380z^2}{0.01 - 4z + 400z^2} = 2180$$

$$21.8 - 8720z + 8.72 \times 10^5 z^2 = 380z^2$$
$$8.72 \times 10^5 z^2 - 380z^2 - 8720z + 21.8 = 0$$
$$8.7162 \times 10^5 z^2 - 8720z + 21.8 = 0$$

An exact value for z can be obtained by using the general solution for quadratic equations. If we use a slide rule, we will not be able to distinguish between 8.7162 and 8.72. As a result, G-1-P(z) will come out to be 5×10^{-3} M, G-6-P(y) will be 95×10^{-3} M, and ADP ($y + z$) will be 0.1 M. Thus, ATP and glucose will be zero. The concentrations of ATP and glucose at equilibrium are indeed very small—but not zero. Our slide rule solution then is not quite correct. We could have arrived at the same approximate (but nearly correct) solution by a far simpler approach: K'_{eq_1} is very large. Therefore, almost all the ATP and glucose will be converted to G-6-P and ADP by hexokinase. Phosphoglucomutase will then convert a portion of the initial G-6-P (0.1 M) to G-1-P.

Let

$$y = M \text{ G-6-P converted to G-1-P}$$
$$\therefore \quad y = M \text{ G-1-P present at equilibrium}$$

and
$$0.1 - y = M \text{ G-6-P remaining at equilibrium}$$

The y can be solved using only the K'_{eq_2} expression.

$$K'_{eq_2} = \frac{[\text{G-1-P}]}{[\text{G-6-P}]} = 0.0526 = \frac{(y)}{(0.1 - y)}$$

$$5.26 \times 10^{-3} - 5.26 \times 10^{-2}y = y$$

$$5.26 \times 10^{-3} = y + 5.26 \times 10^{-2}y = 1.0526y$$

$$y = \frac{5.26 \times 10^{-3}}{1.0526} = 5 \times 10^{-3}$$

$$\therefore \quad [\text{G-1-P}] \simeq 5 \times 10^{-3} \, M$$

and
$$[\text{G-6-P}] \simeq 0.1 - 5 \times 10^{-3} \, M \simeq 95 \times 10^{-3} \, M$$

and
$$[\text{ADP}] \text{ (unchanged by the PGM reaction)} \simeq 0.1 \, M$$

We can also obtain a reasonably accurate estimate of the equilibrium concentrations of ATP and glucose if we assume that all of the glucose used up appears as G-6-P (if we employ only the K'_{eq_1} expression). The assumption is valid because, as shown above, only about 5% of the G-6-P is converted to G-1-P.

Let
$$y = M \text{ glucose used up}$$
$$\therefore \quad y = M \text{ ATP used up}$$

and
$$y = M \text{ G-6-P produced}$$
$$y = M \text{ ADP produced}$$

and
$$(0.1 - y) = M \text{ ATP remaining at equilibrium}$$
$$(0.1 - y) = M \text{ glucose remaining at equilibrium}$$

$$K'_{eq_1} = \frac{[\text{ADP}][\text{G-6-P}]}{[\text{ATP}][\text{glucose}]} = \frac{(y)(y)}{(0.1 - y)(0.1 - y)} = \frac{y^2}{(0.1 - y)^2} = 2180$$

$$\sqrt{\frac{y^2}{(0.1 - y)^2}} = \sqrt{2180}$$

$$\frac{y}{0.1 - y} = 47.7$$

$$y = 4.77 - 47.7y$$

$$48.7y = 4.77$$

$$y = \frac{4.77}{48.7} = 0.098$$

$$\therefore \quad \text{[G-6-P]} = 0.098 \ M$$
$$\text{[ADP]} = 0.098 \ M$$
$$\text{[ATP]} = 0.1 - 0.098 \ M = 0.002 \ M$$
$$\text{[glucose]} = 0.002 \ M$$

From the two approximate solutions presented above we can state that the equilibrium G-6-P and ADP concentrations lie between 0.095 and 0.098 M, while the glucose and ATP concentrations are about 0.002 M and the G-1-P concentration is about 0.005 M.

C. OXIDATION-REDUCTION REACTIONS

Problem IV-17

One mole each of oxalacetate, malate, acetate, and acetaldehyde was dissolved in sufficient water to make 1 liter final volume at 25° C. (a) Write an equation for a thermodynamically favorable oxidation-reduction reaction that could occur (a reaction with a negative $\Delta F'$ value). (b) Which compound is oxidized? Which compound is reduced? Which compound is the oxidizing agent? Which compound is the reducing agent? (c) Calculate $\Delta E_0'$, $\Delta F'$, and K_{eq}' for the reaction.

Solution

(a) From Table 3 in Appendix II, we can obtain the following half-reactions and their standard reduction potentials.

(1)   oxalacetate $+ 2H^+ + 2e^- \rightarrow$ malate $E_0' = -0.102$ v

(2)   acetate $+ 2H^+ + 2e^- \rightarrow$ acetaldehyde $E_0' = -0.600$ v

Reaction 1 has the more positive standard reduction potential. Therefore, the reaction goes as written (as a reduction). Reaction 2 then is driven in reverse (as an oxidation).

(1)   oxalacetate $+ 2H^+ + 2e^- \rightarrow$ malate

(2a)   acetaldehyde $\rightarrow$ acetate $+ 2H^+ + 2e^-$

Sum: (3) oxalacetate $+$ acetaldehyde $\rightarrow$ malate $+$ acetate

(b)   The compound that is oxidized is acetaldehyde (it loses electrons). The compound that is reduced is oxalacetate (it gains electrons).

Acetaldehyde is the *reducing agent*. The reducing agent is the compound that reduces something else and, in the process, becomes oxidized. Oxalacetate is the *oxidizing agent*. The oxidizing agent is the compound that oxidizes something else and, in the process, becomes reduced.

(c) $\Delta E_0' = (E_0'$ of half-reaction containing the oxidizing agent) $-$

$\qquad\qquad (E_0'$ of half-reaction containing the reducing agent)

$\Delta E_0' = (-0.102) - (-0.600) = -0.102 + 0.600$

$$\boxed{\Delta E_0' = +0.498 \text{ v}}$$

$\Delta F' = -n\mathscr{F}\,\Delta E_0'$

where n is the number of electrons transferred per mole

$$\mathscr{F} = 23{,}063 \text{ cal/v-equiv}$$
$$\Delta F' = -(2)(23{,}063)(0.498)$$

$$\boxed{\Delta F' = -23{,}000 \text{ cal/mole}}$$

Note that the ΔE value must be positive in order to obtain a negative ΔF value.

$$\Delta F' = -1363 \log K_{eq}'$$

$$= -23{,}000 \text{ cal/mole}$$

$$\log K_{eq}' = \frac{-23{,}000}{-1363} = 16.85$$

$$K_{eq}' = \text{antilog of } 16.85$$

$$\boxed{K_{eq}' = 7.08 \times 10^{16}}$$

$$F' = -n\mathscr{F}\,\Delta E_0'$$

$$= -2.3RT \log K_{eq}'$$

$$\frac{-n\mathscr{F}\,\Delta E_0'}{-2.3RT} = \log K_{eq}'$$

$$\frac{n\,\Delta E_0'}{0.059} = \log K_{eq}'$$

$$\frac{(2)(0.498)}{(0.059)} = \log K_{eq}' = 16.85$$

$$\boxed{K_{eq}' = 7.08 \times 10^{16}}$$

The fact that the reaction has a very high K_{eq}' (a high negative $\Delta F'$) does not necessarily mean that it will proceed at a measurable *velocity*. The above calculations only tell us the direction in which the reaction *would* go under standard-state conditions and the position of equilibrium, but they tell us nothing about the *rate* of the reaction. The overall reaction, if it did occur in living cells, would very likely proceed in several steps.

Problem IV-18

A solution containing 0.01 M each of fumarate, succinate, riboflavin, and riboflavin-H_2 was prepared at 30° C. Write an equation for a thermo-dynamically favorable reaction that could occur and calculate the ΔE for the reaction.

Solution

The half-reactions involved and their standard reduction potentials are shown below.

(1)   fumarate $+ 2H^+ + 2e^- \rightarrow$ succinate $E_0' = +0.030$ v

(2)   riboflavin $+ 2H^+ + 2e^- \rightarrow$ riboflavin-H_2 $E_0' = -0.200$ v

Reaction 1 has the more positive reduction potential, so it goes as a reduction, thereby forcing reaction 2 to go as an oxidation.

(1)   fumarate $+ 2H^+ + 2e^- \rightarrow$ succinate

(2a)   riboflavin-$H_2 \rightarrow$ riboflavin $+ 2H^+ + 2e^-$

———

┌——————— is oxidized to ———————┐
                                                              ↓
Sum: (3) fumarate $+$ riboflavin-$H_2 \rightarrow$ succinate $+$ riboflavin
          └——————— is reduced to ———————↑

Thus, under standard-state conditions, riboflavin-H_2 reduces fumarate to succinate and, in the process, is itself oxidized to riboflavin. However, the particular solution does not contain standard-state concentrations (1 M) but rather 0.01 M of each component. The reduction potential of a half-reaction in which the oxidized and reduced forms of the substance are present at concentrations other than 1 M may be calculated from the following expression, called the Nernst equation.

$$E = E_0' + \frac{RT}{n\mathscr{F}} \ln \frac{[\text{oxidized form}]}{[\text{reduced form}]}$$

or

$$E = E_0' + \frac{2.3RT}{n\mathscr{F}} \log \frac{[\text{oxidized form}]}{[\text{reduced form}]}$$

where

$R =$ molar gas constant, 1.99 cal/degree

$T =$ absolute temperature (° K)

$\mathscr{F} =$ 23,063 cal/v-equiv

$n =$ number of electrons transferred per mole

At 25° C, the $2.3RT/\mathscr{F}$ term is 0.059. At 30° C, the value is 0.060.

$$E = E_0' + \frac{0.06}{n} \log \frac{[\text{oxidized form}]}{[\text{reduced form}]}$$

We can see that as long as there is only one oxidized and one reduced form involved in a half-reaction, the log term equals zero when [oxidized form] = [reduced form] and $E = E_0'$. Therefore, because the fumarate/succinate and riboflavin/riboflavin-H_2 ratios are unity, E for each half-reaction equals E_0', and $\Delta E = \Delta E_0'$.

$$\Delta E = \Delta E_0' = (E_0' \text{ of half-reaction containing the oxidizing agent}) -$$

$$(E_0' \text{ of half-reaction containing the reducing agent})$$

$$\Delta E_0' = (0.030) - (-0.200) = (0.030) + (0.200)$$

$$\boxed{\Delta E_0' = +0.230 \text{ v}}$$

General Principle

Standard reduction potentials are the voltages of half-reactions when the oxidized and reduced components are present at *unit* activity *or* when the oxidized and reduced components are present at *equal* activities provided there is the same number of oxidized and reduced forms. Under either of the above conditions the standard reduction potential is the voltage of the half-reduced (or half-oxidized) system.

Problem IV-19

A solution containing 0.2 M dehydroascorbate and 0.2 M ascorbate was mixed with an equal volume of a solution containing 0.01 M acetaldehyde and 0.01 M ethanol. (a) Write the equation for a thermodynamically favorable reaction that could occur. (b) Calculate $\Delta E_0'$.

Solution

(a) The half-reactions involved and their standard reduction potentials are shown below.

(1) dehydroascorbate $+ 2H^+ + 2e^- \rightarrow$ ascorbate $E_0' = +0.06$ v

(2)   acetaldehyde $+ 2H^+ + 2e^- \rightarrow$ ethanol $E_0' = -0.163$ v

Under standard-state conditions, dehydroascorbate gains electrons and becomes reduced while ethanol provides the electrons and becomes oxidized.

(1) dehydroascorbate $+ 2H^+ + 2e^- \longrightarrow$ ascorbate

(2a) ethanol $\longrightarrow 2H^+ + 2e^- +$ acetaldehyde

Sum: (3) dehydroascorbate $+$ ethanol $\longrightarrow$ ascorbate $+$ acetaldehyde
(oxidizing agent)   (reducing agent)

Although the dehydroascorbate and ascorbate concentrations are $0.1 M$ in the final solution and the ethanol and acetaldehyde concentrations are only $0.005 M$, the concentration *ratios* for each oxidized-reduced pair are unity. Consequently, the E values for each half-reaction are identical to their corresponding E_0' values. Any advantage gained by having a high concentration of the oxidized form of a substance on one side of the equation is offset by having an equally high concentration of the reduced form on the other side.

(b) $\Delta E_0' = (E_0'$ of half-reaction containing the oxidizing agent) $-$

$(E_0'$ of half-reaction containing the reducing agent)

$\Delta E_0' = (+0.060) - (-0.163) = (+0.06) + (0.163)$

$$\boxed{\Delta E_0' = +0.223 \text{ v}}$$

Problem IV-20

Write the spontaneous reaction that occurs and calculate the ΔF of the reaction when the enzyme lactic dehydrogenase is added to a solution containing pyruvate, lactate, DPN$^+$, and DPNH at the following concentration ratios: (a) lactate/pyruvate $= 1$, DPN$^+$/DPNH $= 1$, (b) lactate/pyruvate $= 150$, DPN$^+$/DPNH $= 150$, and (c) lactate/pyruvate $= 1000$, DPN$^+$/DPNH $= 1000$.

Solution

The half-reactions involved and their standard reduction potentials are shown below.

(1)   pyruvate $+ 2H^+ + 2e^- \longrightarrow$ lactate $E_0' = -0.190$ v

(2)   DPN$^+ + 2H^+ + 2e^- \longrightarrow$ DPNH $+ H^+$ $E_0' = -0.320$ v

(a) Under standard-state conditions, or at lactate/pyruvate and DPN$^+$/DPNH ratios of 1, the pyruvate/lactate half-reaction goes as written while the DPN$^+$/DPNH half-reaction goes in reverse (as an oxidation).

The overall spontaneous reaction is:

Sum: (3)   pyruvate + DPNH + $H^+ \rightarrow$ lactate + DPN^+

$$\Delta E_0' = (-0.190) - (-0.320) = +0.130$$

The DPNH is the reducing agent; pyruvate is the oxidizing agent.

$$\Delta F' = -n\mathscr{F}\,\Delta E_0' = -(2)(23,063)(0.130)$$

$$\boxed{\Delta F' = -5980 \text{ cal/mole}}$$

(b)   If, instead of standard-state conditions, we set up a reaction mixture containing lactate/pyruvate and DPN^+/DPNH ratios of 150/1, the half-reactions have new reduction potentials as shown below.

(1) $E_1 = -0.190 + \dfrac{0.06}{2} \log \dfrac{[\text{pyruvate}]}{[\text{lactate}]}$

$= -0.190 + 0.03 \log \frac{1}{150}$

$= -0.190 - 0.030 \log 150$

$= -0.190 - 0.030(2.176) = -0.190 - 0.065$

$$\boxed{E_1 = -0.255 \text{ v}}$$

(2) $E_2 = -0.320 + \dfrac{0.06}{2} \log \dfrac{[DPN^+]}{[DPNH]}$

$= -0.320 + 0.030 \log 150$

$= -0.320 + 0.030(2.176)$

$= -0.320 + 0.065$

$$\boxed{E_2 = -0.255 \text{ v}}$$

At these concentration ratios the two half-reactions have the same E values; ΔE is zero, hence $\Delta F = 0$. In other words, at the indicated concentration ratios, the reaction is at equilibrium already so no further net change will occur.

(c)   Now suppose that the lactate/pyruvate and DPN^+/DPNH ratios are set at 1000/1. Under this condition the reduction potentials of the two reactions are as follows:

(1)
$$E_1 = -0.190 + 0.030 \log \tfrac{1}{1000}$$
$$= -0.190 - 0.030 \log 1000$$
$$= -0.190 - 0.030(3) = -0.190 - 0.090$$

$$\boxed{E_1 = -0.280 \text{ v}}$$

(2)
$$E_2 = -0.320 + 0.03 \log 1000$$
$$= -0.320 + 0.03(3) = -0.320 + 0.090$$

$$\boxed{E_2 = -0.230 \text{ v}}$$

Now the $DPN^+/DPNH$ half-reaction has the more positive reduction potential. Thus, this half-reaction goes as written (as a reduction), driving the pyruvate/lactate half-reaction backwards (as an oxidation). Lactate is now the reducing agent; DPN^+ is the oxidizing agent. The overall spontaneous reaction is:

$$DPN^+ + \text{lactate} \rightarrow DPNH + H^+ + \text{pyruvate}$$
$$\Delta E = (-0.230) - (-0.280) = -(0.230) + (0.280)$$
$$\Delta E = +0.05 \text{ v}$$
$$\Delta F = -n\mathscr{F}\Delta E = -(2)\,(23{,}063)\,(0.05)$$

$$\boxed{\Delta F = -2306 \text{ cal/mole}}$$

The ΔF under the nonstandard-state conditions can also be calculated directly from $\Delta F'$ and the actual concentration ratios.

$$\Delta F = \Delta F' + 1386 \log \frac{[DPNH][\text{pyruvate}]}{[DPN^+][\text{lactate}]} \qquad (2.3RT = 1386 \text{ at } 30^\circ \text{ C})$$

The $\Delta F'$ for the *original* reaction under standard-state conditions (part a) was -5980 cal/mole. Because the reaction is now written in reverse under our new conditions, the value for $\Delta F'$ becomes positive.

$$\Delta F = +5980 + 1386 \log \tfrac{1}{1000}\big|\tfrac{1}{1000}$$
$$\Delta F = 5980 - 1386 \log 10^6 = 5980 - 1386\,(6)$$
$$\Delta F = 5980 - 8320$$

$$\boxed{\Delta F = -2340 \text{ cal/mole}}$$

The slight difference between the two answers results from rounding off the $2.3RT/n\mathscr{F}$ term to 0.06.

We can see that under standard-state conditions, lactate does not reduce DPN+ (that is, DPN+ does not oxidize lactate). The $\Delta F'$ for the reaction is positive. However, by altering the concentration ratios, we can make the reduction of DPN+ by lactate a favorable reaction. The fact that the reaction can be driven in either direction by manipulating the concentrations of reactants and products should come as no surprise. This is a consequence of the law of mass action. The above calculations allow us to determine exactly in which direction a reaction will proceed under any given concentration conditions. The calculations also point out the difficulty in deciding which way a reaction will proceed *in vivo* without a prior knowledge of the intracellular concentrations of the reaction components.

Problem IV-21

The reaction catalyzed by triosephosphate dehydrogenase (glyceraldehyde-3-phosphate dehydrogenase) is shown below.

(1) $GAP + DPN^+ + P_i \rightleftharpoons 1,3\text{-DiPGA} + DPNH + H^+$

(a) In which direction will the reaction proceed spontaneously under standard-state conditions at 30° C and pH 7? (b) In which direction will the reaction proceed spontaneously under steady-state conditions in which the DPN+/DPNH ratio is maintained at 100/1, the GAP/1,3-DiPGA ratio is maintained at 200/1, and the P_i concentration is maintained at 0.01 M?

Solution

(a) The overall reaction can be expressed as the sum of two half-reactions:

(2) $GAP + P_i \rightarrow 2e^- + 2H^+ + 1,3\text{-DiPGA}$

(3) $DPN^+ + 2e^- + 2H^+ \rightarrow DPNH + H^+$

Sum: (1) $GAP + P_i + DPN^+ \rightarrow 1,3\text{-DiPGA} + DPNH + H^+$

The two half-reactions written as reductions are:

(2)   1,3-DiPGA $+ 2e^- + 2H^+ \rightarrow$ GAP $+ P_i$ $E_0' = -0.29$ v

(3)   DPN$^+ + 2e^- + 2H^+ \rightarrow$ DPNH $+ H^+$ $E_0' = -0.32$ v

When two half-reactions are coupled, the one with the more positive reduction potential goes spontaneously as a reduction while the one with the less positive reduction potential goes spontaneously as an oxidation. Under standard-state conditions the 1,3-DiPGA/GAP half-reaction has the more positive E_0' value. The *spontaneous* half-reactions then are:

(2a)   1,3-DiPGA $+ 2e^- + 2H^+ \rightarrow$ GAP $+ P_i$

(3a)   DPNH $+ H^+ \rightarrow$ DPN$^+ + 2e^- + 2H^+$
_____

Sum: (1a) 1,3-DiPGA $+$ DPNH $+ H^+ \rightarrow$ GAP $+ P_i +$ DPN$^+$

Consequently, under standard-state conditions the triosephosphate dehydrogenase reaction goes spontaneously toward GAP, P_i, and DPN$^+$ formation. DPNH is the reducing agent (and is oxidized); 1,3-DiPGA is the oxidizing agent (and is reduced).

From the E_0' values we can calculate $\Delta F'$ and K_{eq}' values for the overall reaction (1) written in either direction. For reaction 1a:

$$\Delta E_0' = (-0.29) - (-0.32)$$

$$= -0.29 + 0.32$$

$$\boxed{\Delta E_0' = +0.03 \text{ v}}$$

$$\Delta F' = -n\mathscr{F} \, \Delta E_0' = -(2)(23{,}063)(+0.03)$$

$$\boxed{\Delta F_{1a}' = -1381 \text{ cal/mole}}$$

$$\Delta F' = -1386 \log K_{eq}'$$

$$\log K_{eq}' = \frac{-1381}{-1386} = 0.995$$

$$K_{eq}' = \text{antilog of } 0.995$$

$$\boxed{K_{eq_{1a}}' = 9.9}$$

For the reaction written in the direction of 1,3-DiPGA and DPNH formation (1):

$$\Delta F_1' = +1381 \text{ cal/mole}$$

and

$$K_{eq}' = \frac{1}{9.9}$$

$$K_{eq_1}' = 0.101$$

If we knew the K_{eq}' values to begin with, we could decide upon the spontaneous direction under standard-state conditions without calculating $\Delta E_0'$.

Reaction 1	**Reaction 1a**
$K_{eq_1}' = \dfrac{[\text{1,3-DiPGA}][\text{DPNH}]}{[\text{GAP}][\text{DPN}^+][\text{P}_i]}$	$K_{eq_{1a}}' = \dfrac{[\text{GAP}][\text{DPN}^+][\text{P}_i]}{[\text{1,3-DiPGA}][\text{DPNH}]}$
$= 0.101$	$= 9.9$
At standard-state conditions, all concentrations are 1 M.	At standard-state conditions, all concentrations are 1 M.
$\dfrac{(1)(1)}{(1)(1)(1)} > 0.101$	$\dfrac{(1)(1)(1)}{(1)(1)} < 9.9$
The product of the concentrations is greater than K_{eq}'. $\therefore$ The numerator (1,3-DiPGA and DPNH) decreases and the denominator (GAP, DPN$^+$, and P$_i$) increases until the product of the concentrations equals K_{eq}'.	The product of the concentrations is less than K_{eq}'. $\therefore$ The numerator (GAP, DPN$^+$, and P$_i$) increases and the denominator (1,3-DiPGA and DPNH) decreases until the product of the concentrations equals K_{eq}'.

(b)   Under nonstandard-state conditions, we can decide upon the spontaneous direction either by calculating the new E values or the sign of the nonstandard-state ΔF value, or by comparing the product of the concentrations of reaction components with K_{eq}'.

Method 1. Calculating E Values

First calculate the new E values for the two half-reactions.

$$(2a) \quad E_{2a} = E_0' + \frac{0.06}{2}$$

$$\times \log \frac{[1,3\text{-DiPGA}]}{[\text{GAP}][P_i]}$$

Although the inorganic phosphate is not oxidized or reduced in the half-reaction, its concentration is important in establishing the E value.

$$E_{2_a} = -0.29 + \frac{0.06}{2}$$

$$\times \log \frac{(1)}{(200)(0.01)}$$

$$= -0.29 + 0.03 \log \tfrac{1}{2}$$

$$= -0.29 - 0.03 \log 2$$

$$= -0.29 - 0.03(0.3)$$

$$= -0.290 - 0.009$$

$$\boxed{E_{2a} = -0.299 \text{ v}}$$

$$(3) \quad E_3 = E_0' + \frac{0.06}{2} \log \frac{[\text{DPN}^+]}{[\text{DPNH}]}$$

$$= -0.32 + 0.03 \log \tfrac{100}{1}$$

$$= -0.32 + 0.03(2)$$

$$= -0.32 + 0.06$$

$$\boxed{E_3 = -0.260 \text{ v}}$$

Now the DPN$^+$/DPNH half-reaction has the more positive E value. Consequently, the spontaneous half-reactions are:

Method 2. Calculating ΔF

For the reaction written in the direction of GAP, DPN$^+$, and P$_i$ formation:

$$\Delta F = \Delta F' + 1386$$

$$\times \log \frac{[\text{GAP}][\text{DPN}^+][P_i]}{[1,3\text{-DiPGA}][\text{DPNH}]}$$

$$\Delta F = -1381$$

$$+ 1386 \log \frac{(200)(100)(0.01)}{(1)(1)}$$

$$= -1381 + 1386 \log 200$$

$$= -1381 + 1386(2.3)$$

$$= -1381 + 3181$$

$$\boxed{\Delta F = +1800 \text{ cal/mole}}$$

Thus, the reaction 1,3-DiPGA + DPNH $\rightarrow$ GAP + DPN$^+$ + P$_i$ is not favorable. The reaction goes spontaneously in the opposite direction, toward 1,3-DiPGA and DPNH formation. The ΔF for the spontaneous reaction is -1800 cal/mole.

Method 3. Comparing Product and Substrate Concentrations to K_{eq}'

$$K_{eq}' = \frac{[\text{GAP}][\text{DPN}^+][P_i]}{[1,3\text{-DiPGA}][\text{DPNH}]}$$

$$= 9.9$$

$$(200)(100)(0.01) = 200 > 9.9$$

The product of the concentrations is greater than K_{eq}'. Therefore, the spontaneous reaction is that which tends to bring the product of the concentrations to equal K_{eq}'; that is, the numerator (GAP, DPN$^+$,

Method 1—*continued*

(2) $GAP + P_i \rightarrow$

             $2e^- + 2H^+ + 1,3\text{-DiPGA}$

(3) $DPN^+ + 2e^- + 2H^+ \rightarrow$

                                      $DPNH$

Sum: (1) $GAP + P_i + DPN^+ \rightarrow$

             $1,3\text{-DiPGA} + DPNH$

Now DPN^+ is the oxidizing agent, and GAP is the reducing agent. The overall spontaneous reaction now goes in the direction of 1,3-DiPGA and DPNH formation.

Method 3—*continued*

and P_i) decreases and the denominator (1,3-DiPGA and DPNH) increases.

We can cross-check Methods 1 and 2. The ΔF value calculated from the ΔE value obtained in Method 1 should be the same as the ΔF value obtained directly in Method 2.

$$\Delta E = (-0.260) - (-0.299) = -0.260 + 0.299$$

$$\boxed{\Delta E = +0.039 \text{ v}}$$

$$\Delta F = -n\mathscr{F}\Delta E = -(2)(23,063)(0.039)$$

$$\boxed{\Delta F = -1800 \text{ cal/mole}}$$

Problem IV-22

The reaction catalyzed by the enzyme glutamic dehydrogenase is shown below.

(1) $\alpha\text{-ketoglutarate} + TPNH + NH_4^+ \rightarrow glutamate + TPN^+ + H_2O$

     $(\alpha\text{-KG})$ (GA)

If the intracellular ratios of GA/α-KG and TPNH/TPN$^+$ are each 10/1, how would you calculate the minimum intracellular concentration of NH$_4^+$ required to drive the reaction toward glutamate and TPN$^+$ formation at pH 7 and 30° C?

*Solution**

We can solve the problem in several ways: (1) If we know the K'_{eq} for the reaction, we can calculate the concentration of NH_4^+ that would be in equilibrium with the indicated GA/α-KG and TPNH/TPN$^+$ ratios. Any increase in the NH_4^+ concentration would then drive the reaction toward GA and TPN$^+$ formation. The K'_{eq} can be calculated from the E'_0 values of the two component half-reactions. (2) If we know the $\Delta F'$ value, we can calculate the NH_4^+ concentration that would give a nonstandard-state ΔF value of zero. Any increase in the NH_4^+ concentration would then yield a negative ΔF value for the reaction as written, indicating that the reaction would go spontaneously toward GA and TPN$^+$ formation. The $\Delta F'$ can be calculated from K'_{eq}. (3) We could calculate the NH_4^+ concentration that would make the E value for the α-KG + NH_4^+/GA half-reaction equal to the E value for the TPN$^+$/TPNH half-reaction. Any increase in the NH_4^+ concentration would make the E value for the α-KG + NH_4^+/GA half-reaction more positive; this reaction would go spontaneously as a reduction, forcing the TPN$^+$/TPNH half-reaction to go as an oxidation. Therefore, the overall reaction would go toward GA and TPN$^+$ formation.

Method 1 involves the simplest calculations. The reaction can be expressed as the sum of two half-reactions.

(2) $\qquad \alpha\text{-KG} + NH_3 + 2e^- + 2H^+ \rightarrow GA + H_2O$

(3) $\qquad\qquad\qquad TPNH + H^+ \rightarrow TPN^+ + 2e^- + 2H^+$

Sum (1): $\qquad \alpha\text{-KG} + TPNH + NH_4^+ \rightarrow GA + H_2O + TPN^+$

The reactions may be written as reductions.

(2) $\alpha\text{-KG} + NH_3 + 2e^- + 2H^+ \rightarrow GA + H_2O \qquad E'_0 = -0.14 \text{ v}$

(3a) $\qquad TPN^+ + 2e^- + 2H^+ \rightarrow TPNH + H^+ \qquad E'_0 = -0.32 \text{ v}$

Because reaction 2 has the more positive E'_0 value, the overall reaction 1 goes spontaneously as written under standard-state conditions. TPNH is

---

* The reactions and equations can be formulated using either NH_3 or NH_4^+, whichever is the more convenient. Because we are concerned with K'_{eq}, $\Delta F'$, $\Delta E'_0$, E, and ΔF values at pH 7, both symbols, [NH_3] and [NH_4^+], refer to the *total* (analytical) concentration of "ammonia" present (actually ~1% NH_3 plus ~99% NH_4^+).

oxidized and α-KG is simultaneously reduced and aminated.

$$\Delta E_0' = (-0.14) - (-0.32) = -0.14 + 0.32$$

$$\boxed{\Delta E_0' = +0.18 \text{ v}}$$

$$\Delta F' = -n\mathscr{F}\,\Delta E_0' = -(2)(23{,}063)(0.18)$$

$$\boxed{\Delta F' = -8300 \text{ cal/mole}}$$

$$\Delta F' = -1386 \log K_{eq}' = -8300$$

$$\log K_{eq}' = \frac{-8300}{-1386} = 5.98$$

$$K_{eq}' = \text{antilog of } 5.98$$

$$\boxed{K_{eq}' = 9.55 \times 10^5}$$

Method 1:

$$K_{eq}' = \frac{[\text{GA}][\text{TPN}^+]}{[\alpha\text{-KG}][\text{TPNH}][\text{NH}_4^+]} = 9.55 \times 10^5$$

$$\frac{(10)(1)}{(1)(10)[\text{NH}_4^+]} = 9.55 \times 10^5$$

$$[\text{NH}_4^+] = \frac{1}{9.55 \times 10^5} = 0.1046 \times 10^{-5}$$

$$\boxed{[\text{NH}_4^+] = 1.046 \times 10^{-6}\,M}$$

Thus, at about $10^{-6}\,M\;\text{NH}_4^+$ the reaction is at equilibrium. Any increase will drive the reaction toward GA and TPN⁺ formation.

Method 2:

$$\Delta F = \Delta F' + 1386 \log \frac{[\text{GA}][\text{TPN}^+]}{[\alpha\text{-KG}][\text{TPNH}][\text{NH}_4^+]}$$

$$\Delta F = -1386 \log K_{eq}' + 1386 \log \frac{[\text{GA}][\text{TPN}^+]}{[\alpha\text{-KG}][\text{TPNH}][\text{NH}_4^+]}$$

At equilibrium

$$\Delta F = 0$$

$$0 = -1386 \log K_{eq}' + 1386 \log \frac{[\text{GA}][\text{TPN}^+]}{[\alpha\text{-KG}][\text{TPNH}][\text{NH}_4^+]}$$

$$1386 \log K_{eq}' = 1386 \log \frac{[\text{GA}][\text{TPN}^+]}{[\alpha\text{-KG}][\text{TPNH}][\text{NH}_4^+]}$$

or

$$K'_{eq} = \frac{[GA][TPN^+]}{[\alpha\text{-}KG][TPNH][NH_4^+]}$$

Thus, Method 2 breaks down to Method 1.

Method 3: First calculate the E value for the $TPN^+/TPNH$ half-reaction under the nonstandard-state condition.

$$E = E'_0 + \frac{0.06}{2} \log \frac{[TPN^+]}{[TPNH]}$$

$$= -0.32 + 0.03 \log 1/10$$

$$= -0.32 - 0.03 \log 10$$

$$= -0.32 - 0.03(1)$$

$$\boxed{E_{TPN^+/TPNH} = -0.35 \text{ v}}$$

Next, calculate the NH_3 concentration that would make the E value for the α-KG + NH_3/GA half-reaction -0.35 v. At this NH_3 concentration the reaction will be at equilibrium ($\Delta E = 0$, $\therefore$ $\Delta F = 0$).

$$E = E'_0 + \frac{0.06}{2} \log \frac{[\alpha\text{-}KG][NH_3]}{[GA]}$$

$$-0.35 = -0.14 + 0.03 \log \frac{(1)[NH_3]}{(10)}$$

$$-0.35 = -0.14 - 0.03 \log \frac{(10)}{(1)[NH_3]}$$

$$-0.21 = -0.03\left[\log 10 + \log \frac{1}{[NH_3]}\right]$$

$$-0.21 = -0.03 \log 10 - 0.03 \log \frac{1}{[NH_3]}$$

$$-0.21 = -0.03 - 0.03 \log \frac{1}{[NH_3]}$$

$$-0.18 = -0.03 \log \frac{1}{[NH_3]}$$

$$\frac{-0.18}{-0.03} = \log \frac{1}{[NH_3]}$$

$$6 = \log \frac{1}{[NH_3]}$$

$$\frac{1}{[NH_3]} = \text{antilog of } 6 = 10^6$$

$$[NH_3] = 1/10^6$$

$$\boxed{[NH_3] = 10^{-6} \, M}$$

Problem IV-23

The oxidation of the $FADH_2$ in succinic dehydrogenase (via the electron-transport system) yields sufficient energy to drive the synthesis of 2 moles of ATP per mole of $FADH_2$. Calculate the standard reduction potential of the enzyme-bound $FADH_2$, assuming that energy conservation reactions in living cells occur with an efficiency of about 40%. The E_0' for the $\frac{1}{2}O_2 + 2H^+ + 2e^-$ half-reaction is $+0.816$ v.

Solution

The overall reaction is shown below.

(1) $FADH_2 + \frac{1}{2}O_2 + 2ADP + 2P_i \rightarrow FAD + 2H_2O + 2ATP$

The reaction may be considered as the sum of two coupled reactions (an oxidation-reduction reaction and a dehydration reaction).

(2) $FADH_2 + \frac{1}{2}O_2 \rightarrow FAD + H_2O$

(3) $2ADP + 2P_i \rightarrow 2ATP + 2H_2O$ $\Delta F_3' = 2(+8000 \text{ cal/mole})$

Sum: (1) $FADH_2 + \frac{1}{2}O_2 + 2ADP + 2P_i \rightarrow FAD + 3H_2O + 2ATP$

We know that ATP is worth 8000 cal/mole. Consequently, 2(8000 cal/mole) = 16,000 cal are conserved. If this represents 40% of the energy released in reaction 2, then we can calculate ΔF_2 easily. Let y equal energy released in reaction 2.

$$16,000 \text{ cal} = 0.40y$$

$$y = \frac{16,000}{0.40} = 40,000 \text{ cal}$$

$$\boxed{\Delta F_2' = -40,000 \text{ cal/mole}}$$

We can also now calculate the overall $\Delta F'$.

$$\Delta F_1' = \Delta F_2' + \Delta F_3' = (-40,000) + (16,000)$$

$$\boxed{\Delta F_1' = -24,000 \text{ cal/mole}}$$

or

$$\Delta F_1' = 60\% \text{ of } \Delta F_2' = 0.60(-40,000 \text{ cal/mole})$$

$$\boxed{\Delta F_1' = -24,000 \text{ cal/mole}}$$

Reaction 2 is an oxidation-reduction reaction. It can be expressed in terms of two half-reactions.

(4) $$FADH_2 \rightarrow FAD + 2H^+ + 2e^-$$

(5) $$\tfrac{1}{2}O_2 + 2H^+ + 2e^- \rightarrow H_2O$$

Sum: (2) $$FADH_2 + \tfrac{1}{2}O_2 \rightarrow FAD + H_2O$$

The two half-reactions written as reductions are:

(4a) $$FAD + 2H^+ + 2e^- \rightarrow FADH_2 \qquad E_0' = ?$$

(5) $$\tfrac{1}{2}O_2 + 2H^+ + 2e^- \rightarrow H_2O \qquad E_0' = +0.816 \text{ v}$$

Because we already know ΔF_2, we can calculate $\Delta E_0'$.

$$\Delta F_2' = -n\mathscr{F}\,\Delta E_0' = -(2)(23,063)(\Delta E_0') = -40,000 \text{ cal/mole}$$

$$\Delta E_0' = \frac{-40,000}{-(2)(23,063)} = \frac{40,000}{46,126}$$

$$\boxed{\Delta E_0' = +0.870 \text{ v}}$$

Now that we know $\Delta E_0'$ and E_0' for the $\tfrac{1}{2}O_2/H_2O$ half-reaction, we can calculate E_0' for the $FAD/FADH_2$ half-reaction.

$\Delta E_0' = (E_0'$ of the half-reaction containing the oxidizing agent)

$\qquad - (E_0'$ of the half-reaction containing the reducing agent)

The O_2 is the oxidizing agent; $FADH_2$ is the reducing agent.

$$0.870 = (+0.816) - (E_{0_{FAD/FADH2}}')$$

$$E_{0_{FAD/FADH2}}' = 0.816 - 0.870$$

$$\boxed{E_{0_{FAD/FADH2}}' = -0.054 \text{ v}}$$

Problem IV-24

When light strikes a pigment in a photosynthetic organism, an electron is raised to an energy level that is sufficient to reduce a molecule of ferredoxin. The electron can then fall back to the positively charged (oxidized) pigment through a series of electron carriers. The electron carriers and their standard-state reduction potentials are shown below.

Predict the likely oxidation-reduction sites at which the synthesis of ATP may occur.

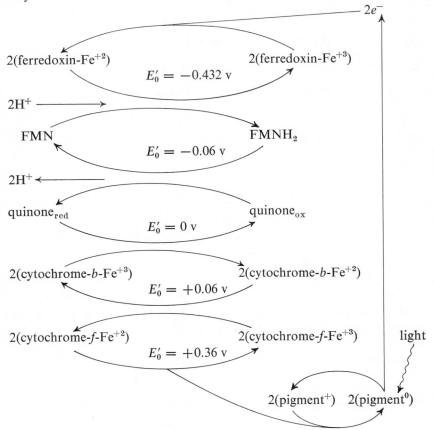

The synthesis of ATP requires about 8000 cal/mole. This corresponds to a $\Delta E_0'$ of 0.178 v as shown below.

$$\Delta F' = -n\mathscr{F}\,\Delta E_0' = -8{,}000 \text{ cal/mole}$$

$$\Delta E_0' = \frac{-8000}{-(2)(23{,}063)} = 0.178 \text{ v}$$

Thus, a $\Delta E_0'$ of *at least* 0.178 v is required (at 100% efficiency). Assuming a reasonable efficiency of 60%, a $\Delta E_0'$ of 0.297 v would be required. In the electron-transport chain shown above, only two oxidation-reduction couples provide sufficient energy: one between ferredoxin and FMN and one between cytochrome b and cytochrome f.

Problem IV-25

Calculate the $\Delta F'$ and K'_{eq} values for the reaction shown below.

(1)

$FADH_2 + 2$ cytochrome-c-$Fe^{+3} \rightleftharpoons FAD + 2$ cytochrome-c-$Fe^{+2} + 2H^+$

Solution

The overall reaction can be expressed as the sum of two half-reactions.

(2) $\qquad\qquad FADH_2 \rightarrow FAD + 2H^+ + 2e^-$

(3) $2[$cytochrome-c-$Fe^{+3} + 1e^- \rightarrow$ cytochrome-c-$Fe^{+2}]$

Sum: (1)

$FADH_2 + 2$ cytochrome-c-$Fe^{+3} \rightarrow FAD + 2$ cytochrome-c-$Fe^{+2} + 2H^+$

The two half-reactions written as reductions are shown below.

(2a) $\qquad FAD + 2H^+ + 2e^- \rightarrow FADH_2 \qquad\qquad E'_0 = -0.06$ v

(3) $2[$cytochrome-c-$Fe^{+3} + 1e^- \rightarrow$ cytochrome-c-$Fe^{+2}] \quad E'_0 = +0.25$ v

The E'_0 value of the cytochrome-c-Fe^{+3}/cytochrome-c-Fe^{+2} half-reaction is more positive than that of the FAD/$FADH_2$ half-reaction. Thus, half-reaction 3 goes as written, forcing half-reaction 2a to go in reverse—as an oxidation. The overall spontaneous reaction (under standard-state conditions) is in the direction shown in reaction 1. $FADH_2$ (the reducing agent) is oxidized to FAD by cytochrome-c-Fe^{+3} (the oxidizing agent).

$\Delta E'_0 = (E'_0$ of the half-reaction containing the oxidizing agent)

$\qquad\qquad - (E'_0$ of the half-reaction containing the reducing agent)

$\qquad \Delta E'_0 = (0.25) - (-0.06) = 0.25 + 0.06$

$$\boxed{\Delta E'_0 = +0.31 \text{ v}}$$

Note that the E'_0 value for the cytochrome-c-Fe^{+3}/cytochrome-c-Fe^{+2} half-reaction is *not* doubled when calculating $\Delta E'_0$. The fact that 2 moles of cytochrome are required per mole of $FADH_2$ is taken into account in calculating $\Delta F'$ and K'_{eq}. (E'_0 values may be thought of as "electron pressures" or volts/electron. As such, they are independent of the *number* of electrons transferred per mole.)

$$\Delta F' = -n\mathscr{F}\Delta E'_0$$

where n equals the number of electrons transferred in the balanced equation.

$$\Delta F' = -(2)(23,063)(0.31)$$

$$\boxed{\Delta F' = -14,300 \text{ cal/mole FADH}_2 = -7150 \text{ cal/mole cytochrome-c}}$$

$$\Delta F' = -1363 \log K'_{eq}$$

$$\log K'_{eq} = \frac{-14,300}{-1363} = 10.5$$

$$K'_{eq} = \text{antilog of } 10.5$$

$$\boxed{K'_{eq} = 3.16 \times 10^{10}}$$

where

$$K'_{eq} = \frac{[\text{FAD}][\text{cytochrome-}c\text{-Fe}^{+2})^2}{[\text{FADH}_2](\text{cytochrome-}c\text{-Fe}^{+3})^2}$$

PRACTICE PROBLEM SET IV

1. The equilibrium constant (at 25° C and pH 7) for the aldolase reaction (fructose-1,6-diphosphate $\rightleftharpoons$ glyceraldehyde-3-phosphate + dihydroxyacetone-phosphate) is 10^{-4}. (a) Calculate the concentration of glyceraldehyde-3-phosphate that is in equilibrium with $2 \times 10^{-3} M$ fructose-1,6-diphosphate and $4 \times 10^{-4} M$ dihydroxyacetone-phosphate. (b) Calculate the $\Delta F'$ for the aldolase reaction written in the direction of fructose-1,6-diphosphate synthesis. (c) Calculate the ΔF for the reaction written in the direction of glyceraldehyde-3-phosphate and dihydroxy acetone-phosphate formation under steady-state conditions in which the concentrations of all components are maintained at $10^{-4} M$.

2. The intracellular concentration of inorganic phosphate is about $10^{-2} M$. The ATP/ADP ratio in a cell is about 10. Calculate the ΔF of hydrolysis of ATP under these steady-state conditions.

3. The hydrolysis of inorganic pyrophosphate proceeds with a $\Delta F'$ of -7000 cal/mole. The K'_{eq} for the hydrolysis of α-glycerolphosphate is 100. From this information, calculate K'_{eq} and $\Delta F'$ for a hypothetical reaction in which α-glycerolphosphate is synthesized from glycerol and inorganic pyrophosphate.

4. Fructose-1,6-diphosphate may be converted to glucose-1-phosphate by three consecutive reactions:

(1) fructose-1,6-diphosphate + H_2O $\xrightarrow{\text{fructose diphosphatase}}$ fructose-6-phosphate + P_i $\Delta F_1' = -3800$ cal/mole

(2) fructose-6-phosphate $\xrightarrow{\text{phosphohexoisomerase}}$ glucose-6-phosphate

$$K_{eq_2}' = 2.0$$

(3) glucose-6-phosphate $\xrightarrow{\text{phosphoglucomutase}}$ glucose-1-phosphate

$$K_{eq_3}' = 0.0526$$

From the above information, calculate the K_{eq}' and $\Delta F'$ for the overall reaction.

5. Calculate the $\Delta F'$ of hydrolysis of phosphenolpyruvate (PEP) to P_i and pyruvate given the following information.

(1) PEP + ADP $\xrightarrow{\text{pyruvic kinase}}$ pyruvate + ATP, $K_{eq_1}' = 3.2 \times 10^3$

(2)   ATP + H_2O $\xrightarrow{\text{ATP-ase}}$ ADP + P_i, $\Delta F_2' = -7700$ cal/mole

6. Estimate the $\Delta F'$ values for the following reactions: (a) acetyl CoA + ethanol $\rightleftharpoons$ ethyl acetate + CoASH, (b) acetyl CoA + $P_i \rightleftharpoons$ acetyl phosphate + CoASH, (c) pyruvate + ATP $\rightleftharpoons$ PEP + AMP + P_i, and (d) acetyl CoA + butyrate $\rightleftharpoons$ acetate + butyryl CoA.

7. Calculate the concentration of glucose-6-phosphate required to force the hexokinase reaction backwards (in the direction of glucose and ATP formation) in the presence of 10^{-5} M glucose, 10^{-3} M ATP, and 10^{-4} M ADP.

8. Calculate the number of moles of ATP that could be synthesized from the complete oxidation of ethanol to CO_2 and H_2O. Assume an energy conservation efficiency of 44%. The $\Delta F'$ for the reaction of glucose $\rightarrow$ 2 ethanol + $2CO_2$ is about $-55,000$ cal/mole. The $\Delta F'$ for the reaction of glucose + $6O_2 \rightarrow 6CO_2 + 6H_2O$ is $-686,000$ cal/mole.

9. Calculate the equilibrium concentrations and concentration ratio of glucose-6-phosphate/glucose-1-phosphate in the phosphoglucomutase reaction when the initial concentration of glucose-6-phosphate is: (a) 1 M, (b) 0.1 M, (c) 10^{-2} M, (d) 10^{-3} M, and (e) 10^{-4} M. The K_{eq}' for the reaction written as glucose-1-phosphate $\rightleftharpoons$ glucose-6-phosphate is 19.

10. Calculate the equilibrium concentrations and concentration ratios of all components of the isocitritase reaction (isocitrate $\rightleftharpoons$ glyoxylate + succinate) when the initial concentration of isocitrate is: (a) 1 M, (b) 0.1 M, (c) 0.01 M, (d) 10^{-3} M, and (e) 10^{-4} M. The $\Delta F'$ for the isocitritase reaction as written is $+2110$ cal/mole.

11. A solution was made $1 M$ in the following compounds: acetoacetate, pyruvate, β-hydroxybutyrate, and lactate. (a) Write an equation for a thermodynamically favorable oxidation-reduction reaction that could occur. (b) Identify the compounds that are oxidized and reduced in the reaction. Identify the oxidizing and reducing agents. (c) Calculate $\Delta E_0'$, $\Delta F'$, and K_{eq}' for the reaction.

12. A solution containing 0.001 M ubiquinone and 0.001 M ubiquinone-H_2 was mixed with an equal volume of a solution containing 0.1 M fumarate and 0.1 M sucinate. (a) Write the equation for a thermodynamically favorable reaction that could occur. (b) Calculate the ΔE, $\Delta F'$, and K_{eq}' values.

13. Calculate the reduction potential of the $FAD/FADH_2$ half-reaction when the concentration (activity) ratios for $FAD/FADH_2$ are: (a) 10^{-3} (b) 0.2, (c) 1, (d) 3, (e) 25, and (f) 400.

14. Determine the direction of the spontaneous reaction that will occur when solutions containing the following concentration ratios of $DPN^+/DPNH$ are added to a solution containing the enzyme malic dehydrogenase and 0.01 M each of malate and oxaloacetate: (a) $DPN^+/DPNH = 10^{-4}$. (b) $DPN^+/DPNH = 0.1$, (c) $DPN^+/DPNH = 0.5$, (d) $DPN^+/DPNH = 2$, (e) $DPN^+/DPNH = 25$, (f) $DPN^+/DPNH = 100$, and (g) $DPN^+/DPNH = 300$.

15. If the intracellular concentrations of succinate and fumarate are $10^{-4} M$ each, calculate the minimum $FADH_2/FAD$ ratio that would be required to drive the reaction in the direction of succinate formation: $E_{0\,FAD/FADH_2}' = -0.06$ v, $E_{0\,fumarate/succinate}' = +0.030$ v.

16. How many moles of ATP could be produced by the oxidation of DPNH by: (a) FAD, (b) cytochrome-a-Fe^{+2}, and (c) oxygen. Assume an efficiency of energy conservation of 40%.

ENZYME KINETICS

V

A. VELOCITY, K_m, [S], AND [P] CALCULATIONS: MICHAELIS-MENTEN EQUATION

Problem V-1

An enzyme that catalyzes the reaction substrate $\rightleftharpoons$ product (S $\rightleftharpoons$ P) was assayed at the initial concentrations of S shown below. The corresponding initial velocities* were recorded. (a) Determine the K_m for the enzyme and the maximum velocity obtainable at the particular enzyme concentration used. (b) What would the initial velocity be at initial S concentrations of $2.5 \times 10^{-5}\ M$ and $5 \times 10^{-5}\ M$? (c) What would the initial velocity be at 0.02 M S? (d) What would the initial velocity be at $10^{-4}\ M$ [S] if the enzyme concentration were doubled? (e) If the initial concentration of S were 0.04 M, what would the product concentration be after 3 minutes?

Solution

(a) K_m and V_{max} can be determined graphically by any of the methods outlined in Appendix III. However, with only the few scattered points given above, it is probably easier to calculate the two values directly.

---

* Velocities are expressed in terms of μmoles/liter-min in this section. In practice, we generally express velocities in terms of μmoles/min in a standard reaction volume. During enzyme purification work the velocities are often recorded as μmoles/min-mg protein which also gives an arbitrary measure of the amount of enzyme present in terms of "units." (One unit of enzyme is defined as that amount of enzyme that catalyzes the production of 1 μmole of product per minute under optimum assay conditions.)

Initial Concentration of S	Initial Velocity
M	μmoles/liter-min
$1 \quad \times 10^{-2}$	75
$1 \quad \times 10^{-3}$	74.9
$1 \quad \times 10^{-4}$	60
7.5×10^{-5}	56.25
6.25×10^{-6}	15.0

Because the velocity barely increases as the concentration is increased from 10^{-3} M to 10^{-2} M, we can assume that V_{max} is 75 μmoles/liter-min. The K_m may be calculated from the Michaelis-Menten equation:

$$\frac{v}{V_{max}} = \frac{[S]}{K_m + [S]}$$

where v is the initial velocity at an initial substrate concentration of S and V_{max} is the maximum velocity at the particular enzyme concentration used.

$$\frac{60}{75} = \frac{10^{-4}}{K_m + 10^{-4}}$$

$$60K_m + 60 \times 10^{-4} = 75 \times 10^{-4}$$

$$60K_m = (75 \times 10^{-4}) - (60 \times 10^{-4}) = 15 \times 10^{-4}$$

$$K_m = \frac{15 \times 10^{-4}}{60} = 0.25 \times 10^{-4}$$

$$\boxed{K_m = 2.5 \times 10^{-5} M}$$

We assumed that the enzyme obeys hyperbolic saturation (Michaelis-Menten) kinetics in performing the above calculation. We should verify this assumption by picking a different set of velocity and concentration values. For example:

$$\frac{v}{V_{max}} = \frac{[S]}{K_m + [S]}$$

$$\frac{56.25}{75} = \frac{7.5 \times 10^{-5}}{K_m + 7.5 \times 10^{-5}}$$

$$56.25K_m + 56.25 \, (7.5 \times 10^{-5}) = 75(7.5 \times 10^{-5})$$

$$56.25K_m + 422 \times 10^{-5} = 562.5 \times 10^{-5}$$

$$56.3K_m = (562.5 \times 10^{-5}) - (422 \times 10^{-5}) = 140.5 \times 10^{-5}$$

$$K_m = \frac{140.5 \times 10^{-5}}{56.3}$$

$$\boxed{K_m = 2.5 \times 10^{-5} M}$$

and

$$\frac{v}{V_{\max}} = \frac{[S]}{K_m + [S]}$$

$$\frac{15}{75} = \frac{6.25 \times 10^{-6}}{K_m + 6.25 \times 10^{-6}}$$

$$15K_m + 15(6.25 \times 10^{-6}) = 75(6.25 \times 10^{-6})$$

$$15K_m + 93.75 \times 10^{-6} = 468.75 \times 10^{-6}$$

$$15K_m = (468.75 \times 10^{-6}) - (93.75 \times 10^{-6}) = 375 \times 10^{-6}$$

$$K_m = \frac{375 \times 10^{-6}}{15} = 25 \times 10^{-6}$$

$$\boxed{K_m = 2.5 \times 10^{-5}\ M}$$

(b)   At [S] = $2.5 \times 10^{-5}\ M$: The K_m is equivalent to the substrate concentration that yields half-maximum velocity. Consequently, at [S] = $2.5 \times 10^{-5}\ M$, [S] = K_m and $v = V_{\max}/2 = \frac{75}{2}$

$$\boxed{v = 37.5\ \mu\text{moles/liter-min}}$$

or, by substitution into the Michaelis-Menten equation:

$$\frac{v}{V_{\max}} = \frac{[S]}{K_m + [S]}$$

$$\frac{v}{75} = \frac{(2.5 \times 10^{-5})}{(2.5 \times 10^{-5}) + (2.5 \times 10^{-5})} = \frac{(2.5 \times 10^{-5})}{(5.0 \times 10^{-5})}$$

$$75(2.5 \times 10^{-5}) = v(5.0 \times 10^{-5})$$

$$v = \frac{75(2.5 \times 10^{-5})}{(5.0 \times 10^{-5})} = \frac{75}{2}$$

$$\boxed{v = 37.5\ \mu\text{moles/liter-min}}$$

At [S] = $5.0 \times 10^{-5}\ M$:

$$\frac{v}{V_{\max}} = \frac{[S]}{K_m + [S]}$$

$$\frac{v}{75} = \frac{(5 \times 10^{-5})}{(2.5 \times 10^{-5}) + (5 \times 10^{-5})}$$

$$\frac{v}{75} = \frac{(5 \times 10^{-5})}{(7.5 \times 10^{-5})}$$

$$v = \frac{75(5 \times 10^{-5})}{(7.5 \times 10^{-5})}$$

$$v = \frac{375 \times 10^{-5}}{7.5 \times 10^{-5}}$$

$$\boxed{v = 50 \ \mu\text{moles/liter-min}}$$

Note that when $[S] = K_m$, $v = V_{max}/2$, but when $[S] = 2K_m$, v is *not* V_{max}. That is, the velocity is *not* the maximum attainable velocity at a substrate concentration equal to twice the K_m value.

(c)   At $[S] = 0.02 \ M$: In general, we can assume that $v = V_{max}$ when the substrate concentration is $100K_m$. Thus, at $0.02 \ M$ $[S]$, $v = V_{max} = 75 \ \mu$moles/liter-min. We can also see from the velocity data given that the enzyme is essentially already saturated at $10^{-3} \ M$, so the velocity will not increase much at higher concentrations.

(d)   The velocity of an enzyme-catalyzed reaction is almost always directly proportional to the enzyme concentration (at least under *in vitro* assay conditions where only catalytic amounts of enzymes are used). If the enzyme concentration were doubled, the initial velocity at *any* given substrate concentration would double.

∴   At $10^{-4} \ M$ $[S]$

$$v_{orig} = 60 \ \mu\text{moles/liter-min}$$

At $2X$ enzyme:

$$\boxed{v_{new} = 120 \ \mu\text{moles/liter-min}}$$

(e)   At $0.04 \ M$ $[S]$ the reaction would be *zero-order*. The conversion of S to P would proceed at essentially a constant velocity until the concentration of S decreased to a region around $100K_m$.

∴   For the first 3 minutes of the reaction:

$$[P] = V_{max} \times \text{time}$$
$$[P] = (75 \ \mu\text{moles/liter-min})(3 \ \text{min})$$
$$[P] = 225 \ \mu\text{moles/liter} = 225 \times 10^{-6} \ M$$

$$\boxed{[P] = 2.25 \times 10^{-4} \ M}$$

Problem V-2

An enzyme catalyzed a reaction at a velocity of 35 μmoles/liter-min when its substrate concentration was 0.01 M. The K_m for the substrate is 2×10^{-5} M. What would the initial velocity be at substrate concentrations of: (a) 3.5×10^{-3} M, (b) 4×10^{-4} M, (c) 2×10^{-4} M, (d) 2×10^{-6} M, and (e) 1.2×10^{-6} M?

Solution

We must assume that the velocity observed at 0.01 M is the maximum velocity ([S] $> 100K_m$). Then all calculations are quite simple.

(a)   At 3.5×10^{-3} M [S], [S] is still greater than $100K_m$. Thus, for all practical purposes, v is still V_{max}.

$$\boxed{v = V_{max} = 35 \text{ μmoles/liter-min}}$$

(b)   At 4×10^{-4} M [S], [S] is between K_m and $100K_m$. Therefore, v will be somewhere between 17.5 μmoles/liter-min ($V_{max}/2$) and 35 μmoles/liter-min (V_{max}).

$$\frac{v}{V_{max}} = \frac{[S]}{K_m + [S]}$$

$$\frac{v}{35} = \frac{4 \times 10^{-4}}{(2 \times 10^{-5}) + (4 \times 10^{-4})}$$

$$\frac{v}{35} = \frac{4 \times 10^{-4}}{(0.2 \times 10^{-4}) + (4 \times 10^{-4})} = \frac{4 \times 10^{-4}}{4.2 \times 10^{-4}}$$

$$v = \frac{35(4 \times 10^{-4})}{(4.2 \times 10^{-4})} = \frac{140 \times 10^{-4}}{4.2 \times 10^{-4}}$$

$$\boxed{v = 33.3 \text{ μmoles/liter-min}}$$

(c)   At 2×10^{-4} M [S]: We can see from the Michaelis-Menten equation that when [S] $= 10K_m$, v will be about 91% of V_{max}.

$$\frac{v}{V_{max}} = \frac{[S]}{K_m + [S]}$$

$$v = \frac{(V_{max})[S]}{K_m + [S]}$$

$$v = \frac{(V_{max})10K_m}{K_m + 10K_m} = \frac{(V_{max})10K_m}{11K_m}$$

$$v = \tfrac{10}{11}V_{max}$$

$$\boxed{v = 0.91 V_{max}}$$

$$\therefore \quad v = (0.91)(35 \ \mu\text{moles/liter-min})$$

$$\boxed{v = 31.82 \ \mu\text{moles/liter-min}}$$

We can also solve for v by substituting the K_m and [S] values into the Michaelis-Menten equation.

$$\frac{v}{V_{\text{max}}} = \frac{[\text{S}]}{K_m + [\text{S}]}$$

$$\frac{v}{35} = \frac{2 \times 10^{-4}}{(2 \times 10^{-5}) + (2 \times 10^{-4})}$$

$$\frac{v}{35} = \frac{2 \times 10^{-4}}{(0.2 \times 10^{-4}) + (2 \times 10^{-4})} = \frac{2 \times 10^{-4}}{2.2 \times 10^{-4}}$$

$$v = \frac{35(2 \times 10^{-4})}{(2.2 \times 10^{-4})}$$

$$\boxed{v = 31.82 \ \mu\text{moles/liter-min}}$$

(d)   At $2 \times 10^{-6} \ M$ [S]: We can also see from the Michaelis-Menten equation that when $[\text{S}] = 0.1K_m$, v will be about 9.1% of V_{max}.

$$\frac{v}{V_{\text{max}}} = \frac{[\text{S}]}{K_m + [\text{S}]}$$

$$v = \frac{(V_{\text{max}})[\text{S}]}{K_m + [\text{S}]} = \frac{(V_{\text{max}})(0.1K_m)}{K_m + 0.1K_m}$$

$$v = \frac{(V_{\text{max}})(0.1K_m)}{1.1K_m}$$

$$v = \tfrac{1}{11}V_{\text{max}}$$

$$\boxed{v = 0.091V_{\text{max}}}$$

$$\therefore \quad v = (0.091)(35 \ \mu\text{moles/liter-min})$$

$$\boxed{v = 3.182 \ \mu\text{moles/liter-min}}$$

We can also determine v by substituting the K_m and [S] values into the Michaelis-Menten equation.

$$\frac{v}{V_{max}} = \frac{[S]}{K_m + [S]}$$

$$\frac{v}{35} = \frac{2 \times 10^{-6}}{(2 \times 10^{-5}) + (2 \times 10^{-6})}$$

$$\frac{v}{35} = \frac{2 \times 10^{-6}}{(20 \times 10^{-6}) + (2 \times 10^{-6})} = \frac{2 \times 10^{-6}}{22 \times 10^{-6}}$$

$$v = \frac{35(2 \times 10^{-6})}{(22 \times 10^{-6})}$$

$$\boxed{v = 3.182 \; \mu\text{moles/liter-min}}$$

(e)   At $1.2 \times 10^{-6} \; M$ [S]:

$$\frac{v}{V_{max}} = \frac{[S]}{K_m + [S]}$$

$$\frac{v}{35} = \frac{(1.2 \times 10^{-6})}{(2 \times 10^{-5}) + (1.2 \times 10^{-6})}$$

$$\frac{v}{35} = \frac{1.2 \times 10^{-6}}{(20 \times 10^{-6}) + (1.2 \times 10^{-6})} = \frac{1.2 \times 10^{-6}}{21.2 \times 10^{-6}}$$

$$v = \frac{35(1.2 \times 10^{-6})}{(21.2 \times 10^{-6})}$$

$$\boxed{v = 1.98 \; \mu\text{moles/liter-min}}$$

Problem V-3

The initial velocity of an enzyme-catalyzed reaction was measured at a series of different initial substrate concentrations. The data are shown in Table V-1. Determine K_m and V_{max} graphically by the Lineweaver-Burk method. (Plot $1/v$ versus $1/[S]$.)

Solution

The concentrations were chosen to yield reciprocals that are easy to plot and also to give a relatively even spread of $1/[S]$ values. If integral values of [S] were chosen (1×10^{-5}, 2×10^{-5}, 3×10^{-5}, etc.), most

TABLE V-1.

[S]	v	$\dfrac{1}{[S]}$	$\dfrac{1}{v}$
M	μmoles/liter-min		
8.35×10^{-6}	13.8	12×10^4	7.24×10^{-2}
1.00×10^{-5}	16.0	10×10^4	6.25×10^{-2}
1.25×10^{-5}	19.1	8×10^4	5.23×10^{-2}
1.67×10^{-5}	23.8	6×10^4	4.20×10^{-2}
2.0×10^{-5}	26.7	5×10^4	3.75×10^{-2}
2.5×10^{-5}	30.8	4×10^4	3.25×10^{-2}
3.3×10^{-5}	36.2	3×10^4	2.76×10^{-2}
5.0×10^{-5}	44.5	2×10^4	2.25×10^{-2}
1.0×10^{-4}	57.2	1×10^4	1.75×10^{-2}
2.0×10^{-4}	66.7	0.5×10^4	1.50×10^{-2}

points would be bunched close to the $1/v$ axis, making it relatively difficult to draw an accurate straight line. The $1/[S]$ values were calculated as shown below. For example, at $[S] = 8.35 \times 10^{-6}$ M,

$$\frac{1}{[S]} = \frac{1}{8.35 \times 10^{-6}} = 0.12 \times 10^6 = 12 \times 10^4$$

The $1/v$ values were calculated similarly. For example, at $[S] = 5.0 \times 10^{-5}$ M,

$$v = 44.5 \ \mu\text{moles/liter-min}$$

$$\frac{1}{v} = \frac{1}{44.5} = 0.0225 = 2.25 \times 10^{-2}$$

The Lineweaver-Burk plot is shown in Figure V-1. Note that the $1/[S]$ axis is labeled $M^{-1} \times 10^{-4}$. The M^{-1} means "the reciprocal of the molarity." The added factor ($\times 10^{-4}$) means that each value of the reciprocal has been multiplied by 10^{-4}. This has the effect of simplifying the numbers used to label the $1/[S]$ axis. Thus, the position labeled 5 represents a $1/[S]$ value of 5×10^4 (and corresponds to an $[S]$ value of $1/(5 \times 10^4)$ $M = 0.2 \times 10^{-4}$ $M = 2 \times 10^{-5}$ M). Another way to label the $1/[S]$ axis would be $1/(M \times 10^4)$ or $(M \times 10^4)^{-1}$. For example, at $[S] = 2 \times 10^{-5}$ M, $M \times 10^4$ equals $(2 \times 10^{-5})(10^4)$ or 0.2, and the reciprocal equals 5.

Similarly, the $1/v$ axis is labeled $(\mu\text{moles/liter-min})^{-1} \times 10^2$, meaning that the $1/v$ value has been multiplied by 100 to simplify the values.

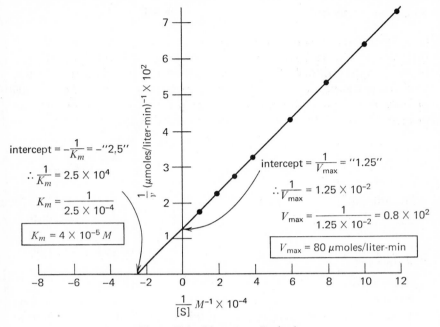

intercept $= -\dfrac{1}{K_m} = -''2.5''$

$\therefore \dfrac{1}{K_m} = 2.5 \times 10^4$

$K_m = \dfrac{1}{2.5 \times 10^{-4}}$

$\boxed{K_m = 4 \times 10^{-5}\,M}$

intercept $= \dfrac{1}{V_{\max}} = ''1.25''$

$\therefore \dfrac{1}{V_{\max}} = 1.25 \times 10^{-2}$

$V_{\max} = \dfrac{1}{1.25 \times 10^{-2}} = 0.8 \times 10^2$

$\boxed{V_{\max} = 80\ \mu moles/liter\text{-}min}$

$\dfrac{1}{v}\ (\mu moles/liter\text{-}min)^{-1} \times 10^2$

$\dfrac{1}{[S]}\ M^{-1} \times 10^{-4}$

Figure V-1   Lineweaver-Burk plot

Another way to label the $1/v$ axis would be $1/(\mu moles/liter\text{-}min \times 10^{-2})$ or $(\mu moles/liter\text{-}min \times 10^{-2})^{-1}$.

B. CALCULATIONS OF AMOUNT OF SUBSTRATE USED AND PRODUCT FORMED

Problem V-4

An enzyme was assayed at an initial substrate concentration of $10^{-5}\ M$. The K_m for the substrate was $2 \times 10^{-3}\ M$. At the end of 1 minute, 2% of the substrate had been converted to product. (a) What percent of the substrate will be converted to product at the end of 3 minutes? What will be the product and substrate concentrations after 3 minutes? (b) If the initial concentration of substrate were $10^{-6}\ M$, what percent of the substrate will be converted to product after 3 minutes? (c) What is the maximum attainable velocity ($V_{\max}$) with the enzyme concentration used? (d) At about what substrate concentration will $V_{\max}$ be observed? (e) At this S (saturating) concentration, what percent of the substrate will be converted to product in 3 minutes?

Solution

(a)   At an initial S concentration of 10^{-5} *M* (which is less than $0.01K_m$), the reaction will be *first-order*—the velocity will be directly proportional to the S concentration. Because [S] keeps decreasing with time, v will also decrease with time. For first-order reactions, a constant *proportion* of the substrate is converted to product per unit time (*not* a constant *amount*). This can be expressed mathematically as shown below:

$$\frac{-ds}{dt} \qquad = \qquad v \qquad = \qquad k \qquad\qquad [S]$$

The amount of S used up per small increment of time . . .	that is, the velocity . . .	is some constant fraction . . .	of the substrate present at that time.

We can solve the problem in two ways. One way is an approximate method; the other is an exact method that employs an equation derived by calculus from the expression shown above. The approximate method assumes that the velocity is constant over the small measured increment of time. The exact method takes into account that the velocity is actually constantly changing. (See Figure V-2.)

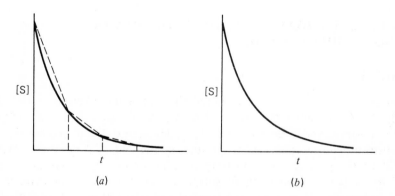

(a)   (b)

Figure V-2   Substrate concentration [S] as a function of time for a reaction that obeys first-order kinetics: (*a*) Approximate method—assumes constant velocity during each interval of time; that is, assumes straight lines between small intervals. (*b*) Exact method—recognizes constantly changing velocity.

Approximate Method

At the end of 1 minute, 2% of the substrate is used, leaving 98%. *During* the second minute, 2% of the remaining 98% will be used up; 2% of 98% is 1.96% more.

∴ By the *end* of 2 minutes, 2% + 1.96% = 3.96% of the substrate will be gone, leaving 100% − 3.96% = 96.04%. During the third minute, 2% of the remaining 96.04% will be used up; 2% of 96.04% is 1.92% more.

∴ By the *end* of the third minute, 3.96% + 1.92% = 5.88% of the substrate will be gone, leaving 100% − 5.88% = 94.12%.

∴ [product]

$$= 5.88\% \text{ of } 10^{-5} M$$

$$= 0.0588 \times 10^{-5}$$

$$\boxed{[\text{product}] = 5.88 \times 10^{-7} M}$$

$$[\text{substrate}] = 94.12\% \text{ of } 10^{-5} M$$

$$= 0.9412 \times 10^{-5} M$$

$$\boxed{[\text{substrate}] = 9.412 \times 10^{-6} M}$$

General Principle

The amount of substrate used in a given time interval (as calculated by the approximate method) will always be larger than the true value. This results from our assumption that the velocity is constant over a short time interval when in fact it is constantly decreasing.

Exact Method

The integrated form of the first-order rate equation is:

$$2.3 \log \frac{S_0}{S_t} = kt$$

where S_0 is the substrate concentration at the beginning of the reaction (zero-time), S_t is the substrate concentration remaining after time t, and k is the first-order rate constant. The units of k are min^{-1}. First calculate k knowing that at the end of 1 minute 98% of the original substrate remains.

$$2.3 \log \frac{S_0}{S_t} = kt$$

Let:

$$S_0 = 100$$

$$S_t = 98$$

$$2.3 \log \tfrac{100}{98} = (k)(1)$$

$$2.3(\log 1.02) = k$$

$$2.3(0.009) = k$$

$$\boxed{k = 0.0207 \text{ min}^{-1}}$$

Next calculate S_t at $t = 3$ min.

$$2.3 \log \frac{100}{S_t} = (0.0207)(3) = 0.0621$$

$$2.3(\log 100 - \log S_t) = 0.0621$$

$$4.6 - 2.3 \log S_t = 0.0621$$

$$-2.3 \log S_t = 0.0621 - 4.60$$

$$= -4.54$$

Approximate Method—*Continued*

The true rate constant, k, in terms of min^{-1} will always be larger than the fraction of the substrate used per minute for the same reason.

Exact Method—*Continued*

$$\log S_t = \frac{-4.54}{-2.3} = 1.975$$

$S_t = $ antilog of 1.975

$S_t = 94.4$

$S_t = 94.4\%$ of original

$S_t = 0.944 \times 10^{-5} \ M$

$$\boxed{S_t = 9.44 \times 10^{-6} \ M}$$

$\therefore \quad$ [product] $= 100\% - 94.4\%$

$\qquad\qquad = 5.6\%$

$\qquad\qquad = 5.6\%$ of $10^{-5} \ M$

$\qquad\qquad = 0.056 \times 10^{-5} \ M$

$$\boxed{\text{[product]} = 5.60 \times 10^{-7} \ M}$$

When using the integrated form of the first-order rate equation, S_0 and S_t can be expressed in terms of percents (100 and 94.4 in the above example), or as decimals (1.0 and 0.944), or as actual concentrations ($10^{-5} \ M$ and $0.944 \times 10^{-5} \ M$).

(b)   If the initial substrate concentration were $10^{-6} \ M$, the reaction would still be first-order. The proportion of the substrate converted to product would still be 5.6% by 3 minutes. The *amount* of product formed would, of course, be less than in part a.

(c) V_{max} can be estimated since we know K_m and an initial velocity (v) at a given substrate [S] concentration. For example, at [S] $= 10^{-5} \ M$,

$$v = 2\% \text{ of } 10^{-5} \ M/min$$

$$2\% \text{ of } 10^{-5} \ M/min = 2\% \times 10^{-5} \text{ moles/liter-min}$$

$$= (0.02)(10^{-5} \text{ moles/liter-min})$$

$$\boxed{v = 2 \times 10^{-7} \text{ moles/liter-min}}$$

or

$$\boxed{v = 0.2 \; \mu\text{moles/liter-min}}$$

$$\frac{v}{V_{\max}} = \frac{[\text{S}]}{K_m + [\text{S}]}$$

$$\frac{2 \times 10^{-7}}{V_{\max}} = \frac{10^{-5}}{(2 \times 10^{-3}) + (10^{-5})}$$

$$\frac{2 \times 10^{-7}}{V_{\max}} = \frac{10^{-5}}{(200 \times 10^{-5}) + (10^{-5})} = \frac{10^{-5}}{201 \times 10^{-5}}$$

$$\frac{V_{\max}}{2 \times 10^{-7}} = \frac{201 \times 10^{-5}}{10^{-5}}$$

$$V_{\max} = \frac{(201 \times 10^{-5})(2 \times 10^{-7})}{(10^{-5})}$$

$$V_{\max} = \frac{402 \times 10^{-12}}{10^{-5}} = 402 \times 10^{-7}$$

$$\boxed{V_{\max} = 4.02 \times 10^{-5} \text{ moles/liter-min}}$$

or

$$\boxed{V_{\max} = 40.2 \; \mu\text{moles/liter-min}}$$

(d) $V_{\max}$ will be observed at about 100 K_m.

$$100(2 \times 10^{-3} \, M) = 2 \times 10^{-1} \, M = 0.2 \, M$$

$$\boxed{[\text{S}] \simeq 0.2 \, M}$$

(e) At 0.2 M, the reaction will be essentially zero-order.

$$[\text{product}] = V_{\max} \times t$$
$$= 4.02 \times 10^{-5} \text{ moles/liter-min} \times 3 \text{ min}$$

$$\boxed{[\text{product}] = 12.06 \times 10^{-5} \, M \text{ at 3 min}}$$

$$\frac{12.06 \times 10^{-5} \, M}{0.2 \, M} \times 100 = \% \text{ conversion of substrate to product}$$

$$\frac{12.06 \times 10^{-3}}{2 \times 10^{-1}} \, \% = 6.03 \times 10^{-2} \, \% = \boxed{0.0603\,\%}$$

Now that we know V_{max} and K_m, we can obtain another estimate of the first-order rate constant, k.

$$k = \frac{V_{max}}{K_m} = \frac{4.02 \times 10^{-5} \text{ moles/liter-min}}{2 \times 10^{-3} \text{ moles/liter}}$$

$$\boxed{k = 2.01 \times 10^{-2} \text{ min}^{-1}}$$

Problem V-5

An enzyme was assayed at an initial substrate concentration of 2×10^{-5} M. In 6 minutes, half of the substrate had been used. The K_m for the substrate is 5×10^{-3} M. Calculate: (a) k, (b) V_{max}, and (c) the concentration of product produced by 15 minutes.

Solution

The [S] is $<0.01 K_m$. $\therefore$ The reaction is first-order.

(a) $$\frac{0.693}{k} = t_{1/2}$$

where $t_{1/2}$ = the half-life—the time required for half of the substrate to be converted to product when the reaction is first-order.

$$\frac{0.693}{k} = 6 \text{ min}$$

$$6k = 0.693$$

$$k = \frac{0.693}{6}$$

$$\boxed{k = 0.115 \text{ min}^{-1}}$$

(b) $$k = \frac{V_{max}}{K_m}$$

$$V_{max} = (k)(K_m)$$
$$V_{max} = (0.115 \text{ min}^{-1})(5 \times 10^{-3} \ M) = 0.575 \times 10^{-4} \ M \text{ min}^{-1}$$
$$= 0.575 \times 10^{-4} \text{ mole/liter-min}$$

$$\boxed{V_{max} = 57.5 \ \mu\text{moles/liter-min}}$$

(c)
$$2.3 \log \frac{S_0}{S_t} = kt$$

$$2.3 \log \frac{2 \times 10^{-5}}{S_t} = (0.115)(15)$$

$$2.3(\log 2 \times 10^{-5} - \log S_t) = 1.725$$

$$\log 2 \times 10^{-5} - \log S_t = \frac{1.725}{2.3} = 0.75$$

$$0.301 - 5 - \log S_t = 0.75$$

$$-\log S_t = 5 + 0.75 - 0.301 = 5.449$$

$$\log \frac{1}{S_t} = 5.449$$

$$\frac{1}{S_t} = 2.81 \times 10^5$$

$$S_t = \frac{1}{2.81 \times 10^5} = 0.356 \times 10^{-5}$$

$$[P] = S_0 - S_t = (2 \times 10^{-5}) - (0.356 \times 10^{-5})$$

$$\boxed{[P] = 1.644 \times 10^{-5} \ M}$$

General Rules

The Michaelis-Menten equation can be used to relate v, $V_{\max}$, [S], and K_m at any S concentration (provided the enzyme obeys hyperbolic saturation kinetics).

At $[S] \geqq 100 K_m$, the reaction is zero-order.

$$[P] = V_{\max} \times t$$

At $[S] \leqq 0.01 K_m$, the reaction is first-order and the integrated first-order rate equation must be used to calculate [S] or [P] at any given t.

$$k = \frac{V_{\max}}{K_m} \qquad t_{1/2} = \frac{0.693}{k}$$

Problem V-6

One mg/liter of a pure enzyme (MW $= 25,000$, $K_m = 2 \times 10^{-5} \ M$) catalyzed a reaction at a rate of 25 mmoles/liter-min when [S] was $3 \times 10^{-3} \ M$. Calculate the *turnover number* (molecular activity) of the enzyme.

Solution

The turnover number of an enzyme is defined as the maximum moles of substrate utilized per mole of enzyme per minute under optimum assay conditions (saturating substrate concentration, optimum pH, etc.). For the enzyme described above, $[S] > 100K_m$; $\therefore$ velocity $= V_{max}$.

$$1 \text{ mole enzyme} = 25{,}000 \text{ g}$$

$$\therefore \quad 1 \text{ mg enzyme} = \frac{1 \times 10^{-3} \text{ g}}{25 \times 10^3 \text{ g/mole}} = \frac{100 \times 10^{-5}}{25 \times 10^3}$$

$$= 4 \times 10^{-8} \text{ moles enzyme}$$

$$25 \text{ mmoles/liter-min} = 25 \times 10^{-3} \text{ moles/liter-min}$$

$$\text{turnover number} = \frac{25 \times 10^{-3} \text{ moles S/liter-min}}{4 \times 10^{-8} \text{ moles enzyme/liter}}$$

---

turnover number $= 6.25 \times 10^5$ moles S/mole enzyme-min

Problem V-7

Calculate the ratio of the substrate concentration required for 90% of V_{max} to the substrate concentration required for 10% of V_{max} (that is, $[S]_{90}/[S]_{10}$) for an enzyme that obeys hyperbolic saturation kinetics.

Solution

$$\frac{v}{V_{max}} = \frac{[S]}{K_m + [S]}$$

$$\frac{0.90}{1} = \frac{[S]_{90}}{K_m + [S]_{90}}$$

$$0.90K_m + 0.90[S]_{90} = [S]_{90}$$

$$0.90K_m = [S]_{90} - 0.90[S]_{90}$$

$$0.90K_m = 0.10[S]_{90}$$

$$[S]_{90} = \frac{0.90K_m}{0.10}$$

$$\boxed{[S]_{90} = 9K_m}$$

That is, when the substrate concentration is $9K_m$, $v = 90\%$ of V_{max}.

$$\frac{v}{V_{max}} = \frac{[S]}{K_m + [S]}$$

$$\frac{0.10}{1} = \frac{[S]_{10}}{K_m + [S]_{10}}$$

$$0.10K_m + 0.10[S]_{10} = [S]_{10}$$

$$0.10K_m = [S]_{10} - 0.10[S]_{10}$$

$$0.10K_m = 0.90[S]_{10}$$

$$[S]_{10} = \frac{0.10K_m}{0.90}$$

$$\boxed{[S]_{10} = 0.111K_m}$$

That is, when the substrate concentration is $0.111K_m$, $v = 10\%$ of V_{max}.

$$\frac{[S]_{90}}{[S]_{10}} = \frac{9K_m}{0.111K_m}$$

$$\boxed{\frac{[S]_{90}}{[S]_{10}} = 81}$$

C. TWO-SUBSTRATE REACTIONS

Problem V-8

The concentration-velocity data shown in Table V-2 were obtained for a two-substrate enzyme that catalyzes the reaction $S_1 + S_2 \rightarrow P$. Determine: (a) V_{max} and (b) the K_m values for each substrate (K_{m_1} and K_{m_2}).

Solution

(a)   From the data obtained in experiments 1–3, we can see that $[S_1]$ and $[S_2]$ can be decreased to 10^{-2} M and 10^{-3} M, respectively, before any decrease in the initial velocity of the reaction is detectable. Consequently, the enzyme must be saturated with S_1 at 10^{-2} M and with S_2 at 10^{-3} M. The velocity observed at these concentrations (and above) must be V_{max}.

$$\boxed{\therefore \quad V_{max} = 20 \ \mu\text{moles/liter-min}}$$

TABLE V-2.

Experiment	Initial Concentration of S_1	Initial Concentration of S_2	Initial Velocity
	M	M	μmoles/liter-min
1	0.1	10^{-2}	20
2	10^{-2}	10^{-2}	20
3	10^{-2}	10^{-3}	20
4	10^{-3}	10^{-2}	18.2
5	10^{-4}	10^{-2}	10
6	2×10^{-5}	10^{-2}	3.33
7	10^{-2}	10^{-4}	18.2
8	10^{-2}	10^{-5}	10
9	10^{-2}	3×10^{-6}	4.62
10	4×10^{-4}	10^{-3}	16.0

(b)   The K_m for S_1 (K_{m_1}) can be obtained graphically by measuring the velocities at different S_1 concentrations in the presence of saturating concentrations of S_2. Similarly, K_{m_2} can be obtained graphically by measuring the velocities at different S_2 concentrations in the presence of saturating concentrations of S_1. We can also calculate K_{m_1} and K_{m_2} directly. To calculate K_{m_1}, use the Michaelis-Menten equation at some nonsaturating concentration of S_1 and a saturating concentration of S_2. For example, in experiment 6:

$$\frac{v}{V_{\max}} = \frac{[S_1]}{K_{m_1} + [S_1]}$$

$$[S_2] = 10^{-2} \, M$$

$$\frac{3.33}{20} = \frac{2 \times 10^{-5}}{K_{m_1} + 2 \times 10^{-5}}$$

$$3.33 K_{m_1} + 3.33(2 \times 10^{-5}) = 20(2 \times 10^{-5})$$

$$3.33 K_{m_1} + 6.66 \times 10^{-5} = 40 \times 10^{-5}$$

$$3.33 K_{m_1} = (40 \times 10^{-5}) - (6.66 \times 10^{-5})$$

$$3.33 K_{m_1} = 33.34 \times 10^{-5}$$

$$K_{m_1} = \frac{33.34 \times 10^{-5}}{3.33}$$

$$K_{m_1} = 10 \times 10^{-5}$$

$$\boxed{K_{m_1} = 10^{-4} \, M}$$

Similarly, from experiment 10:

$$\frac{v}{V_{\max}} = \frac{[S_1]}{K_{m_1} + [S_1]}$$

$$[S_2] = 10^{-3}\ M$$

$$\frac{16.0}{20} = \frac{4 \times 10^{-4}}{K_{m_1} + 4 \times 10^{-4}}$$

$$16K_{m_1} + 16(4 \times 10^{-4}) = 20(4 \times 10^{-4})$$

$$16K_{m_1} + 64 \times 10^{-4} = 80 \times 10^{-4}$$

$$16K_{m_1} = (80 \times 10^{-4}) - (64 \times 10^{-4})$$

$$16K_{m_1} = 16 \times 10^{-4}$$

$$\boxed{K_{m_1} = 10^{-4}\ M}$$

The K_{m_2} may be calculated in an identical manner. For example, in experiment 9:

$$\frac{v}{V_{\max}} = \frac{[S_2]}{K_{m_2} + [S_2]}$$

$$[S_1] = 10^{-2}\ M$$

$$\frac{4.61}{20} = \frac{3 \times 10^{-6}}{K_{m_2} + 3 \times 10^{-6}}$$

$$4.61K_{m_2} + 4.61(3 \times 10^{-6}) = 20(3 \times 10^{-6})$$

$$4.61K_{m_2} + 13.83 \times 10^{-6} = 60 \times 10^{-6}$$

$$4.61K_{m_2} = (60 \times 10^{-6}) - (13.83 \times 10^{-6})$$

$$4.61K_{m_2} = 46.17 \times 10^{-6}$$

$$K_{m_2} = \frac{46.17 \times 10^{-6}}{4.61} = 10 \times 10^{-6}$$

$$\boxed{K_{m_2} = 10^{-5}\ M}$$

The above calculations assume (and the data confirm) that S_1 and S_2 bind independently and in random order to the enzyme. In other words, the binding of one substrate neither facilitates nor hinders the binding of the other and there is no preference as to which substrate binds first. The overall reaction may take place by either of the two paths shown below.

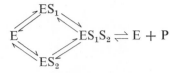

The same treatment can be applied to one-substrate plus one-activator reactions.

Problem V-9

The enzyme glutathione reductase catalyzes the reaction: oxidized glutathione (GSSG) + TPNH → reduced glutathione (2GSH) + TPN⁺ + H⁺. Assuming that $K_{m_{\text{GSSG}}}$ is 2×10^{-5} M and that $K_{m_{\text{TPNH}}}$ is 10^{-6} M, calculate the velocity of the reaction at the following concentrations of substrates: (a) 2×10^{-5} M GSSG and 10^{-6} M TPNH, (b) 2.5×10^{-6} M GSSG and 3×10^{-7} M TPNH, and (c) 10^{-4} M GSSG and 10^{-6} M TPNH. The velocity observed at 5×10^{-3} M GSSG and 2×10^{-4} M TPNH was 450 μmoles/liter-min. Assume random and independent binding of the two substrates.*

Solution

The velocity observed at 5×10^{-3} M GSSG and 2×10^{-4} M TPNH (both substrates at concentrations $>100K_m$) is the maximum attainable velocity for the amount of enzyme used.

$$\therefore \quad V_{\text{max}} = 450 \text{ μmoles/liter-min}$$

The velocity at any given set of S_1 and S_2 concentrations for a two-substrate enzyme is given by the expression shown below. The expression is the product of two Michaelis-Menten expressions.

$$\frac{v}{V_{\text{max}}} = \frac{[S_1]}{K_{m_1} + [S_1]} \times \frac{[S_2]}{K_{m_2} + [S_2]} = \frac{[S_1][S_2]}{([S_1] + K_{m_1})([S_2] + K_{m_2})}$$

(a)   At 2×10^{-5} M GSSG and 10^{-6} M TPNH, both substrates are present at their respective K_m values. The concentration of each substrate is sufficient to provide a velocity of $\frac{1}{2}V_{\text{max}}$ in the presence of an *excess* of the

* It has been shown experimentally that the two substrates do not actually bind randomly and independently. The K_m value for GSSG is in fact dependent on the concentration of TPNH. The actual kinetics suggest that TPNH binds first and converts the enzyme to a modified form before the GSSG binds.

other substrate. When both substrates are present at concentrations equal to K_m,

$$v = \tfrac{1}{2} \times \tfrac{1}{2}V_{max} = \tfrac{1}{4}V_{max}$$
$$\therefore \quad v = \tfrac{450}{4} = 112.5 \ \mu\text{moles/liter-min}$$

We would obtain the same result by substituting the $[S_1]$ and $[S_2]$ values into the two-substrate Michaelis-Menten expression.

$$v = V_{max} \frac{[S_1]}{K_{m_1} + [S_1]} \times \frac{[S_2]}{K_{m_2} + [S_2]}$$

$$= 450 \ \frac{(2 \times 10^{-5})}{(2 \times 10^{-5}) + (2 \times 10^{-5})} \times \frac{(10^{-6})}{(10^{-6}) + (10^{-6})}$$

$$= 450 \frac{(2 \times 10^{-5})}{(4 \times 10^{-5})} \times \frac{(10^{-6})}{(2 \times 10^{-6})} = 450 \times \tfrac{1}{2} \times \tfrac{1}{2} = \tfrac{450}{4}$$

$$\boxed{v = 112.5 \ \mu\text{moles/liter-min}}$$

(b) $\quad v = V_{max} \dfrac{[S_1][S_2]}{([S_1] + K_{m_1})([S_2] + K_{m_2})}$

$$= 450 \ \frac{(2.5 \times 10^{-6})(3 \times 10^{-7})}{[(2.5 \times 10^{-6}) + (2 \times 10^{-5})][(3 \times 10^{-7}) + (10^{-6})]}$$

$$= 450 \ \frac{(7.5 \times 10^{-13})}{[(2.5 \times 10^{-6}) + (20 \times 10^{-6})][(3 \times 10^{-7}) + (10 \times 10^{-7})]}$$

$$= 450 \ \frac{(7.5 \times 10^{-13})}{(22.5 \times 10^{-6})(13 \times 10^{-7})} = 450 \ \frac{(7.5 \times 10^{-13})}{(292 \times 10^{-13})}$$

$$= 450 \ \frac{7.5}{292}$$

$$\boxed{v = 11.55 \ \mu\text{moles/liter-min}}$$

We could just as easily have handled the effect of each substrate on the velocity separately.

The effect of GSSG:

$$\frac{v}{V_{max}} = \frac{[S_1]}{K_{m_1} + [S_1]}$$

$$\frac{v}{V_{max}} = \frac{(2.5 \times 10^{-6})}{(2.0 \times 10^{-5}) + (2.5 \times 10^{-6})}$$

$$\frac{v}{V_{max}} = \frac{(2.5 \times 10^{-6})}{(20 \times 10^{-6}) + (2.5 \times 10^{-6})} = \frac{(2.5 \times 10^{-6})}{(22.5 \times 10^{-6})} = 0.111$$

$$\therefore \quad v = 0.111 V_{max} \quad \text{(contribution of GSSG)}$$

The effect of TPNH:

$$\frac{v}{V_{max}} = \frac{[S_2]}{K_{m_2} + [S_2]}$$

$$\frac{v}{V_{max}} = \frac{(3 \times 10^{-7})}{(10^{-6}) + (3 \times 10^{-7})}$$

$$\frac{v}{V_{max}} = \frac{(3 \times 10^{-7})}{(10 \times 10^{-7}) + (3 \times 10^{-7})} = \frac{(3 \times 10^{-7})}{(13 \times 10^{-7})} = 0.231$$

$$\therefore \quad v = 0.231 V_{max} \quad \text{(contribution of TPNH)}$$

When S_1 and S_2 are considered together:

$$v = 0.111 \times 0.231 V_{max}$$

$$= (0.111)(0.231)(450)$$

$$\boxed{v = \textbf{11.55 } \mu\textbf{moles/liter-min}}$$

(c)   At $10^{-6} M$ TPNH ($[S] = K_m$), the TPNH reduces the velocity to $\frac{1}{2} V_{max}$. We need only determine the effect of $10^{-4} M$ GSSG and multiply the two effects.

$$\frac{v}{V_{max}} = \frac{[S]}{K_m + [S]}$$

$$\frac{v}{V_{max}} = \frac{10^{-4}}{(2 \times 10^{-5}) + (10^{-4})}$$

$$\frac{v}{V_{max}} = \frac{(10^{-4})}{(0.2 \times 10^{-4})(1 \times 10^{-4})} = \frac{(1 \times 10^{-4})}{(1.2 \times 10^{-4})} = 0.833$$

$$v = 0.833 V_{max} \quad \text{(effect of GSSG)}$$

and

$$v = 0.5 V_{max} \quad \text{(effect of TPNH)}$$

$$\therefore \quad v = (0.833)(0.5) V_{max}$$

$$= (0.833)(0.5)(450)$$

$$\boxed{v = \textbf{187 } \mu\textbf{moles/liter-min}}$$

D. EFFECT OF COMPETITIVE INHIBITORS

Problem V-10

An enzyme with a single substrate ($K_m = 2 \times 10^{-4} M$) was assayed in the presence of $2 \times 10^{-4} M$ substrate and $2.5 \times 10^{-3} M$ competitive inhibitor ($K_i = 2.5 \times 10^{-3} M$), The V_{max} (uninhibited) is 55 μmoles/liter-min. Calculate the initial velocity in the presence of the competitive inhibitor.

Solution

In the presence of a competitive inhibitor, v, V_{max}, K_m, K_i, and the concentrations of substrate, [S], and inhibitor, [I], are related by the modified Michaelis-Menten expression shown below.

$$\frac{v_i}{V_{max}} = \frac{[S]}{K_m\left(1 + \frac{[I]}{K_i}\right) + [S]} = \frac{[S]}{K_{m_{app}} + [S]}$$

where

v_i = velocity in presence of inhibitor

V_{max} = maximum velocity in absence of inhibitor

K_i = dissociation constant of the enzyme-inhibitor complex

The competitive inhibitor has the effect of increasing the K_m for the substrate.

$$\frac{v}{55} = \frac{(2 \times 10^{-4})}{(2 \times 10^{-4})\left[1 + \frac{(2.5 \times 10^{-3})}{(2.5 \times 10^{-3})}\right] + (2 \times 10^{-4})}$$

$$v = 55\frac{(2 \times 10^{-4})}{(2 \times 10^{-4})(1 + 1) + (2 \times 10^{-4})}$$

$$= 55\frac{(2 \times 10^{-4})}{(4 \times 10^{-4}) + (2 \times 10^{-4})}$$

$$= 55\frac{(2 \times 10^{-4})}{(6 \times 10^{-4})} = 55 \times \tfrac{2}{6}$$

$$\boxed{v = 18.35 \ \mu\text{moles/liter-min}}$$

Or, in general, when [S] = K_m and [I] = K_i (for a competitive inhibitor), $v_i = \tfrac{1}{3}V_{max}$.

Problem V-11

Calculate K_i for a competitive inhibitor given the following information: $K_m = 6.7 \times 10^{-4}$ M, $V_{\max}$ (minus inhibitor) = 300 μmoles/liter-min, and v_i (in the presence of 10^{-5} M [I]) at 2×10^{-5} M [S] = 1.5 μmoles/liter-min.

Solution

$$\frac{v_i}{V_{\max}} = \frac{[S]}{K_{m_{app}} + [S]}$$

where

$$K_{m_{app}} = K_m\left(1 + \frac{[I]}{K_i}\right)$$

$$\frac{1.5}{300} = \frac{(2 \times 10^{-5})}{K_{m_{app}} + (2 \times 10^{-5})}$$

$$1.5K_{m_{app}} + 1.5(2 \times 10^{-5}) = 300(2 \times 10^{-5})$$

$$1.5K_{m_{app}} + 3 \times 10^{-5} = 600 \times 10^{-5}$$

$$1.5K_{m_{app}} = (600 \times 10^{-5}) - (3 \times 10^{-5})$$

$$1.5K_{m_{app}} = 597 \times 10^{-5}$$

$$K_{m_{app}} = \frac{597 \times 10^{-5}}{1.5} = 398 \times 10^{-5}$$

$$\boxed{K_{m_{app}} = 39.8 \times 10^{-4} \ M}$$

$$K_{m_{app}} = K_m\left(1 + \frac{[I]}{K_i}\right) = 39.8 \times 10^{-4}$$

$$6.7 \times 10^{-4}\left(1 + \frac{10^{-5}}{K_i}\right) = 39.8 \times 10^{-4}$$

$$6.7 \times 10^{-4} + \frac{6.7 \times 10^{-9}}{K_i} = 39.8 \times 10^{-4}$$

$$\frac{6.7 \times 10^{-9}}{K_i} = (39.8 \times 10^{-4}) - (6.7 \times 10^{-4})$$

$$= 33.1 \times 10^{-4}$$

$$33.1 \times 10^{-4}K_i = 6.7 \times 10^{-9}$$

$$K_i = \frac{6.7 \times 10^{-9}}{33.1 \times 10^{-4}} = 0.202 \times 10^{-5}$$

$$\boxed{K_i = 2.02 \times 10^{-6} \ M}$$

Problem V-12

Calculate the degree of inhibition of the enzyme-catalyzed reaction described in Problem V-11 under the following conditions: (a) $[S] = 2 \times 10^{-5} M$ and $[I] = 10^{-5} M$, (b) $[S]$ is increased 10-fold while $[I]$ remains the same, (c) $[S]$ is increased 100-fold while $[I]$ remains the same, (d) $[S]$ remains at $2 \times 10^{-5} M$ and $[I]$ is decreased to $10^{-6} M$, (e) $[S]$ and $[I]$ remain at $2 \times 10^{-5} M$ and $10^{-5} M$, respectively, but I is a poorer inhibitor (K_i is 20-fold higher than calculated in Problem V-11), and (f) the ratio of $[S]/[I]$ remains the same but both $[S]$ and $[I]$ are increased 10-fold (and $K_i = 2.02 \times 10^{-6} M$).

Solution

To calculate the degree of inhibition, we first must calculate what the velocity would be at $2 \times 10^{-5} M$ $[S]$ in the absence of the inhibitor.

$$\frac{v}{V_{\max}} = \frac{[S]}{K_m + [S]}$$

$$\frac{v}{300} = \frac{(2 \times 10^{-5})}{(6.7 \times 10^{-4}) + (2 \times 10^{-5})}$$

$$v = \frac{300(2 \times 10^{-5})}{(67 \times 10^{-5}) + (2 \times 10^{-5})}$$

$$= \frac{600 \times 10^{-5}}{69 \times 10^{-5}}$$

$$\boxed{v = 8.7 \ \mu\text{moles/liter-min}}$$

(a)   relative velocity $= \dfrac{1.5}{8.7} \times 100\% = 17.25\%$ of the uninhibited velocity

degree of inhibition $= 100 - 17.25\%$

$$\boxed{\textbf{degree of inhibition} = \textbf{82.75}\%}$$

(b)   If $[S]$ were increased 10-fold (to $2 \times 10^{-4} M$):

$$\frac{v}{V_{\max}} = \frac{[S]}{K_m + [S]}$$

$$\frac{v}{300} = \frac{(2 \times 10^{-4})}{(6.7 \times 10^{-4}) + (2 \times 10^{-4})}$$

$$v = \frac{300(2 \times 10^{-4})}{8.7 \times 10^{-4}}$$

$$= \frac{600 \times 10^{-4}}{8.7 \times 10^{-4}}$$

$$\boxed{v = 69 \ \mu\text{moles/liter-min}}$$

$$\frac{v_i}{V_{\max}} = \frac{[S]}{K_{m_{\text{app}}} + [S]}$$

$$\frac{v_i}{300} = \frac{(2 \times 10^{-4})}{(39.8 \times 10^{-4}) + (2 \times 10^{-4})}$$

$$v_i = \frac{300(2 \times 10^{-4})}{41.8 \times 10^{-4}} = \frac{600 \times 10^{-4}}{41.8 \times 10^{-4}}$$

$$\boxed{v_i = 14.35 \ \mu\text{moles/liter-min}}$$

$$\text{relative velocity} = \frac{v_i}{v} = \frac{14.35}{69} \times 100\% = 20.8\%$$

$$\text{degree of inhibition} = 100 - 20.8\%$$

$$\boxed{\textbf{degree of inhibition} = \textbf{79.2}\%}$$

(c) If [S] were increased 100-fold (to $2 \times 10^{-3} \ M$):

$$\frac{v}{V_{\max}} = \frac{[S]}{K_m + [S]}$$

$$\frac{v}{300} = \frac{(2 \times 10^{-3})}{(6.7 \times 10^{-4}) + (2 \times 10^{-3})}$$

$$v = \frac{300(2 \times 10^{-3})}{(0.67 \times 10^{-3}) + (2 \times 10^{-3})} = \frac{600 \times 10^{-3}}{2.67 \times 10^{-3}}$$

$$\boxed{v = 224.5 \ \mu\text{moles/liter-min}}$$

$$\frac{v_i}{V_{\max}} = \frac{[S]}{K_{m_{\text{app}}} + [S]}$$

$$\frac{v_i}{300} = \frac{(2 \times 10^{-3})}{(39.8 \times 10^{-4}) + (2 \times 10^{-3})}$$

$$v_i = \frac{300(2 \times 10^{-3})}{(3.98 \times 10^{-3}) + (2 \times 10^{-3})} = \frac{600 \times 10^{-3}}{5.98 \times 10^{-3}}$$

$$\boxed{v_i = 100 \ \mu\text{moles/liter-min}}$$

relative velocity $= \dfrac{v_i}{v} = \dfrac{100}{224.5} \times 100\% = 44.5\%$

degree of inhibition $= 100 - 44.5\%$

$$\boxed{\textbf{degree of inhibition} = \textbf{55.5}\%}$$

Thus, the degree of inhibition decreases as [S] increases.

(d)   If $[S] = 2 \times 10^{-5} \ M$ and $[I] = 10^{-6} \ M$:
The $K_{m_{\text{app}}}$ changes when [I] changes. First calculate the new $K_{m_{\text{app}}}$.

$$K_{m_{\text{app}}} = K_m\left(1 + \frac{[I]}{K_i}\right) = 6.7 \times 10^{-4}\left(1 + \frac{10^{-6}}{2.02 \times 10^{-6}}\right)$$

$$K_{m_{\text{app}}} = 6.7 \times 10^{-4}(1 + 0.495) = 6.7 \times 10^{-4}(1.495)$$

$$K_{m_{\text{app}}} = 10 \times 10^{-4}$$

$$\boxed{K_{m_{\text{app}}} = 10^{-3} \ M}$$

Next, calculate v_i.

$$\frac{v_i}{V_{\max}} = \frac{[S]}{K_{m_{\text{app}}} + [S]}$$

$$\frac{v_i}{300} = \frac{(2 \times 10^{-5})}{(10^{-3}) + (2 \times 10^{-5})}$$

$$v_i = \frac{300(2 \times 10^{-5})}{(100 \times 10^{-5}) + (2 \times 10^{-5})} = \frac{600 \times 10^{-5}}{102 \times 10^{-5}}$$

$$\boxed{v_i = 5.88 \ \mu\text{moles/liter-min}}$$

$$\text{relative velocity} = \frac{v_i}{v} = \frac{5.88}{8.7} \times 100\% = 67.5\%$$

$$\text{degree of inhibition} = 100 - 67.5\%$$

$$\boxed{\text{degree of inhibition} = \mathbf{32.5\%}}$$

(e)   If K_i were $4.04 \times 10^{-5}\ M$: First calculate the new $K_{m_{app}}$. The $K_{m_{app}}$ also depends on the value of K_i.

$$K_{m_{app}} = K_m\left(1 + \frac{[I]}{K_i}\right)$$

$$= 6.7 \times 10^{-4}\left(1 + \frac{10^{-5}}{4.04 \times 10^{-5}}\right)$$

$$= 6.7 \times 10^{-4}(1 + 0.248)$$

$$= 6.7 \times 10^{-4}(1.248)$$

$$\boxed{K_{m_{app}} = \mathbf{8.36 \times 10^{-4}\ M}}$$

$$\frac{v_i}{V_{\max}} = \frac{[S]}{K_{m_{app}} + [S]}$$

$$\frac{v_i}{300} = \frac{(2 \times 10^{-5})}{(8.36 \times 10^{-4}) + (2 \times 10^{-5})}$$

$$v_i = \frac{300(2 \times 10^{-5})}{(83.6 \times 10^{-5}) + (2 \times 10^{-5})} = \frac{600 \times 10^{-5}}{85.6 \times 10^{-5}}$$

$$\boxed{v_i = \mathbf{7.01\ \mu moles/liter\text{-}min}}$$

$$\text{relative velocity} = \frac{v_i}{v} = \frac{7.01}{8.7} \times 100\% = 80.5\%$$

$$\text{degree of inhibition} = 100 - 80.5\%$$

$$\boxed{\text{degree of inhibition} = \mathbf{19.5\%}}$$

(f)   If $[S] = 2 \times 10^{-4} M$ and $[I] = 10^{-4} M$:

$$\frac{v_i}{V_{max}} = \frac{[S]}{K_m\left(1 + \frac{[I]}{K_i}\right) + [S]}$$

$$\frac{v_i}{300} = \frac{(2 \times 10^{-4})}{(6.7 \times 10^{-4})\left(1 + \dfrac{10^{-4}}{2.02 \times 10^{-6}}\right) + (2 \times 10^{-4})}$$

$$v_i = \frac{300(2 \times 10^{-4})}{(6.7 \times 10^{-4})(1 + 0.495 \times 10^2) + (2 \times 10^{-4})}$$

$$= \frac{300(2 \times 10^{-4})}{(6.7 \times 10^{-4})(50.5) + (2 \times 10^{-4})}$$

$$v_i = \frac{(600 \times 10^{-4})}{(338 \times 10^{-4}) + (2 \times 10^{-4})} = \frac{600 \times 10^{-4}}{340 \times 10^{-4}}$$

$$v_i = 1.765 \ \mu\text{moles/liter-min}$$

$$v = 69.0 \ \mu\text{moles/liter-min} \qquad \text{(from part b)}$$

$$\text{relative velocity} = \frac{v_i}{v} = \frac{1.765}{69} \times 100\% = 2.56\%$$

$$\text{degree of inhibition} = 100 - 2.56\%$$

> **degree of inhibition = 97.44%**

Problem V-13

(a)   What concentration of competitive inhibitor ($K_i = 2 \times 10^{-5} M$) is required to yield 50% inhibition at a substrate concentration of $10^{-4} M$ ($K_m = 3 \times 10^{-3} M$)?   (b) To what concentration must the substrate be raised to reduce the degree of inhibition to 25%?

Solution

(a)   Although the V_{max} is not given, we can work the problem in terms of v and V_{max}. In the absence of inhibitor:

$$\frac{v}{V_{max}} = \frac{[S]}{K_m + [S]}$$

$$\frac{v}{V_{\max}} = \frac{10^{-4}}{(3 \times 10^{-3}) + (10^{-4})} = \frac{10^{-4}}{(30 \times 10^{-4}) + (10^{-4})}$$

$$\frac{v}{V_{\max}} = \frac{10^{-4}}{30 \times 10^{-4}}$$

$$v = \frac{10^{-4} V_{\max}}{31 \times 10^{-4}}$$

$$\boxed{v = 0.0323 \ V_{\max}}$$

At 50% inhibition, $v_i/v = 50\%$, $v_i = 0.5v = 0.5(0.0323) V_{\max} = 0.0162 \ V_{\max}$. Calculate the [I] that would yield $0.0162 \ V_{\max}$ at $10^{-4} \ M$ [S].

$$\frac{v_i}{V_{\max}} = \frac{[S]}{K_m \left(1 + \dfrac{[I]}{K_i}\right) + [S]} = 0.0162$$

$$= \frac{10^{-4}}{(3 \times 10^{-3})\left(1 + \dfrac{[I]}{2 \times 10^{-5}}\right) + 10^{-4}} = 0.0162$$

$$= \frac{10^{-4}}{\left(3 \times 10^{-3} + \dfrac{3 \times 10^{-3} \, [I]}{2 \times 10^{-5}}\right) + 10^{-4}} = 0.0162$$

$$10^{-4} = 0.0162\left(3 \times 10^{-3} + \frac{3 \times 10^{-3} \, [I]}{2 \times 10^{-5}}\right) + 0.0162(10^{-4})$$

$$10^{-4} = 4.85 \times 10^{-5} + 2.43[I] + 0.0162 \times 10^{-4}$$

$$10^{-4} = 0.485 \times 10^{-4} + 0.0162 \times 10^{-4} + 2.43[I]$$

$$10^{-4} - 0.501 \times 10^{-4} = 2.43[I]$$

$$0.499 \times 10^{-4} = 2.43[I]$$

$$[I] = \frac{4.99 \times 10^{-5}}{2.43}$$

$$\boxed{[I] = 2.05 \times 10^{-5} \ M}$$

(b) At 25% inhibition $v_i/v = 75\%$, $v_i = 0.75v = 0.75(0.0323) V_{\max} = 0.0242 \ V_{\max}$. (Note that a reduction in the degree of inhibition by a factor of 2 does *not* mean that the velocity has doubled.) Calculate the value of [S] that would yield $v = 0.0242\ V_{\max}$ in the presence of $1.93 \times 10^{-5}\ M$ [I].

$$\frac{v_i}{V_{\max}} = \frac{[S]}{K_m\left(1 + \dfrac{[I]}{K_i}\right) + [S]} = 0.0242$$

$$\frac{[S]}{(3 \times 10^{-3})\left(1 + \dfrac{1.93 \times 10^{-5}}{2 \times 10^{-5}}\right) + [S]} = 0.0242$$

$$\frac{[S]}{(3 \times 10^{-3})(1.965) + [S]} = 0.0242$$

$$[S] = 0.0242(3 \times 10^{-3})(1.965) + 0.0242[S]$$

$$[S] = 14.25 \times 10^{-5} + 0.0242[S]$$

$$0.9758[S] = 1.425 \times 10^{-4}$$

$$[S] = \frac{1.425 \times 10^{-4}}{0.9758}$$

$$\boxed{[S] = \mathbf{1.46 \times 10^{-4}\ M}}$$

General Principles

The degree of inhibition caused by a competitive inhibitor depends on [S], [I], K_m, and K_i.

An increase in [S] at constant [I] decreases the degree of inhibition. An increase in [I] at constant [S] increases the degree of inhibition. The lower the value of K_i, the greater is the degree of inhibition at any given [S] and [I].

A competitive inhibitor acts only to increase the apparent K_m for the substrate. The $V_{\max}$ remains unchanged but, in the presence of a competitive inhibitor, a much greater substrate concentration is required to attain $V_{\max}$.

The v_i may be considered equal to $V_{\max}$ when $[S] \geqq 100\ K_{m_{\mathrm{app}}}$.

E. EFFECT OF NONCOMPETITIVE INHIBITORS

Problem V-14

Calculate (a) the velocity and (b) the degree of inhibition of an enzyme-catalyzed reaction in the presence of $3.5 \times 10^{-5}\ M$ substrate ($K_m = 2 \times 10^{-4}\ M$) and $4 \times 10^{-5}\ M$ noncompetitive inhibitor ($K_i = 2 \times 10^{-5}$

M). The velocity observed at 0.03 M [S] in the absence of inhibitor is 295 μmoles/liter-min.

Solution

(a)   First calculate the velocity at 3.5×10^{-5} M [S] in the absence of inhibitor. The v at 0.03 M [S] $= V_{max} = 295$ μmoles/liter-min.

$$\frac{v}{V_{max}} = \frac{[S]}{K_m + [S]}$$

$$\frac{v}{295} = \frac{(3.5 \times 10^{-5})}{(2 \times 10^{-4}) + (3.5 \times 10^{-5})}$$

$$v = \frac{295(3.5 \times 10^{-5})}{(20 \times 10^{-5}) + (3.5 \times 10^{-5})} = \frac{1030 \times 10^{-5}}{23.5 \times 10^{-5}}$$

$$\boxed{v = 44 \text{ } \mu\text{moles/liter-min}}$$

The relationship between the inhibited velocity (v_i) and the uninhibited velocity (v) in the presence of a noncompetitive inhibitor is shown below.

$$\frac{v_i}{v} = \frac{K_i}{K_i + [I]}$$

$$v_i = \frac{K_i}{K_i + [I]} v$$

$$v_i = \frac{(2 \times 10^{-5})}{(2 \times 10^{-5}) + (4 \times 10^{-5})} 44 = \frac{(2 \times 10^{-5})}{(6 \times 10^{-5})} 44$$

$$\boxed{v_i = 14.65 \text{ } \mu\text{moles/liter-min}}$$

(b) $$\text{relative velocity} = \frac{v_i}{v} = \frac{14.65}{44} \times 100\% = 33.3\%$$

$$\text{degree of inhibition} = 100 - 33.3\%$$

$$\boxed{\text{degree of inhibition} = 66.7\%}$$

General Principle

The *degree of inhibition* in the presence of a noncompetitive inhibitor depends only upon [I] and K_i. The inhibited velocity (v_i) is always a constant fraction of v, regardless of the substrate concentration or the value of K_m. An increase in [S] causes both v and v_i to increase by the same factor. The net effect of a noncompetitive inhibitor is to make it seem as if less enzyme were present.

Problem V-15

What concentration of a noncompetitive inhibitor ($K_i = 4 \times 10^{-6}\ M$) is required to yield 65% inhibition of an enzyme-catalyzed reaction?

Solution

$$\frac{v_i}{v} = \frac{K_i}{K_i + [I]}$$

degree of inhibition = 65%

∴ relative velocity = 35%

That is,

$$\frac{v_i}{v} = 0.35$$

$$0.35 = \frac{4 \times 10^{-6}}{4 \times 10^{-6} + [I]}$$

$$0.35(4 \times 10^{-6}) + 0.35[I] = 4 \times 10^{-6}$$

$$1.4 \times 10^{-6} + 0.35[I] = 4 \times 10^{-6}$$

$$0.35[I] = 4 \times 10^{-6} - 1.4 \times 10^{-6}$$

$$0.35[I] = 2.6 \times 10^{-6}$$

$$[I] = \frac{2.6 \times 10^{-6}}{0.35}$$

$$\boxed{[I] = 7.43 \times 10^{-6}\ M}$$

Check: When $[I] = K_i$, $v_i = \frac{1}{2}v$.

∴ For more than 50% inhibition, [I] must be greater than K_i.

F. SIGMOIDAL KINETICS

Problem V-16

Calculate $[S]_{90}/[S]_{10}$ for an enzyme that obeys sigmoidal kinetics with interaction coefficients of: (a) 2, (b) 3, and (c) 4.

Solution

(a)
$$\frac{v}{V_{\max}} = \frac{[S]^n}{K' + [S]^n}$$

At $[S]_{90}$

$$\frac{v}{V_{max}} = 0.9$$

$$0.9 = \frac{[S]_{90}^2}{K' + [S]_{90}^2}$$

$$0.9K' + 0.9[S]_{90}^2 = [S]_{90}^2$$

$$0.9K' = 0.1[S]_{90}^2$$

$$[S]_{90}^2 = \frac{0.9K'}{0.1} = 9K'$$

$$[S]_{90} = \sqrt{9K'} = \sqrt{9}\sqrt{K'}$$

At $[S]_{10}$

$$\frac{v}{V_{max}} = 0.1$$

$$0.1 = \frac{[S]_{10}^2}{K' + [S]_{10}^2}$$

$$0.1K' + 0.1[S]_{10}^2 = [S]_{10}^2$$

$$0.1K' = 0.9[S]_{10}^2$$

$$[S]_{10}^2 = \frac{0.1K'}{0.9} = \tfrac{1}{9}K'$$

$$[S]_{10} = \sqrt{\tfrac{1}{9}K'} = \sqrt{\tfrac{1}{9}}\sqrt{K'}$$

$$\frac{[S]_{90}}{[S]_{10}} = \frac{\sqrt{9}\sqrt{K'}}{\sqrt{\tfrac{1}{9}}\sqrt{K'}} = \frac{\sqrt{9}}{\sqrt{\tfrac{1}{9}}} = \frac{3}{\tfrac{1}{3}}$$

$$\boxed{\frac{[S]_{90}}{[S]_{10}} = 9 \qquad \text{when} \qquad n = 2}$$

(b)

$$\frac{v}{V_{max}} = \frac{[S]^n}{K' + [S]^n}$$

At $[S]_{90}$

$$\frac{v}{V_{max}} = 0.9$$

$$0.9 = \frac{[S]_{90}^3}{K' + [S]_{90}^3}$$

$$0.9K' + 0.9[S]_{90}^3 = [S]_{90}^3$$

$$0.9K' = 0.1[S]_{90}^3$$

$$[S]_{90}^3 = \frac{0.9K'}{0.1} = 9K'$$

$$[S]_{90} = \sqrt[3]{9K'} = \sqrt[3]{9}\,\sqrt[3]{K'}$$

At $[S]_{10}$

$$\frac{v}{V_{\max}} = 0.1$$

And by the same calculations shown above:

$$[S]_{10} = \sqrt[3]{\tfrac{1}{9}\,K'} = \sqrt[3]{\tfrac{1}{9}}\,\sqrt[3]{K'}$$

$$\frac{[S]_{90}}{[S]_{10}} = \frac{\sqrt[3]{9}\,\sqrt[3]{K'}}{\sqrt[3]{\tfrac{1}{9}}\,\sqrt[3]{K'}} = \frac{\sqrt[3]{9}}{\sqrt[3]{\tfrac{1}{9}}} = \frac{2.08}{0.482}$$

$$\boxed{\frac{[S]_{90}}{[S]_{10}} = 4.325 \qquad \text{when} \qquad n = 3}$$

(c)   When $n = 4$:

$$\frac{[S]_{90}}{[S]_{10}} = \frac{\sqrt[4]{9}}{\sqrt[4]{\tfrac{1}{9}}} = \frac{9^{1/4}}{(0.111)^{1/4}}$$

$$\log \frac{[S]_{90}}{[S]_{10}} = \tfrac{1}{4}(\log 9) - \tfrac{1}{4}(\log 0.111)$$

$$= \frac{0.954}{4} + \tfrac{1}{4}\log \frac{1}{0.111}$$

$$= 0.2385 + 0.2385 = 0.477$$

$$\frac{[S]_{90}}{[S]_{10}} = \text{antilog of } 0.477$$

$$\boxed{\frac{[S]_{90}}{[S]_{10}} = 3.0 \qquad \text{when} \qquad n = 4}$$

Or

$$\left(\frac{[S]_{90}}{[S]_{10}}\right)^2 = \frac{\sqrt{9}}{\sqrt{\tfrac{1}{9}}} = 9$$

$$\frac{[S]_{90}}{[S]_{10}} = \sqrt{9}$$

$$\boxed{\frac{[S]_{90}}{[S]_{10}} = 3}$$

Or

$$\frac{[S]_{90}}{[S]_{10}} = \frac{\sqrt[4]{9}}{\sqrt[4]{\frac{1}{9}}} = \sqrt[4]{\frac{9}{\frac{1}{9}}} = \sqrt[4]{81}$$

$$\boxed{\frac{[S]_{90}}{[S]_{10}} = 3}$$

General Rule

The $[S]_{90}/[S]_{10}$ for an enzyme that obeys sigmoidal kinetics equals the nth root of 81 where n is the interaction coefficient.

Problem V-17

The $[S]_{90}/[S]_{10}$ for an enzyme that obeys sigmoidal kinetics was found to be equal to 6.5. Calculate the interaction coefficient.

Solution

From the previous problem, we can see that $[S]_{90}/[S]_{10}$ equals the nth root of 9 divided by the nth root of $\frac{1}{9}$ or, as shown below, the nth root of 81.

$$\frac{[S]_{90}}{[S]_{10}} = \frac{\sqrt[n]{9}}{\sqrt[n]{\frac{1}{9}}} = \sqrt[n]{\frac{9}{\frac{1}{9}}}$$

$$\boxed{\frac{[S]_{90}}{[S]_{10}} = \sqrt[n]{81}}$$

$$6.5 = \sqrt[n]{81}$$

$$\log 6.5 = \frac{1}{n} \log 81$$

$$\frac{1}{n} = \frac{\log 6.5}{\log 81} = \frac{0.813}{1.908} = 0.426$$

$$n = \frac{1}{0.426}$$

$$\boxed{n = 2.35}$$

Problem V-18

A plot of v versus $[S]$ for an enzyme-catalyzed reaction was found to be sigmoidal rather than hyperbolic. Some of the data are shown below. (a) Estimate the interaction coefficient (n). (b) Estimate the value of K'. (c) Determine the initial velocity at 4.5×10^{-6} M $[S]$.

[S]	v
M	μmoles/liter-min
10^{-2}	250
5×10^{-3}	250
10^{-3}	250
10^{-4}	250
3×10^{-5}	241
2×10^{-5}	222
1.5×10^{-5}	193
10^{-5}	125
8×10^{-6}	85.3
6×10^{-6}	44.8
5×10^{-6}	27.8
4×10^{-6}	15.0

Solution

The best way to determine the constants n and K' is to plot the velocity-concentration data as $\log v/(V_{max} - v)$ versus $\log [S]$, as shown in Appendix III. The slope of the curve equals n. The $[S]_{50}$ is the substrate concentration when $\log v/(V_{max} - v) = 0$. The K' may then be calculated from the linear equation:

$$n \log [S]_{50} = \log K' \quad \text{or} \quad K' = [S]_{50}^{n}$$

We can estimate n and K' from the data given above.

(a) 90% of V_{max} is observed at about $2 \times 10^{-5} M$ [S]; 10% of V_{max} is observed at about $5 \times 10^{-6} M$ [S].

$$\frac{[S]_{90}}{[S]_{10}} = \sqrt[n]{81} \cong \frac{2 \times 10^{-5}}{5 \times 10^{-6}} = 0.4 \times 10 = 4 \cong \sqrt[n]{81}$$

$$\log 4 \cong \frac{\log 81}{n}$$

$$n \cong \frac{\log 81}{\log 4}$$

$$n \cong \frac{1.908}{0.601}$$

$$\boxed{n \cong 3.17}$$

(The graphical method will show that $n = 3.0$.)

(b) $$K' = [S]_{50}^{n}$$

At 10^{-5} M [S]

$$v = \frac{V_{max}}{2}$$

$$\therefore \quad [S]_{50} = 10^{-5} \, M$$

$$K' = (10^{-5})^{3.17}$$

$$\log K' = 3.17 \log 10^{-5} = 3.17(-5)$$

$$\log K' = -15.85$$

$$\boxed{K' = 10^{-15.85}}$$

Or

$$K' = \text{antilog of} -15.85 = \text{antilog of} \, (-16 + 0.15)$$

$$\boxed{K' = 1.41 \times 10^{-16}}$$

(The graphical method will show that K' actually equals 10^{-15}.)

(c)   At $[S] = 4.5 \times 10^{-6} \, M$, assume $n = 3$ and $K' = 10^{-15}$.

$$\frac{v}{V_{max}} = \frac{[S]^n}{K' + [S]^n}$$

$$\frac{v}{250} = \frac{(4.5 \times 10^{-6})^3}{10^{-15} + (4.5 \times 10^{-6})^3}$$

$$v = \frac{250(91.1 \times 10^{-18})}{10^{-15} + 91.1 \times 10^{-18}} = \frac{250(0.091 \times 10^{-15})}{1.091 \times 10^{-15}}$$

$$\boxed{v = 20.8 \, \mu\text{moles/liter-min}}$$

G. ENZYME UNITS AND PURIFICATION FACTORS

Problem V-19

A crude cell-free extract contains 20 mg of protein per milliliter. Ten μl of the extract in a standard reaction volume of 0.5 ml catalyzed the production of 3 μmoles (total) of product in 1 minute under optimum assay conditions (optimum pH and ionic strength, saturating concentrations of all substrates, coenzymes, activators, etc.). Calculate the: (a) initial velocity of the enzyme-catalyzed reaction in terms of μmoles/liter-min, (b) velocity in terms of (total) μmoles/min, (c) velocity in terms of μmoles/liter-min if the same 10 μl of extract were assayed in a standard reaction volume of 1.0 ml, (d) velocity in terms of μmoles/min if the same 10 μl of extract were assayed in a standard reaction volume of 1.0 ml, (e) concentration of enzyme in the extract in terms of units/ml, and (f) specific activity of the extract in terms of units/mg protein.

Solution

(a)
$$v = \frac{3\,\mu\text{moles/min}}{0.5\ \text{ml}} = \frac{3\,\mu\text{moles/min}}{0.5 \times 10^{-3}\ \text{liters}}$$

$$= \frac{3}{0.5 \times 10^{-3}}\,\mu\text{moles/liter-min}$$

$$\boxed{v = 6 \times 10^3\ \boldsymbol{\mu}\textbf{moles/liter-min}}$$

(b)
$$\boxed{v = 3\ \boldsymbol{\mu}\textbf{moles/min}}$$

(c)
$$v = \frac{3\,\mu\text{moles/min}}{1.0\ \text{ml}} = \frac{3\,\mu\text{moles/min}}{1 \times 10^{-3}\ \text{liter}}$$

$$\boxed{v = 3 \times 10^3\ \boldsymbol{\mu}\textbf{moles/liter-min}}$$

Or, if the total volume were doubled, the enzyme concentration would be halved and the velocity in terms of μmoles/liter-min would be halved.

$$\therefore\quad v = \frac{6 \times 10^3\,\mu\text{moles/liter-min}}{2}$$

$$\boxed{v = 3 \times 10^3\ \boldsymbol{\mu}\textbf{moles/liter-min}}$$

(d)
$$\boxed{v = 3\ \boldsymbol{\mu}\textbf{moles/min}}$$

That is, the total amount of product formed would still be the same. The velocity in terms of μmoles/liter-min would be halved but the total volume would be doubled.

Note: Because all of the reactions were conducted with saturating concentrations of all substrates, etc., the velocities are $V_{\max}$ values for each assay considered.

(e) A unit is defined as that amount of enzyme that catalyzes the production of 1 μmole of product per minute under optimum assay conditions.

$\therefore$ A velocity of 3 μmoles/min corresponds to 3 units of enzyme. The 3 units were present in 10μl of extract.

$$[\text{enzyme}] = \frac{3 \text{ units}}{10\mu\text{l}} = \frac{3 \text{ units}}{0.01 \text{ ml}}$$

$$\boxed{[\textbf{enzyme}] = \textbf{300 units/ml}}$$

(f)   Specific activity (S.A.) is defined as units of enzyme per milligram of protein.

$$\therefore \quad \text{S.A.} = \frac{300 \text{ units/ml}}{20 \text{ mg protein/ml}}$$

$$\boxed{\textbf{S.A.} = \textbf{15 units/mg protein}}$$

Problem V-20

An enzyme preparation has a specific activity of 42 units/mg protein and contains 12 mg of protein per ml. Calculate the initial velocity of the reaction in a standard reaction mixture containing: (a) 20 μl, (b) 5 μl, and (c) 50 μl of the preparation.

Solution

First calculate the enzyme concentration in the preparation in terms of units/ml.

$$[\text{enzyme}] = 42 \, \frac{\text{units}}{\text{mg protein}} \times 12 \, \frac{\text{mg protein}}{\text{ml}}$$

$$\boxed{[\textbf{enzyme}] = \textbf{504 units/ml}}$$

Now the velocities in terms of μmoles/ml can be calculated, taking into consideration that the velocities will be directly proportional to the enzyme concentration in the reaction mixture.

(a)   Twenty μl of preparation contain

$$504 \text{ units/ml} \times 0.02 \text{ ml} = 10.08 \text{ units enzyme}$$

$$\boxed{\therefore \quad v = \textbf{10.08 } \mu\textbf{moles/min}}$$

(b) Five μl of preparation contain

$$504 \text{ units/ml} \times 0.005 \text{ ml} = 2.52 \text{ units enzyme}$$

$$\therefore \quad v = 2.52 \text{ } \mu\text{moles/min}$$

(c) Fifty μl of preparation contain

$$504 \text{ units/ml} \times 0.05 \text{ ml} = 25.2 \text{ units enzyme}$$

$$\therefore \quad v = 25.2 \text{ } \mu\text{moles/min}$$

Problem V-21

Fifteen μl of an enzyme preparation catalyzed the production of 0.52 μmole of product in 1 minute under standard optimum assay conditions. (a) How much product will be produced in 1 minute by 150 μl of the preparation under the same reaction conditions? (b) How long will it take 150 μl of the preparation to produce 0.52 μmole of product under the same assay conditions?

Solution

(a) The initial velocity of the reaction was 0.52 μmole/min. Our first tendency is to say that 10 times as much enzyme will produce 10 times as much product in the same period of time. However, this will only be true if the velocity remains constant for the entire minute, i.e., if the reaction remains zero-order over the minute interval. In many instances, however, the decrease in the substrate concentration will cause the velocity of the reaction to drop back out of the zero-order region. The amount of product formed by 150 μl of enzyme in 1 minute will be $(10)(0.52) = 5.2$ μmoles only if the substrate concentration at the end of the 1 minute is still at least 100 K_m and the product of the reaction is not an inhibitor of the enzyme.

(b) Regardless of the initial substrate concentration or the change in substrate concentration during a reaction, the amount of product produced can be directly related to the amount of enzyme present and the incubation time elapsed.

$$[\text{enzyme}](\text{time}) \doteq \text{constant amount of product}$$

or

$$[\text{enzyme}](t) = k$$

where: [enzyme] = concentration or amount of enzyme in terms of volume of preparation or units. If the enzyme concentration is increased 10-fold, then only one-tenth as much time is needed to produce a given amount of product. Therefore the 0.52 μmole of product will be produced by 150 μl of enzyme preparation in one-tenth as much time as it took 15 μl of enzyme preparation—in 0.1 min.

$$t = 0.1 \text{ min} = 6 \text{ sec}$$

Or

$$[\text{enzyme}] \ (t) = k$$
$$(15 \ \mu\text{l}) \ (1 \text{ min}) = k$$
$$k = 15 \ \mu\text{l-min}$$

Then

$$(150 \ \mu\text{l})(t) = 15 \ \mu\text{l-min}$$
$$t = \frac{15 \ \mu\text{l-min}}{150 \ \mu\text{l}}$$

$$t = 0.1 \text{ min} = 6 \text{ sec}$$

Problem V-22

A crude cell-free extract contained 32 mg protein/ml. Ten μl of the extract catalyzed a reaction at a rate of 0.14 μmole/min under standard optimum assay conditions. Fifty ml of the extract were fractionated by ammonium sulfate precipitation. The fraction precipitating between 20% and 40% saturation was redissolved in 10 ml. This solution was found to contain 50 mg protein/ml. Ten μl of this purified fraction catalyzed the reaction at a rate of 0.65 μmole/min. Calculate (a) the percent recovery of the enzyme in the purified fraction, and (b) the degree of purification obtained by the fractionation (the purification factor).

Solution

The crude cell-free extract contained:

$$\frac{0.14 \ \mu\text{mole/min}}{0.01 \text{ ml}} = 14 \ \mu\text{moles/ml-min} = \boxed{\textbf{14 units/ml}}$$

$$14 \text{ units/ml} \times 50 \text{ ml total volume} = \boxed{\textbf{700 total units}}$$

and

$$32 \text{ mg protein/ml} \times 50 \text{ ml total volume} = \boxed{\textbf{1600 mg total protein}}$$

The specific activity of the crude cell-free extract was:

$$\frac{14 \text{ units/ml}}{32 \text{ mg protein/ml}} = \boxed{\textbf{0.4375 unit/mg protein}}$$

The purified fraction contained:

$$\frac{0.65 \,\mu\text{mole/min}}{0.01 \text{ ml}} = 65 \,\mu\text{moles/ml-min} = \boxed{\textbf{65 units/ml}}$$

$$65 \text{ units/ml} \times 10 \text{ ml} = \boxed{\textbf{650 total units}}$$

and

$$50 \text{ mg protein/ml} \times 10 \text{ ml} = \boxed{\textbf{500 mg total protein}}$$

The specific activity of the purified fraction was:

$$\frac{65 \text{ units/ml}}{50 \text{ mg protein/ml}} = \boxed{\textbf{1.30 units/mg protein}}$$

(a) $\text{recovery} = \dfrac{\text{total units in purified fraction}}{\text{total units in crude extract}} \times 100\%$

$$\text{recovery} = \frac{650}{700} \times 100\% = \boxed{\textbf{93.8\%}}$$

(b) $\text{degree of purification} = \dfrac{\text{specific activity of purified fraction}}{\text{specific activity of crude extract}}$

$$\text{purification} = \frac{1.30}{0.4375}$$

$$\boxed{\textbf{purification = 2.97-fold}}$$

PRACTICE PROBLEM SET V

Initial [S]	Initial Velocity
M	μmoles/liter-min
10^{-2}	120
2×10^{-3}	119
10^{-4}	100
2×10^{-6}	20
10^{-6}	10.9

1. The concentration-velocity data shown on page 255 were obtained for an enzyme catalyzing a reaction S $\rightarrow$ P. (a) Calculate K_m. (b) Verify that the enzyme obeys hyperbolic saturation kinetics. (c) Calculate the first-order rate constant for the enzyme concentration employed.

2. An enzyme with a K_m of 2.4×10^{-4} M was assayed at the following substrate concentrations: (a) 2×10^{-7} M, (b) 6.3×10^{-5} M, (c) 10^{-4} M, (d) 2×10^{-3} M, and (e) 0.05 M. The velocity observed at 0.05 M was 128 μmoles/liter-min. Calculate the initial velocities at the other substrate concentrations.

3. If the enzyme concentrations used in Problems V-1 and V-2 were increased 5-fold, what would be the corresponding velocities at each substrate concentration listed?

4. An enzyme with a K_m of 1.2×10^{-4} M was assayed at an initial substrate concentration of 0.02 M. By 30 seconds, 2.7 μmoles/liter of product had been produced. How much product will be present at: (a) 1 minute, (b) 95 seconds, (c) 3 minutes, and (d) 5.3 minutes? (e) What percent of the original substrate will be utilized by the times indicated?

5. An enzyme with a K_m of 2.6×10^{-3} M was assayed at an initial substrate concentration of 0.3 M. The observed velocity was 5.9×10^{-5} moles/liter-min. Calculate the amount of product that will be produced after (a) 5 minutes and (b) 10 minutes if the initial substrate concentration were 2×10^{-5} M.

6. An enzyme with a K_m of 3×10^{-4} M was assayed at an initial substrate concentration of 10^{-6} M. By 1 minute, 0.5% of the substrate had been utilized. (a) What percent of the substrate will be utilized by 5 minutes? (b) If the initial substrate concentration were 8×10^{-7} M, what percent of the substrate will be utilized by 5 minutes? (c) Calculate $V_{\max}$. (d) At 8×10^{-7} M, how long will it take for 50% of the substrate to be utilized? (e) At 10^{-6} M, how long will it take for 75% of the substrate to be utilized?

7. Calculate: (a) $[S]_{95}/[S]_5$, (b) $[S]_{80}/[S]_{20}$, and (c) $[S]_{75}/[S]_{25}$ for an enzyme that obeys hyperbolic saturation kinetics.

8. Calculate the velocity of an enzyme-catalyzed reaction, A + B $\rightarrow$ C + D, at the following concentrations of A and B: (a) [A] = 2×10^{-5} M and [B] = 6.7×10^{-5} M, (b) [A] = 3.5×10^{-4} M and [B] = 2×10^{-2} M, (c) [A] = 3.7×10^{-2} M and [B] = 10^{-4} M. The K_{m_A} = 2.2×10^{-4} M; K_{m_B} = 1.9×10^{-5} M. At 0.03 M [A] and 2×10^{-3} M [B], the observed velocity was 72.7 μmoles/liter-min. Assume random and independent binding of A and B.

9. Calculate v_i and the degree of inhibition caused by a competitive inhibitor under the following conditions: (a) $[S] = 2 \times 10^{-3}\ M$ and $[I] = 2 \times 10^{-3}\ M$, (b) $[S] = 4 \times 10^{-4}\ M$ and $[I] = 2 \times 10^{-3}\ M$, and (c) $[S] = 7.5 \times 10^{-3}\ M$ and $[I] = 10^{-5}\ M$. Assume that $K_m = 2 \times 10^{-3}\ M$, $K_i = 1.5 \times 10^{-4}\ M$, and $V_{max} = 270$ μmoles/liter-min.

10. (a) What concentration of competitive inhibitor is required to yield 75% inhibition at a substrate concentration of $1.5 \times 10^{-3}\ M$ if $K_m = 2.9 \times 10^{-4}\ M$ and $K_i = 2 \times 10^{-5}\ M$? (b) To what concentration must the substrate be increased to reestablish the velocity at the original uninhibited value?

11. Calculate K_i for a noncompetitive inhibitor if $2 \times 10^{-4}\ M$ [I] yields 75% inhibition of an enzyme-catalyzed reaction.

12. Calculate (a) the velocity and (b) the degree of inhibition of an enzyme-catalyzed reaction in the presence of $6 \times 10^{-4}\ M$ substrate $(K_m = 10^{-3}\ M)$ and $2.5 \times 10^{-4}\ M$ noncompetitive inhibitor $(K_i = 3 \times 10^{-5}\ M)$. The $V_{max} = 515$ μmoles/liter-min.

13. The following data were obtained for an enzyme-catalyzed reaction. Determine whether the enzyme obeys hyperbolic or sigmoidal kinetics and calculate or estimate the appropriate kinetic constants (K_m and V_{max}, or K', $[S]_{50}$, n, and V_{max}).

Initial Substrate Concentration	Initial Velocity
$M \times 10^4$	μmoles/liter-min
6.25	1.539
12.5	5.88
25.0	20.0
50.0	50.0
100.0	80.0
200.0	94.12
400.0	98.46
800.0	99.61

14. A cell-free extract of *Escherichia coli* contains 24 mg protein per milliliter. Twenty μl of this extract in a standard incubation volume of 0.1 ml catalyzed the incorporation of glucose-C^{14} from glucose-1-phosphate-C^{14} into glycogen at a rate of 1.6 mμmole/min. Calculate the velocity of the reaction in terms of: (a) μmoles/min, (b) μmoles/liter-min, (c) μmoles/mg protein-min. Also calculate the phosphorylase activity of the extract in terms of (d) units/ml and (e) units/mg protein.

15. Fifty ml of the cell-free extract described above was fractionated by ammonium sulfate precipitation. The fraction precipitating between 30% and 50% saturation was redissolved in a total volume of 10 ml and dialyzed. The solution after dialysis occupied 12 ml and contained 30 mg protein/ml. Twenty μl of the purified fraction catalyzed the phosphorylase reaction at a rate of 5.9 mμmoles/min. Calculate (a) the recovery of the enzyme and (b) the degree of purification obtained in the ammonium sulfate step.

SPECTROPHOTOMETRY

VI

A. SOLUTIONS CONTAINING ONLY ONE ABSORBING COMPOUND

Problem VI-1

A solution containing 2 mg/liter of a light-absorbing substance in a 1 cm cuvette transmits 75% of the incident light of a certain wavelength. Calculate the transmission of a solution containing: (a) 4 mg/liter, (b) 1 mg/liter, (c) 6 mg/liter, and (d) 5.4 mg/liter. (e) If the molecular weight of the compound is 250, calculate a_m.

Solution

(a)
$$\log \frac{I_0}{I} = acl$$

First solve the equation for a which will be a specific extinction coefficient, a_s, because the concentration is given in g/liter.

$$\log \frac{1.00}{0.75} = (a_s)(2)(1)$$

$$\log 1.335 = 2\, a_s$$

$$0.125 = 2\, a_s$$

$$a_s = \frac{0.125}{2}$$

$$\boxed{a_s = 0.0625}$$

Now that we have the specific extinction coefficient, the I values for solutions of any concentration may be calculated.

$$\log \frac{I_0}{I} = a_s cl$$

Calling I_0 100%

$$\log \frac{100}{I} = (0.0625)(4)(1)$$

$\log 100 - \log I = 0.25$

$\log 100 - 0.25 = \log I$

$2 - 0.25 = \log I$

$1.75 = \log I$

$I = $ antilog of 1.75

$\boxed{I = 56.2\% \text{ or } 0.562}$

Calling I_0 1.00

$$\log \frac{1.00}{I} = (0.0625)(4)(1)$$

$\log 1.00 - \log I = 0.25$

$\log I = -0.25$

$= -1 + 0.75$

$I = 5.62 \times 10^{-1}$

$\boxed{I = 0.562}$

We could also solve the problem by simply recalling that if a solution of concentration c has a transmission of I, then a solution of concentration nc has a transmission of I^n. When $c = 2$, $I = 0.75$. $\therefore$ When $c = 4$ (i.e., $2c_{orig}$), $I = 0.75^2 = 0.562$.

(b)

By Formula

$$\log \frac{I_0}{I} = a_s cl$$

$$\log \frac{100}{I} = (0.0625)(1)(1)$$

$\log 100 - \log I = 0.0625$

$2 - \log I = 0.0625$

$2 - 0.0625 = \log I$

$1.9375 = \log I$

$\boxed{I = 86.6\% \text{ or } 0.866}$

By Inspection

When $c = 2$, $I = 0.75$. $\therefore$
When $c = 1$ (i.e., $\frac{1}{2}c_{orig}$), $I = 0.75^{1/2}$.

$$I = \sqrt{0.75}$$

$\boxed{I = 0.866}$

(c)

By Formula	**By Inspection**

By Formula

$$\log \frac{I_0}{I} = a_s c l$$

$$\log \frac{100}{I} = (0.0625)(6)(1)$$

$$\log 100 - \log I = 0.375$$

$$2 - 0.375 = \log I$$

$$1.625 = \log I$$

$$I = \text{antilog of } 1.625$$

$$\boxed{I = 42.2\% \text{ or } 0.422}$$

By Inspection

When $c = 2$, $I = 0.75$. ∴
When $c = 6$ (i.e., $3c_{\text{orig}}$), $I = 0.75^3$.

$$\boxed{I = 0.422}$$

(d) Because 5.4 mg/liter is not an even multiple of 2 mg/liter it is far easier to solve this problem by formula than by inspection.

$$\log \frac{I_0}{I} = a_s c l$$

$$\log \frac{100}{I} = (0.0625)(5.4)(1)$$

$$\log 100 - \log I = 0.3375$$

$$2 - 0.3375 = \log I$$

$$1.6625 = \log I$$

$$I = \text{antilog of } 1.6625$$

$$\boxed{I = 46.0\% \text{ or } 0.46}$$

(e)

$$a_m = a_s \times \text{MW}$$

$$a_m = (0.0625)(250)$$

$$\boxed{a_m = 15.63}$$

Problem VI-2

A solution containing 10^{-5} M ATP has a transmission of 0.702 (70.2%) at 260 mμ in a 1 cm cuvette. Calculate the: (a) transmission of the solution in a 3 cm cuvette, (b) optical density of the solution in the 1 cm and 3 cm cuvettes, and (c) optical density and transmission of a 5×10^{-5} M ATP solution in a 1 cm cuvette.

Solution

(a)   We can calculate the transmission by formula after first calculating a_m, or by inspection.

<table>
<tr><th>By Formula</th><th>By Inspection</th></tr>
<tr><td>

$$\log \frac{I_0}{I} = a_m c l$$

$$\log \frac{100}{70.2} = (a_m)(10^{-5})(1)$$

$$\log 1.425 = 10^{-5} a_m$$

$$0.154 = 10^{-5} a_m$$

$$a_m = \frac{0.154}{10^{-5}}$$

$$= 0.154 \times 10^5$$

$$\boxed{a_m = 1.54 \times 10^4}$$

$$\log \frac{I_0}{I} = a_m c l$$

$$\log \frac{100}{I} = (1.54 \times 10^4)$$

$$\times (10^{-5})(3)$$

$$\log 100 - \log I = 4.62 \times 10^{-1}$$

$$2 - 0.462 = \log I$$

$$1.538 = \log I$$

$$I = \text{antilog of } 1.538$$

$$\boxed{I = 34.5\% \text{ or } 0.345}$$

</td><td>

When $l = 1$ cm, $I = 0.702$.
$\therefore$ When $l = 3$ cm (i.e., 3 l_{orig}),
$I = 0.702^3$.

$$\boxed{I = 0.345}$$

</td></tr>
</table>

(b)

<table>
<tr><th>1 cm Cuvette</th><th>3 cm Cuvette</th></tr>
</table>

1 cm Cuvette	3 cm Cuvette

$$\text{O.D.} = \log \frac{I_0}{I}$$

$$\text{O.D.} = \log \frac{100}{70.2} = \log 1.425$$

$$\boxed{\text{O.D.} = 0.154}$$

Or

$$\text{O.D.} = a_m cl$$

$$\text{O.D.} = (1.54 \times 10^4)(10^{-5})(1)$$

$$\boxed{\text{O.D.} = 0.154}$$

$$\text{O.D.} = \log \frac{I_0}{I}$$

$$\text{O.D.} = \log \frac{100}{34.5} = \log 2.9$$

$$\boxed{\text{O.D.} = 0.462}$$

Or

$$\text{O.D.} = a_m cl$$

$$\text{O.D.} = (1.54 \times 10^4)(10^{-5})(3)$$

$$\boxed{\text{O.D.} = 0.462}$$

Or

$$\text{O.D.}_{\cdot 3cm} = 3 \times \text{O.D.}_{\cdot 1cm}$$

$$\text{O.D.}_{\cdot 3cm} = (3)(0.154)$$

$$\boxed{\text{O.D.} = 0.462}$$

(c)

$$\text{O.D.}_{\cdot 5 \times 10^{-5}M} = 5 \times \text{O.D.}_{\cdot 1 \times 10^{-5}M} \quad \text{or} \quad \text{O.D.} = a_m cl$$

$$\text{O.D.} = (5)(0.154)$$

$$\boxed{\text{O.D.} = 0.77}$$

$$\text{O.D.} = (1.54 \times 10^{-4})(5 \times 10^{-5})(1)$$

$$\boxed{\text{O.D.} = 0.77}$$

$$\log \frac{I_0}{I} = \text{O.D.}$$

$$\log \frac{100}{I} = 0.77$$

$$\log 100 - \log I = 0.77$$

$$2 - 0.77 = \log I$$

$$1.23 = \log I$$

$$I = \text{antilog of } 1.23$$

$$\boxed{I = 17.0\% \text{ or } 0.17}$$

Problem VI-3

The extinction coefficient ($E_{1\%}^{1cm}$) of a glycogen-iodine complex at 450 mμ is 0.20. Calculate the concentration of glycogen in a solution of the iodine complex which has an optical density of 0.38 in a 3 cm cuvette.

Solution

$$\text{O.D.} = E_{1\%}^{1cm} c_\% l_{cm}$$

$$0.38 = (0.20)(c_\%)(3)$$

$$c_\% = \frac{0.38}{(0.2)(3)} = \frac{0.38}{0.6}$$

$$\boxed{c = 0.634\%}$$

Problem VI-4

A suspension of bacteria containing 300 mg dry weight per liter has an optical density of 1.00 in a 1 cm cuvette at 450 mμ. What is the cell density in a suspension that has a transmission of 30% in a 3 cm cuvette?

Solution

First calculate the optical density of the suspension in a 1 cm cuvette.

$$\text{O.D.}_{3cm} = \log \frac{I_0}{I} = \log \frac{100}{30} = \log 3.33$$

$$\text{O.D.}_{3cm} = 0.522$$

$$\text{O.D.}_{1cm} = \frac{\text{O.D.}_{3cm}}{3} = \frac{0.522}{3} = 0.174$$

Because we know that an O.D. of 1.00 is equivalent to 300 mg/liter of bacterial cells, the density equivalent to an O.D. of 0.174 can be determined by simple proportions.

$$\frac{1.00 \text{ O.D. unit}}{300 \text{ mg/liter}} = \frac{0.174 \text{ O.D. unit}}{X \text{ mg/liter}}$$

$$\boxed{X = 52.2 \text{ mg/liter}}$$

An alternate way of solving the problem is to define a specific extinction coefficient for the bacteria.

If an O.D. = 1.00 ≈ 300 mg/liter bacteria, calculate the O.D. of 1 g/liter bacteria.

$$\frac{1.00 \text{ O.D. unit}}{0.30 \text{ g/liter bacteria}} = \frac{a_s}{1.0 \text{ g/liter bacteria}}$$

$$\boxed{a_s = 3.33}$$

Now use the usual formula.

$$\text{O.D.} = (a_s)(c_{g/liter})(l_{cm})$$

$$0.522 = (3.33)(c_{g/liter})(3)$$

$$c_{g/liter} = \frac{0.522}{9.99} = 0.0522 \text{ g/liter}$$

$$\boxed{c = 52.2 \text{ mg/liter}}$$

Problem VI-5

A protein solution (0.3 ml) was diluted with 0.9 ml of water. To 0.5 ml of this diluted solution, 4.5 ml of biuret reagent were added and the color was allowed to develop. The O.D. of the mixture at 540 mμ was 0.18. A standard solution (0.5 ml, containing 4 mg of protein/ml) plus 4.5 ml of biuret reagent gave an O.D. of 0.12 in the same size cuvette. Calculate the protein concentration in the undiluted unknown solution.

Solution

From the O.D. of the standard reaction mixture, we could calculate a specific absorption (extinction) coefficient.

$$\text{O.D.} = a_{1mg/ml} \times c_{mg/ml} \times l_{cm}$$

However, because the light path length is the same for the standard and the unknown, we can neglect the l term in the equation. (In a sense, we are incorporating l into the specific absorption coefficient.) Similarly, because the total volume of both reaction mixtures is 5 ml, we can replace the concentration term with the weight of the protein.

$$\text{O.D.} = a_{1mg} \times \text{wt}_{mg}$$

$$0.12 = a_{1mg} \times 2 \text{ mg}$$

$$\boxed{a_{1mg} = 0.06}$$

That is, 1 mg of protein (in a standard sample size of 0.5 ml) plus 4.5 ml of biuret reagent will yield an O.D. of 0.06 at 540 mμ in the particular cuvette used.

The weight of protein in 0.5 ml of the diluted unknown can now be calculated.

$$\text{O.D.} = a_{1\text{mg}} \times \text{wt}_{\text{mg}}$$

$$\text{wt}_{\text{mg}} = \frac{\text{O.D.}}{a_{1\text{mg}}} = \frac{0.18}{0.06}$$

$$\boxed{\text{wt}_{\text{mg}} = 3 \text{ mg}}$$

Because the O.D. of the reaction mixture is directly proportional to the amount of protein present, we can also solve for the unknown weight by setting up a simple proportion.

$$\frac{\text{O.D.}_{\text{s}}}{\text{wt}_{\text{s}}} = \frac{\text{O.D.}_{\text{u}}}{\text{wt}_{\text{u}}}$$

$$\frac{0.12}{2 \text{ mg}} = \frac{0.18}{\text{wt}_{\text{u}}}$$

$$\text{wt}_{\text{u}} = \frac{0.18}{0.12} \times 2 = 1.5 \times 2$$

$$\boxed{\text{wt}_{\text{u}} = 3 \text{ mg}}$$

where

$$\text{O.D.}_{\text{s}} = \text{O.D. of standard solution}$$

$$\text{O.D.}_{\text{u}} = \text{O.D. of unknown solution}$$

$$\text{wt}_{\text{s}} = \text{weight of protein in standard sample}$$

$$\text{wt}_{\text{u}} = \text{weight of protein in unknown sample}$$

The 3 mg of protein were present in 0.5 ml of diluted unknown. The concentration of protein in the diluted unknown was:

$$\frac{3 \text{ mg}}{0.5 \text{ ml}} = \boxed{6 \text{ mg/ml}}$$

The sample that was analyzed was a 4-fold dilution of the original unknown solution   The concentration of protein in the original solution can be calculated from the dilution factor.

$$c_{\text{orig}} = c_{\text{final}} \times \text{dilution factor}$$

$$c_{\text{orig}} = 6 \text{ mg/ml} \times 4$$

$$\boxed{c_{\text{orig}} = 24 \text{ mg/ml}}$$

Note that the addition of 0.9 ml of water to 0.3 ml of original solution increased the total volume to 1.2 ml—the dilution was 4-fold, *not* 3-fold.

B. SOLUTIONS CONTAINING TWO ABSORBING COMPOUNDS

Problem VI-6

A solution containing DPN⁺ and DPNH had an optical density in a 1 cm cuvette of 0.311 at 340 mμ and 1.2 at 260 mμ. Calculate the proportion of the oxidized and reduced forms of the coenzyme in the solution. Both DPN⁺ and DPNH absorb at 260 mμ, but only DPNH absorbs at 340 mμ. The extinction coefficients are given below.

	a_m^{1cm}	
Compound	260 mμ	340 mμ
DPN+	15,400	~0
DPNH	15,400	6220

Solution

The concentration of each form may be calculated as follows. First calculate the concentration of DPNH from its absorption at 340 mμ where the DPN⁺ does not absorb.

$$O.D. = (a_m)(c)(l)$$

$$0.311 = (6.22 \times 10^3)(c)(1)$$

$$c = \frac{3.11 \times 10^{-1}}{6.22 \times 10^3} = 0.5 \times 10^{-4}$$

$$\boxed{c_{\text{DPNH}} = 5 \times 10^{-5}\ M}$$

Next calculate the O.D. at 260 mμ resulting from the DPNH.

$$O.D. = (a_m)(c)(l)$$

$$O.D. = (15.4 \times 10^3)(5 \times 10^{-5})(1) = 77 \times 10^{-2}$$

$$\boxed{O.D._{260[\text{DPNH}]} = 0.77}$$

The remainder of the O.D. at 260 mμ must result from the DPN$^+$.

$$1.20 \text{ total O.D. at 260 m}\mu$$
$$\underline{-0.77 \text{ O.D. of DPNH at 260 m}\mu}$$
$$0.43 \text{ O.D. of DPN}^+ \text{ at 260 m}\mu$$

Finally, from the O.D. of the DPN$^+$ at 260 mμ, calculate the concentration of DPN$^+$.

$$\text{O.D.} = (a_m)(c)(l)$$
$$0.43 = (15.4 \times 10^3)(c)(1)$$
$$c = \frac{4.3 \times 10^{-1}}{15.4 \times 10^3} = 0.279 \times 10^{-4}$$

$$\boxed{c_{\text{DPN}^+} = \mathbf{2.79 \times 10^{-5}\ M}}$$

Because the molar extinction coefficients of DPN$^+$ and DPNH are identical at 260 mμ, we could proceed a little differently. We could first calculate the total coenzyme concentration from the O.D. at 260 mμ and then subtract the DPNH concentration calculated from the O.D. at 340 mμ.

$$\text{O.D.} = (a_m)(c)(l)$$
$$1.20 = (15.4 \times 10^3)(c)(1)$$
$$c_{\text{total}} = \frac{12 \times 10^{-1}}{15.4 \times 10^{-3}} = 0.779 \times 10^{-4}$$

$$\boxed{c_{\text{total}} = \mathbf{7.79 \times 10^{-5}\ M}}$$

$$c_{\text{DPN}^+} = (7.79 \times 10^{-5}) - (5 \times 10^{-5})$$

$$\boxed{c_{\text{DPN}^+} = \mathbf{2.79 \times 10^{-5}\ M}}$$

The solution contains:

$$\frac{5 \times 10^{-5}}{7.79 \times 10^{-5}} \times 100\% = \boxed{\mathbf{64.2\%\ DPNH}}$$

$$\frac{2.79 \times 10^{-5}}{7.79 \times 10^{-5}} \times 100\% = \boxed{\mathbf{35.8\%\ DPN^+}}$$

Problem VI-7

Ten g of butter were saponified; the nonsaponifiable fraction was extracted into 25 ml of chloroform. The optical density of the chloroform solution in a 1 cm cuvette was 0.53 at 328 mμ and 0.48 at 458 mμ. Calculate the carotene and vitamin A content of the butter. The extinction coefficients for carotene and vitamin A at the above two wavelengths are given below.

Compound	$E_{1\%}^{1cm}$ in $CHCl_3$	
	328 mμ	458 mμ
Carotene	340	2200
Vitamin A	1550	$\sim$0

Solution

The concentration of carotene may be obtained from the optical density at 458 mμ where the vitamin A has no absorption.

$$O.D. = (E_{1\%})(c_{g/100ml})(l)$$

$$0.48 = (2200)(c)(1)$$

$$c = \frac{0.48}{2200} = \frac{4.8 \times 10^{-1}}{2.2 \times 10^{3}}$$

$$\boxed{c_{\text{carotene}} = \mathbf{2.18 \times 10^{-4} \text{ g/100 ml}}}$$

The total carotene content of the chloroform extract (25 ml) is

$$\frac{2.18 \times 10^{-4} \text{ g/100 ml}}{4} = 0.545 \times 10^{-4} \text{ g} = 0.0545 \text{ mg}$$

The carotene content of the butter is

$$\frac{0.0545 \text{ mg}}{10 \text{ g}} = 5.45 \times 10^{-3} \text{ mg carotene/g butter}$$

or

$$\boxed{\mathbf{5.45 \text{ µg carotene/g butter}}}$$

The vitamin A content of the butter may be calculated from the optical density of the chloroform solution at 328 mμ after correction of the optical density for the carotene present.

$$\text{O.D.}_{328_{\text{carotene}}} = (E_{1\%})(c_{\text{g/100ml}})(l)$$
$$\text{O.D.} = (340)(2.18 \times 10^{-4})(1) = 741 \times 10^{-4}$$
$$\text{O.D.}_{328_{\text{carotene}}} = 0.0741$$

0.530 total O.D. at 328 mμ

−0.074 O.D. of carotene at 328 mμ

_____

0.456 O.D. of vitamin A at 328 mμ

$$\text{O.D.}_{\text{vit.A}} = (E_{1\%})(c_{\text{g/100ml}})(l)$$
$$0.456 = (1550)(c)(1)$$
$$c = \frac{0.456}{1550} = \frac{45.6 \times 10^{-2}}{15.5 \times 10^{2}}$$

$$\boxed{c_{\text{vit.A}} = 2.94 \times 10^{-4} \text{ g/100 ml}}$$

The vitamin A content of the chloroform extract is

$$\frac{2.94 \times 10^{-4} \text{ g/100 ml}}{4} = 0.735 \times 10^{-4} \text{ g} = 0.0735 \text{ mg}$$

The vitamin A content of the butter is

$$\frac{73.5 \ \mu\text{g}}{10 \text{ g}} = \boxed{7.35 \ \mu\text{g vitamin A/g butter}}$$

Problem VI-8

A solution containing two substances, A and B, has an optical density in a 1 cm cuvette of 0.36 at 350 mμ and 0.225 at 400 mμ. The molar extinction coefficients of A and B at the two wavelengths are given below. Calculate the concentrations of A and B in the solution.

Compound	a_m	
	350 mμ	400 mμ
A	15,000	3000
B	7,000	6500

Solution

Because both compounds absorb at both wavelengths, we can set up two simultaneous equations.

$$O.D._{350m\mu} = (a_{m_{350[A]}} \times c_A) + (a_{m_{350[B]}} \times c_B)$$

$$0.36 = (15 \times 10^3 c_A) + (7 \times 10^3 c_B)$$

$$O.D._{400m\mu} = (a_{m_{400[A]}} \times c_A) + (a_{m_{400[B]}} \times c_B)$$

$$0.225 = (3 \times 10^3 c_A) + (6.5 \times 10^3 c_B)$$

Next, solve for c_A in terms of c_B or vice versa.

$$0.36 = 15 \times 10^3 c_A + 7 \times 10^3 c_B$$

$$0.36 - 7 \times 10^3 c_B = 15 \times 10^3 c_A$$

$$c_A = \frac{0.36 - 7 \times 10^3 c_B}{15 \times 10^3}$$

Next, substitute the above value for c_A into the second O.D. expression.

$$0.225 = (3 \times 10^3 c_A) + (6.5 \times 10^3 c_B)$$

$$0.225 = (3 \times 10^3) \frac{0.36 - 7 \times 10^3 c_B}{15 \times 10^3} + 6.5 \times 10^3 c_B$$

$$0.225 = \frac{1.08 \times 10^3 - 21 \times 10^6 c_B}{15 \times 10^3} + \frac{6.5 \times 10^3 c_B}{1}$$

Multiply the numerator and denominator of the second right-hand term by 15×10^3, and collect terms:

$$0.225 = \frac{1.08 \times 10^3 - 21 \times 10^6 c_B}{15 \times 10^3} + \frac{(15 \times 10^3)(6.5 \times 10^3 c_B)}{15 \times 10^3}$$

$$0.225 = \frac{1.08 \times 10^3 - 21 \times 10^6 c_B + 97.5 \times 10^6 c_B}{15 \times 10^3}$$

$$3.38 \times 10^3 = 1.08 \times 10^3 - 21 \times 10^6 c_B + 97.5 \times 10^6 c_B$$

$$2.30 \times 10^3 = 76.5 \times 10^6 c_B$$

$$c_B = \frac{2.3 \times 10^3}{76.5 \times 10^6} = \frac{23 \times 10^2}{7.65 \times 10^7} = 3 \times 10^{-5}$$

$$\therefore \quad c_B = 3 \times 10^{-5} \, M$$

Now that c_B is known, c_A may be calculated by substituting the value for c_B into either O.D. expression.

$$0.36 = (15 \times 10^3 c_A) + (7 \times 10^3 c_B)$$

$$0.36 = 15 \times 10^3 c_A + (7 \times 10^3)(3 \times 10^{-5})$$

$$0.36 = 15 \times 10^3 c_A + 0.21$$

$$0.15 = 15 \times 10^3 c_A$$

$$c_A = \frac{0.15}{15 \times 10^3} = \frac{15 \times 10^{-2}}{15 \times 10^3} = 1 \times 10^{-5}$$

$$\therefore \quad c_A = 1 \times 10^{-5} \, M$$

C. COUPLED REACTIONS

Problem VI-9

To 2.0 ml of a glucose solution, 1.0 ml of solution containing excess ATP, TPN$^+$, MgCl$_2$, hexokinase, and glucose-6-phosphate dehydrogenase was added. The optical density of the final solution (in a 1 cm cuvette) increased to 0.91 at 340 mμ. Calculate the concentration of glucose in the original solution.

Solution

The reactions that take place are shown below.

$$\text{glucose} + \text{ATP} \xrightarrow[\text{Mg}^{++}]{\text{hexokinase}} \text{glucose-6-phosphate}$$

$$\text{glucose-6-phosphate} + \text{TPN}^+ \xrightarrow{\text{glucose-6-phosphate dehydrogenase}}$$

$$\text{6-phosphogluconic acid-}\delta\text{-lactone} + \text{TPNH} + \text{H}^+$$

Although glucose has no absorption at 340 mμ, TPNH does. Because the K_{eq} values of the hexokinase reaction and the glucose-6-phosphate dehydrogenase reaction lie very far to the right, and excess ATP and TPN$^+$ are present, 1 mole of TPNH will be produced for every mole of glucose originally present. From the O.D. at 340 mμ, we can calculate the concentration of TPNH present. Consequently, after correction for dilution, we can calculate the concentration of glucose in the original solution.

$$\text{O.D.} = a_m c l$$

$$0.91 = (6.22 \times 10^3)(c)(1)$$

$$c_{\text{TPNH}} = \frac{0.91}{6.22 \times 10^3} = \frac{9.1 \times 10^{-1}}{6.22 \times 10^3}$$

$$c_{\text{TPNH}} = 1.465 \times 10^{-4} \ M \ \text{TPNH}$$

$$c_{\text{glucose orig}} = (1.465 \times 10^{-4}) \times \text{dilution factor}$$

$$= (1.465 \times 10^{-4})\tfrac{3}{2}$$

$$\boxed{c_{\text{glucose orig}} = 2.2 \times 10^{-4} \ M}$$

Problem VI-10

Describe an assay based on light absorption at 340 mμ by which the concentrations of glucose, glucose-6-phosphate, glucose-1-phosphate, and fructose-6-phosphate in a mixture may be determined.

Solution

All four compounds can give rise to a stoichiometric amount of TPNH by the reactions shown below.

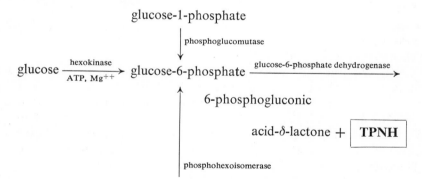

(a)   First add glucose-6-phosphate dehydrogenase and a large excess of TPN$^+$ and MgCl$_2$. The glucose-6-phosphate present is converted to 6-phosphogluconic acid-δ-lactone and a stoichiometric amount of TPNH appears. Measure the O.D. at 340 mμ.

(b)   When no further increase in O.D. occurs, add phosphoglucomutase. This enzyme catalyzes the conversion of the glucose-1-phosphate to glucose-6-phosphate which, in turn, produces more TPNH in an amount equivalent to the amount of glucose-1-phosphate originally present. Measure the increase in O.D.

(c) When no further increase in O.D. occurs, add phosphohexoisomerase. This enzyme catalyzes the conversion of the fructose-6-phosphate to glucose-6-phosphate which, in turn, yields another increment of TPNH. Measure the increase in O.D.

(d) Finally, add hexokinase and ATP. The glucose present yields a stoichiometric amount of TPNH as described previously. Again measure the increase in O.D.

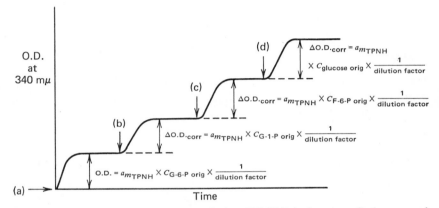

Figure VI-1 Optical density changes during TPNH-linked assay of glucose and derivatives

Although the glucose-6-phosphate dehydrogenase, phosphoglucomutase, and phosphohexoisomerase reactions do not have large K_{eq} values, the overall conversion to 6-phosphogluconic acid-δ-lactone and TPNH can be forced far to the right by using a large excess of TPN$^+$. Furthermore, if any of the enzymes are contaminated with 6-phosphogluconolactonase, the overall reaction sequence is essentially irreversible. The O.D. time course of the assay is shown in Figure VI-1.

In calculating the TPNH concentrations at any point from the ΔO.D. values, corrections must be made for the dilution of any preexisting TPNH. Also, in calculating the concentrations of glucose, glucose-6-phosphate, etc., in the original solution from the ΔTPNH values, the total dilution of the assay mixture must be taken into account.

Problem VI-11

Calculate the concentrations of fumaric acid and L-malic acid in a mixture, given the following information. (a) After the addition of 1.0 ml of a solution containing excess DPN$^+$ and the enzyme malic dehydrogenase to

1.0 ml of the original solution, the optical density at 340 mμ in a 3 cm cuvette increased to 0.227. (b) After the addition of the enzyme fumarase in 1.0 ml to 2.0 ml of solution obtained above, the optical density increased again to 0.72.

Solution

The reactions involved are shown below.

$$\text{fumaric acid} + H_2O \xrightarrow{\text{fumarase}} \text{L-malic acid}$$

$$\text{L-malic acid} + DPN^+ \xrightarrow{\text{malic dehydrogenase}} \text{oxalacetic acid} + DPNH + H^+$$

(a) $$O.D. = a_m \times c_{DPNH} \times l$$

$$0.227 = (6.22 \times 10^3)(c)(3)$$

$$c_{DPNH} = \frac{0.227}{(6.22 \times 10^3)(3)} = \frac{0.227}{18.6 \times 10^3} = \frac{2.27 \times 10^{-1}}{1.86 \times 10^4}$$

$$c_{DPNH} = 1.22 \times 10^{-5}\ M$$

$$c_{\text{L-malic acid orig}} = (1.22 \times 10^{-5}) \times \text{dilution factor}$$

$$c_{\text{L-malic acid orig}} = (1.22 \times 10^{-5})(2)$$

$$\boxed{c_{\text{L-malic acid orig}} = \mathbf{2.44 \times 10^{-5}\ M}}$$

(b) $$\Delta O.D._{corr} = O.D._{final} - O.D._{orig\ corr}$$

$$O.D._{orig\ corr} = (0.227)(\tfrac{2}{3}) = 0.151$$

$$\Delta O.D._{corr} = 0.72 - 0.151 = 0.569$$

$$\Delta O.D._{corr} = a_m \times \Delta c_{DPNH} \times l$$

$$0.569 = (6.22 \times 10^3)(\Delta c_{DPNH})(3)$$

$$\Delta c_{DPNH} = \frac{0.569}{(6.22 \times 10^3)(3)} = \frac{5.69 \times 10^{-1}}{1.86 \times 10^4}$$

$$\boxed{\Delta c_{DPNH} = \mathbf{3.05 \times 10^{-5}\ M}}$$

At this point the original solution has been diluted 3-fold.

$$c_{\text{fumaric acid orig}} = (3.05 \times 10^{-5})(3)$$

$$\boxed{c_{\text{fumaric acid orig}} = \mathbf{9.15 \times 10^{-5}\ M}}$$

Problem VI-12

To 1.5 ml of a solution containing a mixture of glucose-6-phosphate and glucose-1-phosphate, 1.5 ml of a solution containing excess TPN$^+$, MgCl$_2$, and glucose-6-phosphate dehydrogenase were added. The optical density in a 1 cm cuvette increased to 0.57 at 340 mμ. When no further increase in optical density was observed, an additional 2.0 ml of a phosphoglucomutase solution were added. The optical density then decreased to 0.50. Calculate the: (a) glucose-6-phosphate and (b) glucose-1-phosphate concentrations in the original solution.

Solution

(a)
$$\text{O.D.} = a_m \times c_{\text{TPNH}} \times l$$

$$0.57 = (6.22 \times 10^3)(c_{\text{TPNH}})(1)$$

$$c_{\text{TPNH}} = \frac{0.57}{6.22 \times 10^3} = \frac{57 \times 10^{-2}}{6.22 \times 10^3}$$

$$c_{\text{TPNH}} = 9.17 \times 10^{-5}\,M$$

$$c_{\text{G-6-P}} = (9.17 \times 10^{-5}) \times \text{dilution factor}$$

$$c_{\text{G-6-P}} = (9.17 \times 10^{-5})\frac{3.0}{1.5} = 18.34 \times 10^{-5}\,M$$

$$\boxed{c_{\text{G-6-P}} = \mathbf{1.834 \times 10^{-4}\,M}}$$

(b)
$$\Delta\text{O.D.}_{\text{corr}} = \text{O.D.}_{\text{final}} - \text{O.D.}_{\text{orig corr}}$$

$$\text{O.D.}_{\text{orig corr}} = (0.57)\tfrac{3}{5} = 0.342$$

$$\Delta\text{O.D.}_{\text{corr}} = 0.50 - 0.342 = 0.158$$

$$\Delta\text{O.D.} = a_m \times \Delta c_{\text{TPNH}} \times l$$

$$0.158 = (6.22 \times 10^3) \times (\Delta c_{\text{TPNH}}) \times (1)$$

$$\Delta c_{\text{TPNH}} = \frac{0.158}{6.22 \times 10^3} = \frac{15.8 \times 10^{-2}}{6.22 \times 10^3}$$

$$\Delta c_{\text{TPNH}} = 2.54 \times 10^{-5}\,M$$

$$c_{\text{G-1-P orig}} = 2.54 \times 10^{-5} \times \text{dilution factor}$$

$$c_{\text{G-1-P orig}} = (2.54 \times 10^{-5})\frac{5}{1.5}$$

$$\boxed{c_{\text{G-1-P orig}} = \mathbf{8.47 \times 10^{-5}\,M}}$$

Problem VI-13

Calculate the change in optical density that would be observed when 2.0 ml of a solution containing lactic dehydrogenase and 10^{-4} M pyruvic acid are added to a 1 cm cuvette containing 3.0 ml of a 2×10^{-4} M DPNH solution.

Solution

The equilibrium of the reaction catalyzed by lactic dehydrogenase lies far in favor of DPNH oxidation and pyruvate reduction. Consequently, for every mole of pyruvic acid present, 1 mole of DPNH disappears.

$$\text{O.D.}_{\text{orig}} = a_m \times c_{\text{DPNH}} \times l$$
$$\text{O.D.}_{\text{orig}} = (6.22 \times 10^3)(2 \times 10^{-4})(1) = 12.44 \times 10^{-1}$$

$$\boxed{\text{O.D.}_{\text{orig}} = 1.244}$$

Neglecting the enzyme-catalyzed reaction, the addition of 2 ml of solution to 3 ml of 2×10^{-4} M DPNH reduces the "original" optical density proportionately.

$$\text{O.D.}_{\text{dil}} = (1.244)\tfrac{3}{5} = 0.746$$

The cuvette at this point contains 0.003 liter $\times$ $(2 \times 10^{-4} M) = 6 \times 10^{-7}$ moles of DPNH, and 0.002 liter $\times 10^{-4}$ $M = 2 \times 10^{-7}$ moles of pyruvic acid.

Pyruvic acid is limiting. After the reaction takes place, the cuvette contains essentially no pyruvic acid and 4×10^{-7} moles of excess DPNH in 5.0 ml.

$$c_{\text{DPNH final}} = \frac{4 \times 10^{-7}\,\text{moles}}{5 \times 10^{-3}\,\text{liters}} = 0.8 \times 10^{-4}\,M = 8 \times 10^{-5}\,M$$

$$\text{O.D.}_{\text{final}} = a_m \times c_{\text{DPNH}} \times l$$
$$\text{O.D.}_{\text{final}} = (6.22 \times 10^3)(8 \times 10^{-5})(1) = 49.7 \times 10^{-2}$$

$$\boxed{\text{O.D.}_{\text{final}} = 0.497}$$

PRACTICE PROBLEM SET VI

1. Calculate the transmission and optical density at 260 mμ and 340 mμ of the following solutions: (a) 2.2×10^{-6} M TPNH in a 1 cm cuvette, (b) 2.8×10^{-5} M DPNH in a 3 cm cuvette, (c) 1.5×10^{-4} M TPN$^+$ in a

1 cm cuvette and (d) $4.2 \times 10^{-5} M$ ATP plus $7.0 \times 10^{-6} M$ DPNH in a 1 cm cuvette.

2. Calculate the transmission and optical density at 260 mμ and 340 mμ of each solution listed in the above problem after dilutions of: (a) 2-fold, (b) 3-fold, and (c) 4-fold.

3. Calculate the concentration of ATP and TPNH in solutions with optical densities (in a 1 cm cuvette) of: (a) 0.9 at 260 mμ and 0.15 at 340 mμ, (b) 0.75 at 260 mμ and 0 at 340 mμ, and (c) 0.545 at 260 mμ and 0.22 at 340 mμ.

4. Calculate the concentrations of two absorbing compounds, A and B, if the optical density of this solution in a 3 cm cuvette is 0.62 at 450 mμ and 0.54 at 485 mμ. Compound A has an a_m of 12,000 at 450 mμ and 4000 at 485 mμ. Compound B has an a_m of 5000 at 450 mμ and 11,600 at 485 mμ.

5. Suppose that 1.5 ml of a $2 \times 10^{-4} M$ solution of TPNH were added to 1.5 ml of a solution containing an unknown concentration of oxidized glutathione and a catalytic amount of the enzyme glutathione reductase. The final O.D. of the solution at 340 mμ was 0.25 in a 1 cm cuvette. Calculate the concentration of oxidized glutathione in the original 1.5 ml. The reaction catalyzed by glutathione reductase is: GSSG + TPNH + $H^+ \rightarrow$ 2GSH + TPN$^+$, and goes essentially to completion.

6. Devise a spectrophotometric assay based on light absorption at 340 mμ by which the concentrations of fructose-1,6-diphosphate, glyceraldehyde-3-phosphate, and dihydroxyacetone phosphate in a mixture may be determined.

7. Calculate the concentrations of citric acid and isocitric acid in a mixture, given the following information: (a) After the addition of 1.5 ml of solution containing excess DPN$^+$ and the enzyme isocitric dehydrogenase to 2.0 ml of the original solution, the optical density at 340 mμ in a 1 cm cuvette increased to 0.48. (b) After an additional 3.5 ml containing the enzyme aconitase were added, the optical density remained constant at 0.48.

ISOTOPES IN BIOCHEMISTRY

VII

A. DECAY RATES, DECAY CONSTANTS, AND SPECIFIC ACTIVITY

Problem VII-1

Ca^{45} has a half-life of 163 days. Calculate (a) the decay constant (λ) in terms of days^{-1} and sec^{-1}, and (b) the percent of the initial radioactivity remaining in a sample after 90 days.

Solution

(a)
$$\lambda = \frac{0.693}{t_{1/2}} = \frac{0.693}{163 \text{ days}} = \frac{6.93 \times 10^{-1}}{1.63 \times 10^{2}} \text{ days}^{-1}$$

$$\boxed{\lambda = 4.26 \times 10^{-3} \text{ days}^{-1}}$$

$$\lambda = \frac{0.693}{163 \text{ days} \times 24 \text{ hr/day} \times 60 \text{ min/hr} \times 60 \text{ sec/min}}$$

$$= \frac{0.693}{163 \text{ days} \times 86{,}400 \text{ sec/day}} = \frac{0.693}{(1.63 \times 10^{2})(8.64 \times 10^{4})}$$

$$= \frac{69.3 \times 10^{-2}}{14.1 \times 10^{6}} \text{ sec}^{-1}$$

$$\boxed{\lambda = 4.92 \times 10^{-8} \text{ sec}^{-1}}$$

(b)
$$2.3 \log \frac{N_0}{N} = \lambda t$$

Let $N_0 = 100\%$.

$$2.3 \log \frac{100}{N} = (4.26 \times 10^{-3})(90) = 0.384$$

$$\log \frac{100}{N} = \frac{0.384}{2.3} = 0.167$$

$$\log 100 - \log N = 0.167$$

$$\log N = \log 100 - 0.167$$

$$\log N = 2.000 - 0.167 = 1.833$$

$$\boxed{N = 68.1\%}$$

Problem VII-2

C^{14} has a half-life of 5700 years. Calculate (a) the fraction of the C^{14} atoms that decays per year and (b) the specific activity of pure C^{14} (i.e., the specific activity in terms of DPM/g, curies/g, and curies/g-atom).

Solution

(a)   First calculate λ:

$$\lambda = \frac{0.693}{5700 \text{ yr}} = \frac{6.93 \times 10^{-1}}{5.7 \times 10^{3}} \text{ yr}^{-1}$$

$$\boxed{\lambda = 1.215 \times 10^{-4} \text{ yr}^{-1}}$$

$$\lambda = \frac{dN/N}{dt}$$

$$\frac{dN}{N} = \lambda \, dt$$

$$\frac{dN}{N} = (1.215 \times 10^{-4} \text{ yr}^{-1})(1 \text{ yr}) = 1.215 \times 10^{-4}$$

That is, 1.215×10^{-4} atoms per atom decays per year or 1 atom out of $1/1.215 \times 10^{-4}$ atoms decays per year.

$$\frac{1}{1.215 \times 10^{-4}} = 0.823 \times 10^{4} = 8.23 \times 10^{3}$$

$$\boxed{\therefore \quad \textbf{1 out of every 8230 atoms initially present decays per year}}$$

(b)   Before we can calculate the decay rate in terms of disintegrations per minute, we must calculate λ in terms of min^{-1}.

$$\lambda = \frac{0.693}{5700 \times 365 \times 24 \times 60}$$

$$= \frac{0.693}{5700 \times 525,000 \text{ min/yr}} = \frac{6.93 \times 10^{-1}}{2.99 \times 10^{9}}$$

$$\boxed{\lambda = 2.32 \times 10^{-10} \text{ min}^{-1}}$$

$$\text{DPM} = -\frac{dN}{dt} = \lambda N$$

where N = the number of atoms in 1 g of C^{14}.

$$1 \text{ g } C^{14} = \frac{1 \text{ g}}{14 \text{ g/g-atom}} = 0.0714 \text{ g-atom}$$

$$N = 0.0714 \text{ g-atom} \times 6.023 \times 10^{23} \text{ atoms/g-atom}$$

$$\boxed{N = 4.3 \times 10^{22} \text{ atoms}}$$

$$\text{DPM} = 2.32 \times 10^{-10} \text{ min}^{-1} \times 4.3 \times 10^{22} \text{ atoms}$$

$$\boxed{\textbf{specific activity} = \textbf{9.99} \times \textbf{10}^{12} \textbf{ DPM/g}}$$

or

$$\frac{9.99 \times 10^{12} \text{ DPM/g}}{2.22 \times 10^{12} \text{ DPM/curie}} = \boxed{\textbf{4.45 curies/g}}$$

or

$$4.45 \text{ curies/g} \times 14 \text{ g/g-atom} = \boxed{\textbf{62.4 curies/g-atom}}$$

Problem VII-3

C^{14} is produced continuously in the upper atmosphere by the bombardment of N^{14} with neutrons of cosmic radiation. The reaction is $_{7}N^{14} + {_{0}n^{1}} \rightarrow {_{6}C^{14}} + {_{1}H^{1}}$. As a result, all carbon-containing compounds currently being biosynthesized on the earth contain sufficient C^{14} to yield 14 DPM/g carbon. After death of an organism the C^{14} decays with a half-life of 5700 years. Calculate (a) the abundance of C^{14} in the carbon that is participating in the carbon cycle on the surface of the earth today, and (b) the age of a sample of biological material that contains 3 DPM/g carbon.

Solution

(a)
$$\text{DPM/g} = -\frac{dN}{dt} = \lambda N$$

where N = the number of C^{14} atoms per g of carbon and λ = the decay constant, 2.32×10^{-9} min^{-1} for C^{14}.

$$14 = 2.32 \times 10^{-9}N$$

$$N = \frac{14}{2.32 \times 10^{-9}} = \boxed{\textbf{6.03} \times \textbf{10}^9 \text{ atoms } \textbf{C}^{14}/\textbf{g carbon}}$$

1 g of carbon contains

$$\frac{1 \text{ g}}{12 \text{ g/g-atom}} \times 6.02 \times 10^{23} \text{ atoms/g-atom}$$

$$= \boxed{\textbf{0.501} \times \textbf{10}^{23} \text{ total atoms of carbon}}$$

$$\text{abundance} = \frac{6.03 \times 10^9 \text{ atoms } C^{14}}{0.501 \times 10^{23} \text{ total atoms carbon}} \times 100\%$$

$$= \boxed{\textbf{1.2} \times \textbf{10}^{-11}\%}$$

(b)
$$2.3 \log \tfrac{14}{3} = \lambda t$$
where

$$\lambda = 1.215 \times 10^{-4} \text{ yr}^{-1} \text{ for } C^{14}.$$

$$2.3 \log 4.67 = 1.215 \times 10^{-4} \, t$$

$$(2.3)(0.668) = 1.215 \times 10^{-4} \, t$$

$$t = \frac{(2.3)(0.668)}{1.215 \times 10^{-4}} = \frac{1.54}{1.215 \times 10^{-4}} \text{ yr}$$

$$\boxed{\textbf{age} = t = \textbf{12,650 yr}}$$

Check: The C^{14} activity has decayed to about $\frac{1}{5}$ of its original level. After 5700 years, it would decay to $\frac{1}{2}$. After 11,400 years (2 half-lives), it would decay to $\frac{1}{4}$. After 17,100 years (3 half-lives), it would decay to $\frac{1}{8}$. Thus, the sample is between 2 and 3 half-lives old.

Problem VII-4

(a) What is the theoretical maximum specific activity (mc/mmole) at which L-phenylalanine-C^{14} (uniformly labeled) could be prepared? (b) What proportion of the molecules is actually labeled in a preparation of L-phenylalanine-C^{14} that has a specific activity of 200 mc/mmole?

Solution

(a) L-phenylalanine contains 9 g-atoms of carbon per mole. As shown in Problem VII-2, pure C^{14} has a specific activity of 62.4 curies/g-atom.

$\therefore$ maximum specific activity $= 9$ g-atoms/mole $\times$ 62.4 curies/g-atom

$$= \boxed{\textbf{562 curies/mole}}$$

(b) $\dfrac{200 \text{ curie/mole}}{562 \text{ curies/mole}} \times 100 = \%$ of C^{14} labeled molecules

$$= \boxed{\textbf{35.6\%}}$$

Problem VII-5

Calculate (a) the number of radioactive atoms and (b) the weight in grams of phosphorous in 1 curie of P^{32}. The half-life of P^{32} is 14.3 days.

Solution

(a)   1 curie $= 2.22 \times 10^{12}$ DPM

First calculate λ in terms of min^{-1}.

$$\lambda = \frac{0.693}{14.3 \times 24 \times 60} = \frac{0.693}{2.06 \times 10^4} = \frac{6.93 \times 10^{-1}}{2.06 \times 10^4} \text{ min}^{-1}$$

$$\boxed{\lambda = \textbf{3.37} \times \textbf{10}^{-5} \textbf{ min}^{-1}}$$

$$\text{DPM} = -\frac{dN}{dt} = \lambda N$$

$$2.22 \times 10^{12} = 3.37 \times 10^{-5} N$$

$$N = \frac{2.22 \times 10^{12}}{3.37 \times 10^{-5}} = 0.66 \times 10^{17}$$

$$\boxed{N = \textbf{6.6} \times \textbf{10}^{16} \textbf{ atoms/curie}}$$

(b)   1 g-atom of P^{32} (i.e., 32 g) contains 6.023×10^{23} atoms. $\therefore$ 6.6×10^{16} atoms weigh

$$\frac{6.6 \times 10^{16}}{6.023 \times 10^{23}} \times 32\ g = \boxed{3.51 \times 10^{-6}\ g}$$

Problem VII-6

K^{40} ($t_{1/2} = 1.3 \times 10^{10}$ yr) constitutes 0.012% of the potassium in nature. The human body contains about 0.35% potassium by weight. Calculate the total radioactivity resulting from K^{40} decay in a 75 kg human.

Solution

$$\text{total } K^{40} = 0.012\% \times 0.35\% \times 75 \times 10^3\ g$$

$$= (1.2 \times 10^{-4})(3.5 \times 10^{-3})(7.5 \times 10^4)$$

$$= 3.15 \times 10^{-2}\ g$$

$$\text{number of } K^{40} \text{ atoms} = \frac{3.15 \times 10^{-2}\ g}{40\ g/g\text{-atom}} \times 6.023 \times 10^{23} \text{ atoms/g-atom}$$

$$= 4.75 \times 10^{20} \text{ atoms}$$

$$\lambda = \frac{0.693}{1.3 \times 10^{10} \times 365 \times 24 \times 60} = \frac{6.93 \times 10^{-1}}{6.83 \times 10^{14}}\ \text{min}^{-1}$$

$$\lambda = 1.014 \times 10^{-15}\ \text{min}^{-1}$$

$$\text{DPM} = -\frac{dN}{dt} = \lambda N$$

$$\text{DPM} = (1.014 \times 10^{-15})(4.75 \times 10^{20}) = \boxed{4.82 \times 10^5\ \text{DPM}}$$

or

$$\frac{4.82 \times 10^5\ \text{DPM}}{2.22 \times 10^6\ \text{DPM}/\mu c} = \boxed{0.217\ \mu c}$$

Problem VII-7

A bottle contains 1 mc of L-phenylalanine-C^{14} (uniformly labeled) in 2.0 ml of solution. The specific activity of the labeled amino acid is given as 150 mc/mmole. Calculate (a) the concentration of L-phenylalanine in the solution and (b) the activity of the solution in terms of CPM/ml at a counting efficiency of 80%.

Solution

(a)   If 1 mmole is equivalent to 150 mc, calculate the number of mmoles that corresponds to 1 mc:

$$\frac{1 \text{ mmole}}{150 \text{ mc}} = 0.00667 \text{ mmole/mc}$$

The 1 mc is dissolved in 2.0 ml.

$$\therefore \quad \text{concentration} = \frac{6.67 \times 10^{-3} \text{ mmoles}}{2.0 \text{ ml}} = 3.335 \times 10^{-3} \text{ mmoles/ml}$$

$$\boxed{\text{concentration} = 3.335 \times 10^{-3} \; M}$$

(b) $$1 \text{ curie} = 2.22 \times 10^{12} \text{ DPM}$$
$$\therefore \quad 1 \text{ mc} = 2.22 \times 10^{9} \text{ DPM}$$
$$\text{total activity} = (0.80)(2.22 \times 10^{9}) \text{ CPM}$$
$$= 1.775 \times 10^{9} \text{ CPM in 2.0 ml}$$

$$\frac{1.775 \times 10^{9} \text{ CPM}}{2.0 \text{ ml}} = 0.888 \times 10^{9} \text{ CPM/ml}$$

$$\boxed{\text{activity} = 8.88 \times 10^{8} \text{ CPM/ml}}$$

Problem VII-8

A solution of L-glutamic acid-C^{14} (uniformly labeled) contains 1.0 mc and 0.25 mg of glutamic acid per ml. Calculate the specific activity of the labeled amino acid in terms of: (a) mc/mg, (b) mc/mmole, (c) DPM/μmole, and (d) CPM/μmole of carbon at a counting efficiency of 70%.

Solution

(a) $$\text{S.A.} = \frac{1.0 \text{ mc}}{0.25 \text{ mg}} = \boxed{4.0 \text{ mc/mg}}$$

(b) $$\text{S.A.} = 4.0 \text{ mc/mg} \times 147.1 \text{ mg/mmole}$$

$$\boxed{\text{S.A.} = 588 \text{ mc/mmole}}$$

(c) $$\text{S.A.} = \frac{588 \text{ mc}}{1 \text{ mmole}} = \frac{588 \text{ mc}}{1000 \; \mu\text{mole}} = 0.588 \text{ mc/}\mu\text{mole}$$

$$\text{S.A.} = 0.588 \text{ mc/}\mu\text{mole} \times 2.22 \times 10^{9} \text{ DPM/mc}$$

$$\boxed{\text{S.A.} = 1.305 \times 10^{9} \text{ DPM/}\mu\text{mole}}$$

(d)   One μmole of L-glutamic acid contains 5 μmoles of carbon.

$$\therefore \quad \text{S.A.} = \frac{1.305 \times 10^9}{5} \text{ DPM}/\mu\text{mole carbon}$$

$$= 0.261 \times 10^9 \text{ DPM}/\mu\text{mole carbon}$$

At 70% efficiency:

$$\text{S.A.} = (0.70)(2.61 \times 10^8)$$

$$\boxed{\text{S.A.} = 1.83 \times 10^8 \text{ CPM}/\mu\text{mole carbon}}$$

Problem VII-9

Describe the preparation of 100 ml of a 10^{-2} M solution of L-methionine-S^{35} in which the amino acid has a specific activity of 1.5×10^5 DPM/μmole. Assume that you have available a 0.1 M solution of unlabeled L-methionine and a stock solution of L-methionine-S^{35} (30 mc/mmole and 1 mc/ml).

Solution

First calculate the amount of radioactivity needed.

$$10^{-2} M = 10 \ \mu\text{moles/ml}$$

$$10 \ \mu\text{moles/ml} \times 100 \text{ ml} = 1000 \ \mu\text{moles}$$

$$1000 \ \mu\text{moles} \times 1.5 \times 10^5 \text{ DPM}/\mu\text{mole} = \boxed{1.5 \times 10^8 \text{ DPM}}$$

Next calculate the amount of the radioactive stock solution that is needed to provide 1.5×10^8 DPM.

$$1 \text{ mc/ml} \times 2.22 \times 10^9 \text{ DPM/mc} = 2.22 \times 10^9 \text{ DPM/ml}$$

$$\frac{1.5 \times 10^8 \text{ DPM}}{2.22 \times 10^9 \text{ DPM/ml}} = 0.675 \times 10^{-1} \text{ ml} = \boxed{67.5 \ \mu\text{l}}$$

Thus, 67.5 μl of the radioactive stock provide the radioactivity required. Next, calculate whether 67.5 μl also provide any significant amount of L-methionine. Stock solution:

$$\frac{1 \text{ mmole}}{30 \text{ mc}} = 0.0333 \text{ mmoles/mc}$$

Because the stock contains 1 mc/ml, its concentration is 0.0333 mmole/ml or 33.3 μmoles/ml. In the 67.5 μl, we have:

$$0.0675 \text{ ml} \times 33.3 \ \mu\text{moles/ml} = 2.25 \ \mu\text{moles}$$

For most applications the amount of L-methionine added from the radioactive stock (2.25 μmoles) is so small compared to the total (1000 μmoles) that it can be ignored. The radioactive stock is treated as if it were "carrier-free"—as if it contained only radioactivity and no mass.

> ∴ **Take 67.5 μl of radioactive L-methionine-S³⁵ solution, add 10.0 ml of 0.1 M (1000 μmoles) nonradioactive L-methionine solution, and then add sufficient water to make 100 ml final volume. If exact concentrations are important, then use only 9.978 ml of unlabeled L-methionine solution.**

Problem VII-10

Ten ml of a 10^{-3} M unlabeled L-methionine solution and 100 μl of the L-methionine-S³⁵ stock solution described in the previous problem were mixed and diluted to a final volume of 100 ml. What are (a) the concentration and (b) the specific activity of the L-methionine-S³⁵ in the final solution?

Solution

(a) The unlabeled L-methionine solution contains:

$$0.010 \text{ liter} \times 0.001 \ M = 10^{-5} \text{ moles} = 10 \ \mu\text{moles}$$

The radioactive solution contains:

$$33.3 \ \mu\text{moles/ml} \times 0.1 \text{ ml} = 3.33 \ \mu\text{moles}$$

The final solution will contain:

$$10 + 3.33 = 13.33 \ \mu\text{moles}/100 \text{ ml} = 133.3 \ \mu\text{moles/liter}$$
$$= 133.3 \times 10^{-6} \ M$$

$$= \boxed{1.333 \times 10^{-4} \ M}$$

(b) The radioactive stock solution provides:

$$1 \text{ mc/ml} \times 0.1 \text{ ml} = 0.1 \text{ mc}$$

$$\text{S.A.} = \frac{0.1 \text{ mc}}{13.33 \ \mu\text{moles}} = \frac{100 \ \mu\text{c}}{13.33 \ \mu\text{moles}}$$

$$\boxed{\text{S.A.} = 7.5 \ \mu\text{c}/\mu\text{mole (or 7.5 mc/mmole or 7.5 curie/mole)}}$$

In terms of DPM:

$$S.A. = 7.5 \ \mu c/\mu mole \times 2.22 \times 10^6 \ DPM/\mu c$$

$$\boxed{S.A. = 16.65 \times 10^6 \ DPM/\mu mole}$$

Note that in the above problem the amount of L-methionine provided by the radioactive stock solution was significant compared to the amount provided by the nonradioactive solution.

Problem VII-11

Glucose-1-phosphate-C^{14} (uniformly labeled, specific activity 16,000 CPM/μmole) was incubated with glycogen in the presence of a cell-free extract containing the enzyme glycogen phosphorylase. Radioactivity was incorporated into the glycogen primer at an initial velocity of 2550 CPM/min. (a) Calculate the rate of the enzymic reaction in terms of μmoles glucose incorporated/min. (b) Calculate the rate in terms of μmoles/liter-min if the reaction volume was 0.2 ml. (c) Calculate the rate in terms of μmoles/mg protein-min if the incubation mixture contained 0.35 mg of protein.

Solution

(a) $$v = \frac{2550 \ CPM/min}{16,000 \ CPM/\mu mole} = \frac{2.55 \times 10^3}{1.6 \times 10^4} \ \mu moles/min$$

$$v = 1.59 \times 10^{-1} \ \mu moles/min$$

$$\boxed{v = 0.159 \ \mu mole/min}$$

(b) $$v = \frac{0.159 \ \mu mole/min}{0.2 \ ml} = \frac{15.9 \times 10^{-2} \ \mu moles/min}{2 \times 10^{-4} \ liters}$$

$$\boxed{v = 795 \ \mu moles/liter\text{-}min}$$

(c) $$v = \frac{0.159 \ \mu mole/min}{0.35 \ mg \ protein}$$

$$\boxed{v = 0.454 \ \mu mole/mg \ protein\text{-}min}$$

Problem VII-12

A microorganism was grown in a synthetic medium containing $S^{35}O_4^=$ as the sole sulfur source. The initial concentration of $S^{35}O_4^=$ in the medium was $7 \times 10^{-3} \ M$. One ml of the medium contained 2×10^6 CPM of radioactivity. After several days of growth the cells were harvested, washed, and extracted with boiling water. The extract was fractionated by ion-exchange chromatography. One g wet weight of cells contained 53,000 CPM of S^{35} in the L-methionine fraction. Calculate the intracellular concentration of L-methionine in the organism, assuming that the 1 g wet weight contained 0.2 g of dry cell constitutents and 0.8 ml of intracellular water.

Solution

To convert CPM to moles, we must first know the specific activity of the S^{35}. We know that the original medium contained $7 \times 10^{-3}\ M$ $S^{35}O_4^=$ (i.e., 7 μmoles/ml) and 2×10^6 CPM/ml.

$$\text{S.A.} = \frac{2 \times 10^6\ \text{CPM/ml}}{7\ \mu\text{moles/ml}} = 0.286 \times 10^6\ \text{CPM}/\mu\text{mole}$$

$$\boxed{\text{S.A.} = 2.86 \times 10^5\ \textbf{CPM}/\mu\textbf{mole}}$$

All sulfur compounds in the organism are derived from the $S^{35}O_4^=$. Consequently, the specific activity of all sulfur compounds containing 1 atom of sulfur per molecule also is 2.86×10^5 CPM/μmole.

$$\therefore \quad \text{amount of L-methionine} = \frac{53,000\ \text{CPM}}{2.86 \times 10^5\ \text{CPM}/\mu\text{mole}}$$

$$= \frac{5.3 \times 10^4}{2.86 \times 10^5}\ \mu\text{mole}$$

$$= 0.185\ \mu\text{mole}$$

The L-methionine came from 0.8 ml of intracellular water.

$$\therefore \quad \text{intracellular concentration of L-methionine} = \frac{1.85 \times 10^{-1}\ \mu\text{moles}}{0.8\ \text{ml}}$$

$$= 2.31 \times 10^{-1}\ \mu\text{moles/ml}$$

$$= 2.31 \times 10^{-7}\ \text{moles/ml}$$

$$= 2.31 \times 10^{-4}\ \text{moles/liter}$$

$$\boxed{\textbf{intracellular concentration of L-methionine} = \textbf{2.31} \times \textbf{10}^{-4}\ \textbf{\textit{M}}}$$

Problem VII-13

Ten μc of C^{14}-labeled inulin were added to 15.0 ml of a yeast suspension. The suspension was then centrifuged and the supernatant fluid carefully drawn off. The pellet of packed yeast occupied 0.2 ml and contained

10,000 CPM. The counting efficiency was 25%. Calculate the proportion of the packed yeast pellet that is interstitial space, assuming that the yeast cells were completely impermeable to the inulin and that the inulin did not adsorb to the cell surface.

Solution

The "specific activity" of the suspension is 10 μc/15 ml or 0.667 μc/ml. Because the yeast cells occupy such a small proportion of the total volume, we could assume that each milliliter of extracellular fluid contains 0.667 μc. However, for a more exact calculation, we can assume that the volume of the suspension occupied by the yeast cells is at least the same volume as the packed yeast pellet. The "specific activity" of the extracellular fluid then becomes:

$$\frac{10\ \mu c}{15.0 - 0.2\ \text{ml}} = \frac{10}{14.8} = \boxed{\textbf{0.676 } \mu\textbf{c/ml}}$$

Under the given counting conditions (for example, 0.2 ml of packed yeast cells spread out and dried on a planchet or suspended in a given volume of scintillation fluid), the efficiency of counting is 25%.

∴   One ml of extracellular fluid is equivalent to:

$$0.676\ \mu c/\text{ml} \times 2.22 \times 10^6\ \text{DPM}/\mu c \times 0.25 = \boxed{\textbf{3.75} \times \textbf{10}^5\ \textbf{CPM}}$$

$$∴\quad \text{interstitial volume} = \frac{10 \times 10^3\ \text{CPM}}{3.75 \times 10^5\ \text{CPM/ml}} = \boxed{\textbf{2.67} \times \textbf{10}^{-2}\ \textbf{ml}}$$

The packed yeast pellet contains:

$$\frac{0.0267\ \text{ml interstitial space}}{0.2\ \text{ml total volume}} = \boxed{\textbf{0.134 ml interstitial space/ml packed cells}}$$

or

$$\boxed{\textbf{13.4\% interstitial volume}}$$

B. CARRIER DILUTION, ISOTOPE DILUTION, AND RADIOACTIVE DERIVATIVE ANALYSES

Problem VII-14

Carrier-free $S^{35}O_4^=$ (5 × 10^8 CPM) was added to a sample containing an unknown amount of unlabeled sulfate. After equilibration, a small sample of the sulfate was reisolated as $BaS^{35}O_4$. A 2 mg sample of the $BaS^{35}O_4$ contained 2.9 × 10^5 CPM. Calculate M_u, the amount of unlabeled $SO_4^=$ in the sample as Na_2SO_4.

Solution

The specific activity of the reisolated $BaS^{35}O_4(S.A._r)$ is:

$$S.A._r = S.A._{BaS^{35}O_4} = \frac{2.9 \times 10^5 \text{ CPM}}{2 \text{ mg}} = \boxed{1.45 \times 10^5 \text{ CPM/mg}}$$

or

$$1.45 \times 10^5 \text{ CPM/mg} \times 233.5 \text{ mg/mmole} = \boxed{3.42 \times 10^7 \text{ CPM/mmole}}$$

One mmole of $BaSO_4$ contains 1 mmole of $SO_4^=$. Therefore, the specific activity of the $SO_4^=$ is also 3.42×10^7 CPM/mmole. We can now calculate the amount of unlabeled $SO_4^=$ in the sample.

$$M_u = \frac{A_o}{S.A._r}$$

where

$$A_o = \text{total activity added} = 5 \times 10^8 \text{ CPM}$$
$$S.A._r = 3.42 \times 10^7 \text{ CPM/mmole}$$
$$M_u = \frac{5 \times 10^8 \text{ CPM}}{3.42 \times 10^7 \text{ CPM/mmole}}$$

$$\boxed{M_u = 14.6 \text{ mmoles}}$$

Expressed as Na_2SO_4:

$$M_u = 14.6 \text{ mmoles} \times 142 \text{ mg/mmole} = 2080 \text{ mg}$$

$$\boxed{M_u = 2.08 \text{ g}}$$

Problem VII-15

A solution containing 1.3 mg of cystine-S^{35} (30 mc/mmole) was added to an HCl solution containing a mixture of amino acids, including an unknown amount of unlabeled cystine. A small amount of labeled cystine was then reisolated from the mixture and purified to constant specific activity. The reisolated cystine had a specific activity of 1750 CPM/mg. How much cystine was in the HCl solution?

Solution

In this problem the tracer cystine is not carrier-free, but we can see from a quick comparison of the specific activity of the added labeled cystine-S^{35} (S.A._o) with the specific activity of the reisolated cystine-S^{35}

(S.A.$_r$) that the amount of unlabeled cystine in the mixture was far greater than the amount of tracer cystine-S^{35} added.

$$S.A._o = 30 \text{ mc/mmole} = \frac{30 \text{ mc}}{240.3 \text{ mg/mmole}} = 0.125 \text{ mc/mg}$$

$$= 0.125 \text{ mc/mg} \times 2.22 \times 10^9 \text{ CPM/mc}$$

$$\boxed{S.A._o = 2.77 \times 10^8 \text{ CPM/mg}}$$

while

$$S.A._r = 1.75 \times 10^3 \text{ CPM/mg}$$

∴ The labeled cystine was diluted by more than 10^5-fold i.e., the amount of unlabeled cystine in the mixture (M_u) was greater than one-hundred thousand times the amount of labeled cystine added (M_o). Consequently, we can treat the labeled tracer cystine as if it were carrier-free.

$$M_u = \frac{A_o}{S.A._r}$$

$$A_o = 2.77 \times 10^8 \text{ CPM/mg} \times 1.3 \text{ mg} = 3.6 \times 10^8 \text{ CPM}$$

$$M_u = \frac{3.6 \times 10^8 \text{ CPM}}{1.75 \times 10^3 \text{ CPM/mg}} = 2.06 \times 10^5 \text{ mg} = 2.06 \times 10^2 \text{ g}$$

$$\boxed{M_u = 206 \text{ g}}$$

We could use the unsimplified expression which is valid regardless of the ratio of M_u/M_o.

$$M_u = \left[\frac{S.A._o}{S.A._r} - 1\right]M_o \qquad M_u = \left[\frac{2.77 \times 10^8}{1.75 \times 10^3} - 1\right]1.3 \text{ mg}$$

$$= (1.585 \times 10^5 - 1)1.3$$

Obviously, the 1 is insignificant compared to the 1.585×10^5 and may be discarded.

∴ $M_u = (1.585 \times 10^5)1.3 = 2.06 \times 10^5 \text{ mg} = 2.06 \times 10^2 \text{ g}$

$$\boxed{M_u = 206 \text{ g}}$$

Problem VII-16

Twenty mg of C^{14}-labeled glycogen (6.7×10^4 CPM/mg) were added to a solution containing an unknown amount of unlabeled glycogen. A small

amount of glycogen was then reisolated from the solution and repre-
cipitated with ethanol to constant specific activity (2.8×10^4 CPM/mg).
Calculate the amount of unlabeled glycogen in the solution.

Solution

Because S.A.$_r$ is of the same order of magnitude as S.A.$_o$, the amount of
unlabeled glycogen in the solution obviously was significant compared to
the amount of labeled material added.

$$M_u = \left[\frac{\text{S.A.}_o}{\text{S.A.}_r} - 1\right] M_o \qquad M_u = \left[\frac{6.7 \times 10^4}{2.8 \times 10^4} - 1\right] 20 \text{ mg}$$

$$= (2.92 - 1)20 = (1.92)(20)$$

$$\boxed{M_u = 38.4 \text{ mg}}$$

Problem VII-17

A plant was grown in an atmosphere containing $C^{14}O_2$ (3×10^8 CPM/
μmole). After several weeks a leaf extract was prepared for glucose-1-
phosphate determination via reverse isotope dilution analysis. To 20 ml
of the extract, 1.5 mmoles of unlabeled dipotassium glucose-1-phosphate
were added. A small amount of dipotassium glucose-1-phosphate was
reisolated from the extract and recrystallized to constant specific activity
from aqueous ethanol. The recrystallized salt had a specific activity of
2.6×10^5 CPM/μmole. Calculate the concentration of labeled glucose-1-
phosphate in the extract.

Solution

All carbon compounds in the plant have the same specific activity (on
a per mole of carbon basis) as the $C^{14}O_2$ provided. Glucose-1-phosphate
contains 6 g-atoms of carbon per mole.

$$\therefore \quad \text{S.A.}_u = \text{specific activity of the unknown amount of G-1-P}$$

$$= 6 \times 3 \times 10^8 = 18 \times 10^8 \text{ CPM}/\mu\text{mole}$$

$$M_u = \frac{\text{S.A.}_r}{\text{S.A.}_u - \text{S.A.}_r} M_o$$

where:

$$M_o = \text{amount of nonradioactive carrier added to the extract}$$

$$= 1.5 \text{ mmoles}$$

$$M_u = \frac{2.6 \times 10^5}{1.8 \times 10^9 - 2.6 \times 10^5} 1.5 \text{ mmoles}$$

The 2.6×10^5 is insignificant compared to the 1.8×10^9 and may be discarded in the denominator. The fact that $S.A._r$ is so small compared to $S.A._u$ immediately shows that the amount of carrier added was very large compared to the amount of radioactive glucose-1-phosphate in the extract.

$$M_u = \frac{2.6 \times 10^5}{1.8 \times 10^9} 1.5 = 2.165 \times 10^{-4} \text{ mmoles}$$

$$\boxed{M_u = 0.2165 \text{ } \mu\text{mole}}$$

$$\text{concentration} = \frac{0.2165 \text{ } \mu\text{mole}}{20 \text{ ml}} = \frac{21.65 \times 10^{-2}}{20}$$

$$= 1.083 \times 10^{-2} \text{ } \mu\text{moles/ml}$$

$$\boxed{\text{concentration} = 1.083 \times 10^{-5} M}$$

Problem VII-18

A mixture of amino acids was reacted with p-iodobenzene sulfonyl chloride ("pipsyl" chloride) labeled with I^{131} to produce the radioactive pipsyl derivatives of the individual amino acids. The specific activity of the pipsyl chloride was 4.23×10^5 CPM/μmole. After the reaction, 250 mg of unlabeled pipsyl derivative of leucine were added to the mixture. A small amount of the leucine derivative was reisolated and purified to a constant specific activity of 1700 CPM/μmole. Calculate the amount of unlabeled leucine in the original mixture.

Solution

After the reaction with pipsyl chloride the pipsyl derivatives of all the amino acids present have the same specific activity (4.23×10^5 CPM/μmole).

$$M_u = \frac{S.A._r}{S.A._u - S.A._r} M_o \qquad M_u = \frac{1700}{4.23 \times 10^5 - 1700} 250 \text{ mg}$$

$$= \frac{17 \times 10^2}{4.23 \times 10^5 - 0.017 \times 10^5} 250 \text{ mg}$$

$$= \frac{17 \times 10^2}{4.21 \times 10^5} 250 = (4.17 \times 10^{-3})(250) \text{ mg}$$

$$\boxed{M_u = 1.04 \text{ mg}}$$

C. DOUBLE-LABEL ANALYSES

Problem VII-19

A microorganism was grown in a synthetic medium containing $S^{35}O_4^=$ (2.9×10^8 CPM/μmole) and P^{32} inorganic phosphate (5.2×10^8 CPM/μmole) as sole sulfur and phosphorous sources, respectively. A cell-free extract of the organism contained a sulfur- and phosphorous-containing compound which could be detected by paper chromatography. The activity of the chromatographic peak was determined periodically over a 4-week interval. The results are shown below; calculate the P/S ratio in the compound.

Time (days)	Activity (CPM)
0	39,400
14	31,225
21	28,310
28	25,890

Solution

The half-lives of P^{32} and S^{35} are 14.3 days and 87.1 days, respectively. Thus, the activity of each isotope decays at a different rate. We can calculate the fraction of each isotope remaining at any given time from the relationship:

$$2.3 \log \frac{N_0}{N} = \lambda t = \frac{0.693}{t_{1/2}} t$$

where N/N_0 is the fraction remaining.

We can also determine the fraction of each isotope remaining at any time from a two-point semilog plot of amount remaining versus time where 100% is taken as the amount at zero-time and 50% is taken as the amount present at t = half-life.

On a slide rule containing e^{-x} scales, we need only line up the half-life (on the C-scale) beneath 0.50 (on the $LL/2$ scale). The fraction remaining at any other time (on the C-scale) can be read off e^{-x} scales directly. Using any of these methods, we obtain the results shown in Table VII-1.

TABLE VII-1

Isotope	Fraction Remaining After:			
	0 day	14 days	21 days	28 days
P^{32}	1.00	0.5075	0.362	0.258
S^{35}	1.00	0.8948	0.8461	0.8005

If we pick any two times, we can solve for the original P^{32} and S^{35} activities by solving two simultaneous equations. For example, picking 0 day and 28 days:

total activity

$$= \text{(original } P^{32} \text{ activity)(fraction of } P^{32} \text{ activity remaining)}$$
$$+ \text{(original } S^{35} \text{ activity)(fraction of } S^{35} \text{ activity remaining)}$$

$$\text{total activity}_{0 \text{ day}} = (P^{32})(1.00) + (S^{35})(1.00) = 39{,}400 \text{ CPM}$$

$$\text{total activity}_{28 \text{ days}} = (P^{32})(0.258) + (S^{35})(0.8005) = 25{,}890 \text{ CPM}$$

Solving for S^{35} in terms of P^{32}:

$$S^{35} = 39{,}400 - P^{32}$$

Substituting into the 28-day equation:

$$P^{32}(0.258) + (39{,}400 - P^{32})(0.8005) = 25{,}890$$

Now solving for P^{32}:

$$0.258 \ P^{32} + (0.8005)(39{,}400) - 0.8005 \ P^{32} = 25{,}890$$
$$0.258 \ P^{32} + 31{,}600 - 0.8005 \ P^{32} = 25{,}890$$
$$31{,}600 - 25{,}890 = 0.8005 \ P^{32} - 0.258 \ P^{32}$$
$$5710 = 0.5425 \ P^{32}$$
$$P^{32} = \frac{5710}{0.5425}$$

$$\boxed{P^{32} = 10{,}500 \text{ CPM}}$$

$$\therefore \quad S^{35} = 39{,}400 - 10{,}500$$

$$\boxed{S^{35} = 28{,}900 \text{ CPM}}$$

In terms of μmoles:

$$P^{32} = \frac{10{,}500 \text{ CPM}}{5.2 \times 10^8 \text{ CPM}/\mu\text{mole}} = \frac{10.5 \times 10^3}{5.2 \times 10^8} \mu\text{moles}$$

$$\boxed{P^{32} = 2.02 \times 10^{-5} \ \mu\text{moles}}$$

$$S^{35} = \frac{28{,}900 \text{ CPM}}{2.9 \times 10^8 \text{ CPM}/\mu\text{mole}} = \frac{28.9 \times 10^3}{2.9 \times 10^8} \mu\text{moles}$$

$$\boxed{S^{35} = 0.997 \times 10^{-5} \ \mu\text{moles}}$$

$$\frac{P}{S} \text{ ratio} = \frac{2.02 \times 10^{-5}}{0.997 \times 10^{-5}} = 2.025$$

$$\boxed{\frac{P}{S} \text{ ratio} = 2}$$

Note: Because the beta particles of P^{32} and S^{35} have significantly different energies, the two isotopes may be determined simultaneously in a two-channel liquid scintillation counter.

D. BIOLOGICAL HALF-LIFE

Problem VII-20

An animal was given a single injection of Na^{24}-labeled NaCl. Periodically, blood samples were withdrawn and analyzed immediately for radioactivity. From the data shown below, calculate (a) the biological half-life of Na^{24} in the bloodstream, and (b) the specific activity of each sample if all samples had been counted at 24 hours.

Time After Injection (hours)	Specific Activity (CPM/ml)
1	3600
3	2760
5	2380
10	1415
16	758
24	330

Solution

(a) The decrease in the specific activity of the blood results from a combination of normal radioactive decay of the short-lived isotope and removal of the isotope from the bloodstream by excretion and by transport of the Na into various tissues. First calculate the percent of the original radioactivity that would be present if none of the isotope were removed from the bloodstream. This involves correcting only for radioactive decay. The half-life of Na^{24} is 15 hours.

The data at the top of page 298 can be compiled using the usual equation for first-order radioactive decay or the e^{-x} scales on a slide rule, as described in Problem VII-19.

We can now calculate the specific activities that the samples would have if no radioactive decay occurred; that is, if the radioactivity decreased only because the isotope was being removed from the bloodstream. For

Time After Injection (hours)	Original Na^{24} Present (percent)
0	100
1	95.48
3	82.05
5	79.4
10	63.0
16	47.7
24	33.0

example, at 5 hours the specific activity is 2380 CPM/ml. The NaCl at this point contains only 79.4% of the original Na^{24}. Therefore, if no radioactive decay occurred the blood sample would count $2380/0.794 =$ 3000 CPM/ml. Similarly, at 16 hours the "corrected" count would be $758/0.478 = 1585$ CPM/ml. These "corrected" values are a measure of the Na actually present. The biological half-life may now be calculated using the "corrected" values.

$$2.3 \log \frac{N_0}{N} = \lambda t = \frac{0.693}{t_{1/2}} t$$

where t is the time elapsed between N_0 and N, i.e., $16 - 5 = 11$ hours, and $t_{1/2}$ is the biological half-life.

$$2.3 \log \frac{3000}{1585} = \frac{0.693}{t_{1/2}} 11$$

$$2.3 \log 1.892 = \frac{(0.693)(11)}{t_{1/2}}$$

$$(2.3)(0.278) = \frac{7.63}{t_{1/2}}$$

$$t_{1/2} = \frac{7.63}{0.639} \text{ hr}$$

$$\boxed{t_{1/2} = 12 \text{ hr}}$$

(b) If all the samples were counted simultaneously, at 24 hours for example, the radioactive decay factor would be the same for each sample (0.33). The decrease in specific activity of the blood samples would then reflect solely the rate at which Na was removed from the bloodstream. First calculate the percent of the original Na injected that would be present if no radioactive decay occurred—if the decrease in radioactivity resulted solely from removal of the isotope from the bloodstream with a half-life of 12 hours.

Time After Injection (hours)	Original Na^{24} Present (percent)
0	100
1	94.4
3	84.1
5	75
10	56.2
16	39.7
24	25

We can now see that the 3600 CPM/ml in the 1-hour sample was 94.4% of 95.48% of the zero-time specific activity. Similarly, the 1415 CPM/ml in the 10-hour sample was 56.2% of 63.0% of the zero-time specific activity. The zero-time specific activity must have been:

$$\frac{3600}{0.944 \times 0.9548} = 4000 \text{ CPM/ml}$$

or

$$\frac{1415}{0.562 \times 0.63} = 4000 \text{ CPM/ml}$$

If all the samples were counted at 24 hours, their specific activities would be as follows:

Sample	Specific Activity	
1 hr	$0.944 \times 0.33 \times 4000 =$	**1245 CPM/ml**
3 hr	$0.84 \times 0.33 \times 4000 =$	**1100 CPM/ml**
5 hr	$0.75 \times 0.33 \times 4000 =$	**990 CPM/ml**
10 hr	$0.562 \times 0.33 \times 4000 =$	**742 CPM/ml**
16 hr	$0.397 \times 0.33 \times 4000 =$	**524 CPM/ml**
24 hr	$0.25 \times 0.33 \times 4000 =$	**330 CPM/ml**

PRACTICE PROBLEM SET VII

1. An isotope has a half-life of 4 years. Calculate: (a) the decay constant, λ, in terms of yr^{-1}, $days^{-1}$, hr^{-1}, min^{-1}, and sec^{-1}, and (b) the fraction of the original activity remaining after 13 months.

2. I^{131} has a half-life of 8 days. Calculate (a) the fraction of the I^{131} atoms that decays per day and per minute and (b) the specific activity of 1 g of pure I^{131} in terms of DPM/g, curies/g, and curies/g-atom.

3. A sample of organic matter from a stream near a petroleum-refining plant contains a level of C^{14} sufficient to provide a count of 10 DPM/g carbon. What fraction of the carbon in the stream is contamination from the plant? (Note: The organic matter in petroleum is millions of years old; essentially all of the C^{14} has decayed.)

4. (a) What is the theoretical maximum specific activity (mc/mmole) at which glucose-6-phosphate-P^{32} could be prepared? (b) What proportion of the molecules is actually labeled in a sample of glucose-6-phosphate-P^{32} that has a specific activity of 2×10^6 DPM/μmole?

5. Calculate the weight in grams of calcium in 1 mc of Ca^{45}. The half-life of Ca^{45} is 163 days.

6. A bottle of serine-C^{14} (uniformly labeled) contains 2.0 mc in 3.5 ml of solution. The specific activity is given as 160 mc/mmole. Calculate (a) the concentration of serine in the solution and (b) the activity of the solution in terms of CPM/ml at a counting efficiency of 68%.

7. A solution of L-lysine-C^{14} (uniformly labeled) contains 1.2 mc and 0.77 mg of L-lysine per ml. Calculate the specific activity of the lysine in terms of: (a) mc/mg, (b) mc/mmole, (c) DPM/μmole, and (d) CPM/μmole of carbon at a counting efficiency of 80%.

8. Describe the preparation of 75 ml of a 10^{-2} M solution of L-cysteine-S^{35} hydrochloride in which the amino acid has a specific activity of 3.92×10^4 DPM/μmole. Assume that you have available solid unlabeled L-cysteine hydrochloride and a stock solution of L-cysteine-S^{35} (14 mc/mmole and 1.2 mc/ml).

9. Describe the preparation of 50 ml of a 10^{-3} M solution of glucose-C^{14} in which the sugar has a specific activity of 3000 DPM/μmole. Assume that you have available a 10^{-2} M stock solution containing 0.02 μc/ml and solid glucose.

10. An aliquot of a cell-free extract of *Neurospora crassa*, containing 0.72 mg of protein, was incubated with 0-acetylhomoserine and labeled methylmercaptan ($C^{14}H_3$-SH, specific activity 2.4×10^6 CPM/μmole) in a total volume of 1.5 ml. The C^{14} was enzymatically incorporated into L-methionine at a rate of 2240 CPM/min. Calculate the rate of the reaction in terms of: (a) μmoles/min, (b) μmoles/liter-min, and (c) μmoles/mg protein-min.

11. A 10 ml suspension of Cr^{51}-labeled red blood cells, containing 3×10^8 CPM total radioactivity, was injected into a subject. After 10 minutes a small blood sample was taken and found to contain 5×10^4 CPM/ml. Calculate the total blood volume of the individual.

12. A solution containing 0.5 mg of D-mannose-C^{14} (uniformly labeled, specific activity 3.3×10^6 CPM/μmole) was added to 50 ml of a solution containing an unknown amount of unlabeled mannose. After mixing, the D-mannose was reisolated as the osazone. The osazone had a specific activity of 14,280 CPM/μmole. Calculate the concentration of unlabeled D-mannose in the original solution.

13. Fifty-six μg of Co^{60}-labeled vitamin B_{12}, containing 7.39×10^5 CPM, were added to a sample containing an unknown amount of unlabeled vitamin B_{12}. The sample was then extracted and the vitamin B_{12} purified by chromatography. The final product contained 49 μg of vitamin B_{12} and 1.58×10^5 CPM of radioactivity. Calculate the amount of unlabeled vitamin B_{12} in the sample.

14. A yeast culture was grown in a synthetic medium containing $S^{35}O_4^=$ (specific activity 4.78×10^7 CPM/μmole) as the sulfur source. After several days of growth, the cells were harvested and extracted. To 50 ml of extract, 500 mg of unlabeled reduced glutathione were added. Glutathione was then reisolated from the mixture. The reisolated compound had a specific activity of 6.97×10^6 CPM/μmole. Calculate the concentration of glutathione in the extract.

15. A mixture of fatty acids was treated with C^{14}-labeled diazomethane (specific activity 1.93×10^5 CPM/μmole) to produce the methyl-C^{14}-labeled esters of each acid present. Unlabeled methyl stearate (2 mmoles) was then added to the mixture. A small amount of methyl stearate was reisolated from the mixture and found to have a specific activity of 4.87×10^3 CPM/μmole. Calculate the amount of stearic acid in the mixture.

16. An isotope with a half-life of 10 hours was injected into the bloodstream of an animal. Blood samples were taken periodically and counted immediately. The specific activities of the samples are shown below. From the data, calculate the biological half-life of the isotope in the bloodstream.

Sample Time (hours)	Specific Activity (CPM/ml)
2	9400
4	5730
6	3960
10	1890
18	431

17. A sample of C^{14}- and P^{32}-labeled AMP was prepared with a specific activity of 9500 CPM/μmole (75% of the activity results from decay of P^{32}, 25% from decay of C^{14}). Calculate the specific activity of the sample after: (a) 5 days, (b) 10 days, and (c) 25 days.

MISCELLANEOUS
CALCULATIONS

VII

A. IONIC STRENGTH

Ionic strength (I) is a measure of the electrical environment of ions in a solution. I is given by the relationship:

$$I = \tfrac{1}{2} \sum M_i Z_i^2$$

where:

M = the molarity of the ion

Z_i = the total charge of the ion (regardless of sign)

$\sum$ = a symbol meaning "the sum of"

Problem VIII-1

Calculate the ionic strengths of the following solutions: (a) 0.1 M NaCl, (b) 0.1 M K_2SO_4, (c) 0.3 M K_3PO_4, and (d) 0.2 M $Fe_2(SO_4)_3$.

(a)
$$I = \tfrac{1}{2} \sum M_i Z_i^2 = \tfrac{1}{2}[M_{Na^+} Z_{Na^+}^2 + M_{Cl^-} Z_{Cl^-}^2]$$

$$I = \frac{(0.1)(1)^2 + (0.1)(1)^2}{2} = \frac{0.2}{2}$$

The 0.1 M NaCl yields 0.1 M Na^+ and 0.1 M Cl^-.

$$\boxed{I = 0.1}$$

(b)
$$I = \tfrac{1}{2} \sum M_i Z_i^2 = \tfrac{1}{2}[M_{K^+} Z_{K^+}^2 + M_{SO_4^{-2}} Z_{SO_4^{-2}}^2]$$

The 0.1 M K_2SO_4 yields 0.2 M K^+ and 0.1 M SO_4^{-2}.

$$I = \frac{(0.2)(1)^2 + (0.1)(-2)^2}{2} = \frac{(0.2) + (0.4)}{2} = \frac{(0.6)}{2}$$

$$\boxed{I = 0.3}$$

(c) $I = \frac{1}{2}\sum M_i Z_i^2 = \frac{1}{2}[M_{K^+} Z_{K^+}^2 + M_{PO_4^{-3}} Z_{PO_4^{-3}}^2]$

The 0.3 M K_3PO_4 yields 0.9 M K^+ and 0.3 M PO_4^{-3}.

$$I = \frac{(0.9)(1)^2 + (0.3)(-3)^2}{2} = \frac{(0.9) + (2.7)}{2} = \frac{3.6}{2}$$

$$\boxed{I = 1.8}$$

(d) $I = \frac{1}{2}\sum M_i Z_i^2 = \frac{1}{2}[M_{Fe^{+3}} Z_{Fe^{+3}}^2 + M_{SO_4^{-2}} Z_{SO_4^{-2}}^2]$

The 0.02 M $Fe_2(SO_4)_3$ yields 0.04 M Fe^{+3} and 0.06 M SO_4^{-2}.

$$I = \frac{(0.04)(3)^2 + (0.06)(-2)^2}{2} = \frac{(0.04)(9) + (0.06)(4)}{2}$$

$$I = \frac{(0.36) + (0.24)}{2} = \frac{0.60}{2}$$

$$\boxed{I = 0.30}$$

Problem VIII-2

Calculate the ionic strength of a 0.1 M solution of butyric acid; $K_a = 1.5 \times 10^{-5}$.

Solution

Butyric acid is only partially ionized. The undissociated molecules have no effect on the ionic strength of the solution. First calculate M_{H^+} and $M_{butyrate^{-1}}$ as shown in Chapter II.

$$K_a = \frac{[H^+][butyrate^{-1}]}{[butyric\ acid]} = \frac{(X)(X)}{0.1 - X} = \frac{X^2}{0.1} = 1.5 \times 10^{-5}$$

$$X^2 = 1.5 \times 10^{-6}$$

$$X = \sqrt{150 \times 10^{-8}} = 12.25 \times 10^{-4}$$

$$[\text{H}^+] = 1.225 \times 10^{-3} \, M, \quad [\text{butyrate}^{-1}] = 1.225 \times 10^{-3} \, M$$

$$I = \tfrac{1}{2} \sum M_i Z_i^2 = \tfrac{1}{2}[M_{\text{H}^+}Z_{\text{H}^+}^2 + M_{\text{butyrate}^{-1}}Z_{\text{butyrate}^{-1}}^2]$$

$$I = \frac{(1.225 \times 10^{-3})(1)^2 + (1.225 \times 10^{-3})(-1)^2}{2}$$

$$\boxed{I = 1.225 \times 10^{-3}}$$

Problem VIII-3

Calculate the ionic strength of a 0.06 M phosphate buffer, pH 7.5 ($pK_{a_2} = 7.2$).

Solution

First calculate $M_{\text{HPO}_4^=}$ and $M_{\text{H}_2\text{PO}_4^-}$.

$$\text{pH} = pK_{a_2} + \log \frac{\text{conjugate base}}{\text{conjugate acid}}$$

$$7.5 = 7.2 + \log \frac{[\text{HPO}_4^=]}{[\text{H}_2\text{PO}_4^-]}$$

$$0.3 = \log \frac{[\text{HPO}_4^=]}{[\text{H}_2\text{PO}_4^-]}$$

$$\frac{[\text{HPO}_4^=]}{[\text{H}_2\text{PO}_4^-]} = \text{antilog of } 0.3 = 2$$

$$\therefore \quad \tfrac{2}{3} \times 0.06 \, M = 0.04 \, M \, \text{HPO}_4^=$$

and

$$\tfrac{1}{3} \times 0.06 \, M = 0.02 \, M \, \text{H}_2\text{PO}_4^-$$

$$I = \tfrac{1}{2} \sum M_i Z_i^2 = \tfrac{1}{2}[M_{\text{HPO}_4^=} Z_{\text{HPO}_4^=}^2 + M_{\text{H}_2\text{PO}_4^-} Z_{\text{H}_2\text{PO}_4^-}^2]$$

$$I = \frac{(0.04)(-2)^2 + (0.02)(-1)^2}{2} = \frac{(0.04)(4) + (0.02)(1)}{2}$$

$$I = \frac{(0.16) + (0.02)}{2} = \frac{(0.18)}{2}$$

$$\boxed{I = 0.09}$$

Problem VIII-4

Calculate the ionic strength of a 0.03 M lysine buffer, pH 2.18.

Solution

The ionic species present in a lysine buffer at pK_{a_1} are AA^{+2} and AA^{+1} as shown below.

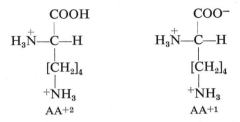

Because the pH exactly equals pK_{a_1}, $[AA^{+2}] = [AA^{+1}] = 0.015\ M$. The ionic strength of the buffer can be calculated as usual, taking into account the *net* charge on each species.

$$I = \tfrac{1}{2} \sum M_i Z_i^2 = \tfrac{1}{2}[M_{AA^{+2}}Z_{AA^{+2}}^2 + M_{AA^{+1}}Z_{AA^{+1}}^2]$$

$$I = \frac{(0.015)(2)^2 + (0.015)(1)^2}{2} = \frac{(0.015)(4) + (0.015)(1)}{2}$$

$$I = \frac{(0.015)(5)}{2} = \frac{0.075}{2}$$

$$\boxed{I = 0.0325}$$

B. OPTICAL ROTATION

Optical rotation refers to the ability of certain compounds (solid or in solution) to rotate the plane of polarized light. Such "optically active" compounds contain at least one asymmetric center and no plane of symmetry. Specific rotation $[\alpha]$ is given by the relationship:

$$[\alpha]_\lambda^T = \frac{A° \times 100}{l \times C}$$

where:

$A°$ = the observed rotation in degrees

l = the light path length (through the solution) in decimeters (dm)

C = the concentration of the solution % w/v (g/100 ml)

T = the temperature of the solution (generally 20°–25° C)

λ = the wavelength of the polarized light (generally the D-line of sodium, 5893 Å)

The relationship is also given as:

$$[\alpha]_\lambda^T = \frac{A^\circ}{l \times C}$$

where now C = concentration in g/ml. The "molar rotation" (M) is $[\alpha] \times$ MW.

Optical rotation is measured with an instrument called a polarimeter. The essential parts of a polarimeter are shown in Figure VIII-1.

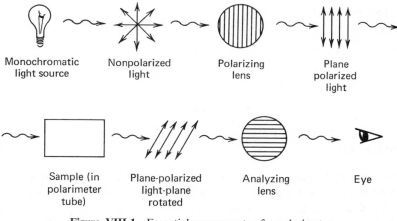

Figure VIII-1   Essential components of a polarimeter

The polarizing and analyzing lenses both transmit light that is plane-polarized. In the absence of an optically active sample, the analyzing lens can be adjusted so that the light intensity as seen by the viewer is minimal. (This is accomplished by rotating the analyzing lens until its transmission plane is perpendicular to the transmission plane of the polarizing lens.) An optically active sample rotates the plane of the polarized light. The analyzing lens then has to be rotated to a new position in order to minimize the light intensity. The angle through which the analyzing lens is rotated is the observed rotation (in degrees) from which $[\alpha]$ may be calculated. If the analyzing lens is rotated clockwise, the substance is said to be dextrorotary (d or $+$). If the analyzing lens is rotated counterclockwise, the substance is said to be levorotary (l or $-$). The symbols d or l are not to be confused with the symbols D or L.

Problem VIII-5

A solution of L-leucine (3.0 g/50 ml of 6 N HCl) had an observed rotation of $+1.81°$ in a 20 cm polarimeter tube. Calculate (a) the specific rotation $[\alpha]$ and (b) the molar rotation (M) of L-leucine (in 6 N HCl).

Solution

(a) $[\alpha] = \dfrac{A°}{l_{dm} \times C_{g/ml}}$ $l = 20 \text{ cm} = 2 \text{ dm}$ $C = 3 \text{ g}/50 \text{ ml} = 0.06 \text{ g/ml}$

$[\alpha] = \dfrac{+1.81}{2 \times 0.06}$

$$\boxed{[\alpha] = +15.1°}$$

(b) $(M) = [\alpha] \times MW$

$(M) = (+15.1°)(131.2)$

$$\boxed{(M) = +1980°}$$

Problem VIII-6

A solution of L-arabinose (containing an equilibrium mixture of α and β forms) has an observed rotation of $+23.7°$ in a 10 cm polarimeter tube at 25° C. Calculate the concentration of L-arabinose in the solution. The $[\alpha]_D^{25}$ for an equilibrium mixture of α and β L-arabinose is $+105°$.

Solution

$$[\alpha]_D^{25} = \frac{A°}{l \times C} \qquad (+105°) = \frac{+23.7°}{(1)(C)} \qquad C = \frac{23.7}{105} \text{ g/ml}$$

$$\boxed{C = 0.225 \text{ g/ml}}$$

Problem VIII-7

An equilibrium mixture of α- and β-D-glucose has an $[\alpha]_D^{25}$ of $+52.7°$. Pure α-D-glucose has an $[\alpha]_D^{25}$ of $+112°$. Pure β-D-glucose has an $[\alpha]_D^{25}$ of $+18.7°$. Calculate the proportions of α- and β-D-glucose in the equilibrium mixture.

Solution

The contribution of each anomer towards the total $[\alpha]_D^{25}$ is directly proportional to the concentrations of each anomer present. Let:

$$X = \% \beta$$
$$\therefore \quad (100 - X) = \% \alpha$$
$$\therefore \quad (+18.7)(X) + (+112)(100 - X) = (100)(+52.7)$$
$$18.7X + 11,200 - 112X = 5270$$
$$5930 = 93.3X$$

$$X = \frac{5930}{93.3} \times 100 = \% \beta$$

$$\boxed{\beta = 63.5\%}$$

$$\therefore \quad \alpha = 100 - 63.5\%$$

$$\boxed{\alpha = 36.5\%}$$

Problem VIII-8

Three g of a polysaccharide containing only D-mannose and D-glucose were acid hydrolyzed. The hydrolysate was diluted to 100 ml. The observed rotation of the solution was $+90.7°$ in a 10 cm polarimeter tube. Calculate the ratio of D-mannose/D-glucose in the polysaccharide. The specific rotations of α/β-D-glucose and α/β-D-mannose are $+52.7°$ and $+14.5°$, respectively.

Solution

After acid hydrolysis, the two sugars are present as the equilibrium mixture of their α and β forms. Furthermore, the total weight of monosaccharides after hydrolysis is 3.33 g as a result of adding 1 mole of water (18 g) per mole of monosaccharide (180 g).

Let

$$X = g \text{ D-glucose present}$$
$$\therefore \quad (3.33 - X) = g \text{ D-mannose present}$$

$$[\alpha] = \frac{A°}{l \times C}$$

$$A^\circ_{\text{glucose}} = [\alpha](l)(C) = (52.7)(1)(X) = 52.7X$$

$$A^\circ_{\text{mannose}} = [\alpha](l)(C) = (14.5)(1)(3.33 - X) = 48.3 - 14.5X$$

$$A^\circ_{\text{glucose}} + A^\circ_{\text{mannose}} = A^\circ_{\text{total}}$$

$$(52.7X) + (48.3 - 14.5\,X) = 90.7$$

$$38.2\,X = 42.4$$

$$X = g\ \text{D-glucose} = \frac{42.4}{38.2}$$

$$\boxed{\text{D-glucose} = 1.11\ \text{g}}$$

$$\therefore\quad \text{D-mannose} = 3.33 - 1.11\ \text{g}$$

$$\boxed{\text{D-mannose} = 2.22\ \text{g}}$$

Because mannose and glucose have the same molecular weights, the molar ratio of the two sugars is the same as the weight ratio.

$$\boxed{\text{D-mannose/D-glucose} = 2}$$

Problem VIII-9

Two distinct crystalline compounds were isolated from an acid hydrolysate of a polysaccharide. Qualitative tests suggested that one of the compounds (compound A) was a ketose while the other (compound B) was an aldose. A freshly prepared solution of compound A had a specific rotation of $-134°$ which rapidly changed to $-92°$ upon standing. A freshly prepared solution of compound B had a specific rotation of $+151°$ which rapidly changed to $+80.2°$ upon standing. What are the possible identities of the two sugars?

Solution (Consult Table AVI-1.)

> **Compound A may be D-fructose (the β form was isolated).**
> **Compound B may be D-galactose (the α form was isolated).**

C. EMPIRICAL FORMULA AND MOLECULAR WEIGHT CALCULATIONS

Problem VIII-10

A compound was found to contain 40.4% carbon, 7.87% hydrogen, and 15.72% nitrogen. The only other element present was oxygen. Calculate

(a) the simplest empirical formula and (b) the minimum molecular weight of the compound.

Solution

The simplest empirical formula of a compound may be calculated from elemental composition data. The calculations are based on the fact that a molecule of the compound must contain at least 1 atom of every element shown to be present. Once the simplest empirical formula is known, the minimal molecular weight can be calculated as the sum of the atomic weights of all elements present.

(a)   First, divide the percent composition values by the atomic weights of each element present. This yields the number of gram-atoms of each element per 100 g of compound.

$$C = \frac{40.4}{12} \qquad H = \frac{7.87}{1} \qquad N = \frac{15.72}{14} \qquad O \text{ (by difference)} = \frac{36.01}{16}$$

$$C = 3.37 \qquad H = 7.87 \qquad N = 1.15 \qquad\qquad\qquad O = 2.25$$

Next, divide each of the relative molar amounts shown above by the smallest relative value.

$$C = \frac{3.37}{1.15} \qquad H = \frac{7.87}{1.15} \qquad N = \frac{1.15}{1.15} \qquad O = \frac{2.25}{1.15}$$

$$C = 2.93 \qquad H = 6.83 \qquad N = 1 \qquad O = 1.95$$

The simplest empirical formula is probably

$$\boxed{C_3H_7NO_2}$$

(b)   The minimum molecular weight is:

$$\text{No. C atoms} \times 12 \text{ g/g-atom} = 3 \times 12 = 36$$
$$+ \text{No. H atoms} \times 1 \ \text{ g/g-atom} = 7 \times 1 \ = 7$$
$$+ \text{No. N atoms} \times 14 \text{ g/g-atom} = 1 \times 14 = 14$$
$$+ \text{No. O atoms} \times 16 \text{ g/g-atom} = 2 \times 16 = 32$$

$$\boxed{\textbf{minimum MW} = \textbf{89}}$$

The actual molecular weight could be any integral multiple of 89.

Problem VIII-11

Hemoglobin contains 0.34% iron. Calculate the minimum molecular weight of hemoglobin.

Solution

At least 1 atom of Fe must be present per molecule of hemoglobin. A gram-atom of Fe weighs 55.85 g. Therefore, the minimum molecular weight is that amount of protein in which 55.85 g represent 0.34% of the (molecular) weight.

$$0.34\% \times \text{MW}_{min} = 55.85$$

$$3.4 \times 10^{-3}\,\text{MW}_{min} = 55.85$$

$$\text{MW}_{min} = \frac{55.85}{3.4 \times 10^{-3}}$$

$$\boxed{\text{MW}_{min} = 16{,}420}$$

Problem VIII-12

A protein was found to contain only 0.58% tryptophan. Calculate the minimum molecular weight of the protein.

Solution

At least 1 mole of tryptophan must be present per mole of protein. Therefore, the minimum molecular weight is that weight of protein in which 1 mole of tryptophan (204 g) represents 0.58% of the (molecular) weight.

$$0.58\% \times \text{MW}_{min} = 204$$

$$5.8 \times 10^{-3}\,\text{MW}_{min} = 204$$

$$\text{MW}_{min} = \frac{204}{5.8 \times 10^{-3}}$$

$$\boxed{\text{MW}_{min} = 35{,}200}$$

Of course, if 2 moles of tryptophan were actually present per mole of protein, the true molecular weight would be 70,400.

Problem VIII-13

A 12 mg sample of an unidentified organic compound was dissolved in 0.8 g of pure camphor. The freezing point of the solution was 175.9° C. The freezing point of pure camphor is 178.4° C. The molal freezing point depression constant for camphor is 37.7°. Calculate the apparent molecular weight of the compound.

Solution

The degree to which the freezing point of any solution is depressed (compared to that of the pure solvent) is directly proportional to the molality of the solute present. One mole of solute dissolved in 1000 g of solvent (to produce a 1 molal solution) depresses the freezing point by the molal freezing point depression constant (K_D).

$$\frac{MW_g \text{ solute}/1000 \text{ g solvent}}{K_D} = \frac{wt_g \text{ solute}/wt_g \text{ solvent}}{\Delta FP}$$

$$\frac{MW/1000}{37.7} = \frac{0.012/0.8}{178.4 - 175.9} = \frac{15/1000}{2.5}$$

$$\frac{MW}{37.7} = \frac{15}{2.5}$$

$$MW = \frac{(15)(37.7)}{2.5}$$

$$\boxed{MW_{solute} = 226}$$

The freezing point depression is a colligative property—it depends on the number of particles of solute present. If the solute dissociates in the solvent, the calculation yields a low value for the molecular weight.

Problem VIII-14

An aqueous solution containing 0.5 g of a neutral polysaccharide per 100 ml exerted an osmotic pressure of 25.82 mm Hg at 25° C. Calculate the molecular weight of the polysaccharide.

Solution

When a solution is separated from the pure solvent by a membrane that is permeable to the solvent but not to the solute, molecules of solvent migrate through the membrane into the solution compartment. The pressure that must be exerted to prevent the passage of solvent molecules is known as the osmotic pressure (π). The osmotic pressure of a solution depends on the concentration of solute and the temperature of the solution. The relationship (shown below) is identical to the P.V.T. relationship of gases.

$$\boxed{\pi V = nRT}$$

where:

$\pi =$ osmotic pressure in atm

$V =$ volume of the solution in liters

$n =$ number of moles of solute.

$R =$ gas constant, 0.0821 liter-atm/mole-° K

$T =$ the absolute temperature.

$$\pi = \frac{n}{V} RT$$

$$\boxed{\therefore \quad \pi = MRT}$$

where $M =$ molarity of the solution.

The molecular weight of the solute may be determined from measurement of π.

$$\pi V = \frac{\text{wt}_g}{MW} RT$$

$$MW = \frac{\text{wt}_g}{V} \frac{RT}{\pi}$$

$$\boxed{MW = \frac{CRT}{\pi}}$$

where $C =$ concentration in g/liter.

Thus, for the polysaccharide described above:

$$MW = \frac{CRT}{\pi} = \frac{(5)(0.0821)(298)}{25.82/760 \text{ atm}}$$

$$MW = \frac{(5)(0.0821)(298)(760)}{(25.82)}$$

$$\boxed{MW = 3600}$$

PRACTICE PROBLEM SET VIII

1. Calculate the ionic strengths of the following solutions: (a) 0.25 M KBr, (b) 0.36 M $ZnCl_2$, (c) 0.15 M $AlCl_3$, (d) 0.4 M Na_2SO_3, (e) 0.01 M H_2SO_4, (f) 0.08 M $Cr_2(SO_4)_3$, (g) 0.016 M acetic acid, (h) 0.30 M citrate buffer, pH 5.65, and (i) 0.025 M glutamate buffer, pH 5.1.

2. A solution of D-histidine (4 g/100 ml of 1 M HCl) had an observed rotation of $-0.41°$ in a 20 cm polarimeter tube. Calculate (a) the specific rotation $[\alpha]$ and (b) the molar rotation (M) of D-histidine in 1 M HCl.

3. A solution of L-ribulose (containing an equilibrium mixture of α and β forms) has an observed rotation of $-3.75°$ in a 10 cm polarimeter tube. Calculate the concentration of L-ribulose in the solution. The $[\alpha]_D^{25}$ for an equilibrium mixture of α- and β-L-ribulose is $-16.6°$.

4. An equilibrium mixture of α- and β-D-mannose has an $[\alpha]_D^{25}$ of $+14.5°$. Pure α-D-mannose has an $[\alpha]_D^{25}$ of $+29.3°$. Pure β-D-mannose has an $[\alpha]_D^{25}$ of $-16.3°$. Calculate the proportions of α- and β-D-mannose in the equilibrium mixture.

5. A compound was found to contain 55% carbon, 9.93% hydrogen, and 10.7% nitrogen. The only other element present was oxygen. Calculate (a) the simplest empirical formula and (b) the minimum molecular weight of the compound.

6. Egg albumin contains 1.62% sulfur in addition to 52.75% carbon, 7.10% hydrogen, and 15.51% nitrogen. Calculate the minimum molecular weight of egg albumin.

7. An enzyme was found to contain 0.8% cysteine. Calculate the minimum molecular weight of the enzyme.

8. A 15.7 mg sample of an unidentified organic compound was dissolved in 0.3 g of pure phenol. The freezing point of the solution was 39.6° C. The freezing point of pure phenol is 42.0° C. The molal freezing point depression constant for phenol is 7.27°. Calculate the apparent molecular weight of the unidentified compound.

9. An aqueous solution containing 0.3 g of isoelectric polyalanine per 100 ml exerted an osmotic pressure of 7.85 mm Hg at 25° C. Calculate the apparent molecular weight of the polypeptide.

Appendix: ACIDS AND BASES

I

A. ABBREVIATIONS AND DEFINITIONS

MW = molecular weight	1 g-mole (or 1 mole) = MW in grams
EW = equivalent weight	1 g-equivalent (or 1 equivalent) = EW in grams
AW = atomic weight	1 g-atom = AW in grams
	1 g-ion = AW of ion in grams

M = "molarity" = number of moles of solute per liter of solution

N = "normality" = number of equivalents of solute per liter of solution

EW = weight of substance that contains 1 g-atom (1 "mole") of replaceable hydrogen, or 1 g-ion ("mole") of replaceable hydroxyl

a = activity (effective or apparent concentration)

γ = activity coefficient (fraction of actual concentration that is active or effective)

a_{H+} = activity of hydrogen ion

a_{OH-} = activity of hydroxyl ion

$[H^+]$ = concentration of hydrogen ion (molarity)

$[OH^-]$ = concentration of hydroxyl ion (molarity)

Avogadro's number = number of molecules per g-mole

= number of atoms per g-atom

= number of ions per g-ion

In general practice, 1 Avogadro's number worth of particles (i.e., 1 g-mole or 1 g-atom or 1 g-ion) is frequently called a "mole" regardless of whether the substance is ionic, monoatomic, or molecular in nature. For example, 35.5 g of Cl^- ions may be called a "mole" instead of a "gram-ion."

density = weight per unit volume

specific gravity = density relative to water; since the density of water is 1 g/ml, specific gravity is numerically equal to density

B. USEFUL FORMULAE AND RELATIONSHIPS

The formulae listed below are useful in solving many problems. Note that many of the formulae and relationships may be obtained simply by rearranging the above definitions.

$$\text{molarity of a solution} = \frac{\text{total number of moles of solute}}{\text{total volume of solution in liters}}$$

$$M = \frac{\text{number of moles}}{\text{liters}} \qquad \text{liters} \times M = \text{number of moles}$$

$$\text{normality of a solution} = \frac{\text{total number of equivalents of solute}}{\text{total volume of solution in liters}}$$

$$N = \frac{\text{number of equivalents}}{\text{liters}} \qquad \text{liters} \times N = \text{number of equivalents}$$

$$EW = \frac{\text{MW in grams}}{\text{number of replaceable hydrogens or hydroxyls per molecule}}$$

$$N = M \times \text{number of replaceable hydrogens or hydroxyls}$$

$$\text{number of moles} = \frac{\text{weight in grams of compound}}{\text{MW}}$$

$$\text{number of equivalents} = \frac{\text{weight in grams of compound}}{\text{EW}}$$

wt_g (of a solution or solid) = $volume_{ml} \times density_{g/ml}$

wt_g (of a pure substance in dilute solution) = volume $\times$ density $\times$ % (as a decimal of pure substance by weight)

In dilutions of solutions or preparation of solutions from solids:

number of moles of substance in one form = number of moles of same substance in other form

$$\therefore \quad \text{liters}_a \times M_a = \text{liters}_b \times M_b$$

$$\text{liters} \times M = \frac{\text{wt}_g}{\text{MW}}$$

In neutralization of acids with bases and vice versa:

number of equivalents of H^+ = number of equivalents of OH^-

$$\text{liters} \times N_{H^+} = \text{liters} \times N_{OH^-}$$

$$\text{liters} \times N_{H^+} = \frac{\text{wt}_g \text{ of solid base}}{\text{EW}}$$

$$\text{liters} \times N_{OH^-} = \frac{\text{wt}_g \text{ of solid acid}}{\text{EW}}$$

C. BRONSTED CONCEPT OF CONJUGATE ACID-CONJUGATE BASE PAIRS

The most useful way of discussing acids and bases in general biochemistry is to define an "acid" as a substance that donates protons (hydrogen ions) and a "base" as a substance that accepts protons. This concept is generally referred to as the Bronsted concept of acids and bases. When a Bronsted acid loses a proton, a Bronsted base is produced. The original acid and resulting base are referred to as a conjugate acid-conjugate base pair. The substance that accepts the proton is a different Bronsted base; by accepting the proton, another Bronsted acid is produced. Thus, in every ionization of an acid or base, two conjugate acid-conjugate base pairs are involved.

$$\text{HA} \quad + \quad \text{B} \quad \rightleftharpoons \quad \text{A}^- \quad + \quad \text{HB}$$

[conjugate acid]$_1$ [conjugate base]$_2$ [conjugate base]$_1$ [conjugate acid]$_2$

It is sometimes convenient (especially in inorganic chemistry) to define a base as a substance that yields OH^- ions in solution.

D. IONIZATION OF WATER

The ionization of water can be considered in two ways: (1) as a simple dissociation to yield H^+ and OH^- ions and (2) in terms of Bronsted conjugate acid-conjugate base pairs. In either case, it is obvious that water is amphoteric—it yields both H^+ and OH^- ions; it can both donate and accept protons.

The ionization of water can be described by a "dissociation constant," K_d, an "ionization constant," K_i, and a specific constant for water, K_w, as shown below.

Simple Dissociation	Conjugate Acid-Conjugate Base
$HOH \rightleftharpoons H^+ + OH^-$	$HOH + HOH \rightleftharpoons H_3O^+ + OH^-$
$K_d = \dfrac{[H^+][OH^-]}{[HOH]}$	$K_i = \dfrac{[H_3O^+][OH^-]}{[HOH]^2}$

Note that water produces two conjugate acid-conjugate base pairs: HOH/OH^- and H_3O^+/HOH.

For every mole of H^+ (or H_3O^+), 1 mole of OH^- is produced. In pure water $[H^+] = 10^{-7}\ M$. $\therefore$ $[OH^-] = 10^{-7}\ M$. The molarity of HOH can be calculated as follows:

$$M = \frac{\text{number of moles } H_2O}{\text{liter}}$$

$$\text{number of moles} = \frac{wt_g}{MW}$$

A liter of water weighs 1000 g. The MW of H_2O is 18.

$$\therefore \quad M = \frac{1000/18\text{ g}}{1\text{ liter}} = 55.6$$

The M of H_2O is actually 55.6 M (original concentration), minus $10^{-7}\ M$ (the amount that ionized). However, this amount is so close to 55.6 that we may neglect the 10^{-7} in the denominator. We can now substitute the above values into the K_d and K_i expressions.

$K_d = \dfrac{(10^{-7})(10^{-7})}{55.6}$	$K_i = \dfrac{(10^{-7})(10^{-7})}{(55.6)^2}$
$= \dfrac{10^{-14}}{55.6}$	$= \dfrac{10^{-14}}{3.09 \times 10^3}$
$\boxed{K_d = 1.8 \times 10^{-16}}$	$\boxed{K_i = 3.24 \times 10^{-18}}$

The molarity of H_2O is essentially constant in the dilute solutions considered in most biochemical problems. Consequently, we can define a new constant for the dissociation or ionization of water, K_w, which combines the two constants (K_d and M_{H_2O} or K_i and $(M_{H_2O})^2$).

$$K_w = K_d \times [H_2O]$$

$$K_w = (1.8 \times 10^{-16})(55.6)$$

$$\boxed{K_w = 1 \times 10^{-14}}$$

$$\boxed{[H^+][OH^-] = 10^{-14}}$$

$$K_w = K_i \times [H_2O]^2$$

$$K_w = (3.24 \times 10^{-18})(55.6)^2$$

$$K_w = (3.24 \times 10^{-18})(3.09 \times 10^3)$$

$$\boxed{K_w = 1 \times 10^{-14}}$$

$$\boxed{[H_3O^+][OH^-] = 1 \times 10^{-14}}$$

E. IONIZATION OF ACIDS

In an aqueous solution an acid ionizes as follows:

$$HA \quad + \quad H_2O \quad \rightleftharpoons \quad H_3O^+ \quad + \quad A^-$$

[conjugate acid]$_1$ [conjugate base]$_2$ [conjugate acid]$_2$ [conjugate base]$_1$

The proton released from HA is accepted by water to form the hydronium ion H_3O^+. The hydronium ion is the actual species of "hydrogen ion" present in an aqueous solution. "Strong acids" are those that ionize almost completely; "weak acids" are those that ionize only partially.

The ionization of a weak acid can be represented by an equilibrium or ionization constant, K_{eq} or K_i.

$$K_i = \frac{(a_{H_3O^+})(a_{A^-})}{(a_{HA})(a_{H_2O})} = \frac{\gamma_{H_3O^+}[H_3O^+]\,\gamma_{A^-}[A^-]}{\gamma_{HA}[HA]\,\gamma_{H_2O}[H_2O]}$$

The a represents the "activity" or *effective* or *apparent* concentration. The γ represents the fraction of the actual concentration that is effective or active. For dilute aqueous solutions, we may assume that $\gamma = 1$; hence, concentrations may be substituted for activities.

$$\therefore \quad K_i = \frac{[H_3O^+][A^-]}{[HA][H_2O]}$$

Because $[H_2O]$ is itself a constant, we can define a new constant, K_a, that combines K_i and $[H_2O]$.

$$K_i[H_2O] = K_a = \frac{[H_3O^+][A^-]}{[HA]}$$

Because $[H_3O^+]$ is the same as the "hydrogen ion concentration," $[H^+]$, the K_a expression is usually written as shown below.

$$K_a = \frac{[H^+][A^-]}{[HA]}$$

It is not surprising that the above K_a expression is identical to the K_d expression that we would obtain if we assume that the weak acid dissociates directly to yield H^+ and A^-.

$$HA \rightleftharpoons H^+ + A^-$$

$$K_d = \frac{[H^+][A^-]}{[HA]}$$

F. IONIZATION OF BASES

In an aqueous solution, inorganic bases yield OH^- ions directly by dissociation.

$$KOH \rightarrow K^+ + OH^-$$

Organic bases such as amines $R\text{-}NH_2$, contain no OH to dissociate. However, if we assume that the $R\text{-}NH_2$ reacts with H_2O to form the OH-containing substance "$R\text{-}NH_3OH$," then we can consider the "dissociation" of organic bases to yield OH^- ions directly just as we do for inorganic bases. In fact, this is frequently done when we consider aqueous ammonia, NH_3; we assume that the dissociable substance present is "NH_4OH."

$$\text{"}R\text{-}NH_3OH\text{"} \rightleftharpoons R\text{-}NH_3^+ + OH^-$$

$$\text{"}NH_4OH\text{"} \rightleftharpoons NH_4^+ + OH^-$$

We should bear in mind that "$R\text{-}NH_3OH$" refers to the sum of $R\text{-}NH_2$ plus any small amount of $R\text{-}NH_3OH$ that might exist.

It usually makes little difference whether we consider the ionization of an acid as a simple dissociation or as the true ionization involving water as a conjugate base. However, in dealing with organic bases, it is far more fruitful to consider the ionization as it actually occurs.

$$R\text{-}NH_2 \quad + \quad HOH \quad \rightleftharpoons \quad R\text{-}NH_3^+ \quad + \quad OH^-$$
[conjugate base]$_1$ [conjugate acid]$_2$ [conjugate acid]$_1$ [conjugate base]$_2$

The two conjugate acid-conjugate base pairs involved are $R\text{-}NH_3^+/$ $R\text{-}NH_2$ and HOH/OH^-. The ionization can be described by an ionization constant, K_i

$$K_i = \frac{[R\text{-}NH_3^+][OH^-]}{[R\text{-}NH_2][H_2O]}$$

Again, we can define a new constant, K_b, that combines the two constants, K_i and $[H_2O]$.

$$K_i[H_2O] = K_b = \frac{[R\text{-}NH_3^+][OH^-]}{[R\text{-}NH_2]}$$

The K_b expression is exactly the same as that which we obtain if we assume that $R\text{-}NH_2$ is actually $R\text{-}NH_3OH$ which dissociates directly to yield $R\text{-}NH_3^+$ and OH^-.

G. pH AND pOH

The pH is a shorthand way of designating the hydrogen ion activity of a solution. By definition, pH is "the negative logarithm of the hydrogen ion activity." The pOH is "the negative logarithm of the hydroxyl ion activity."

$$pH = -\log a_{H^+} = \log \frac{1}{a_{H^+}} \qquad pOH = -\log a_{OH^-} = \log \frac{1}{a_{OH}}$$

$$= -\log \gamma_{H^+}[H^+] \qquad\qquad = -\log \gamma_{OH^-}[OH^-]$$

$$= \log \frac{1}{\gamma_{H^+}[H^+]} \qquad\qquad = \log \frac{1}{\gamma_{OH^-}[OH^-]}$$

In dilute solutions of acids and bases and in pure water, the activities of H^+ and OH^- may be considered to be the same as their concentrations.

$$pH = -\log [H^+] = \log \frac{1}{[H^+]} \qquad pOH = -\log [OH^-] = \log \frac{1}{[OH^-]}$$

In all aqueous solutions the equilibrium for the ionization of water must be satisfied, i.e., $[H^+][OH^-] = K_w = 10^{-14}$. Thus, if $[H^+]$ is known, we can easily calculate $[OH^-]$. Furthermore, we can derive the following relationship between pH and pOH:

$$[H^+][OH^-] = K_w$$

Taking logarithms:

$$\log [H^+] + \log [OH^-] = \log K_w$$
$$-\log [H^+] - \log [OH^-] = -\log K_w$$
$$-\log [H^+] = pH \qquad -\log [OH^-] = pOH \qquad -\log K_w = pK_w$$
$$\therefore \quad pH + pOH = pK_w$$
$$K_w = 10^{-14}$$
$$\therefore \quad pK_w = -\log 10^{-14} = +14$$

$$\boxed{\therefore \quad \textbf{pH} + \textbf{pOH} = \textbf{14}}$$

Thus, if any one of the values $[H^+]$, $[OH^-]$, pH, or pOH is known, the other three can be calculated easily. At concentrations of H^+ and OH^- greater than 0.1 M, the activity coefficients begin to decrease from unity.

TABLE AI-1. RELATIONSHIP BETWEEN MOLARITIES
OF H^+ AND OH^- AND pH AND pOH

$[H^+]$	pH	$[OH^-]$	pOH
10^{-13}	13	10^{-1}	1
10^{-12}	12	10^{-2}	2
10^{-11}	11	10^{-3}	3
10^{-10}	10	10^{-4}	4
10^{-9}	9	10^{-5}	5
10^{-8}	8	10^{-6}	6
10^{-7}	7	10^{-7}	7
10^{-6}	6	10^{-8}	8
10^{-5}	5	10^{-9}	9
10^{-4}	4	10^{-10}	10
10^{-3}	3	10^{-11}	11
10^{-2}	2	10^{-12}	12
10^{-1}	1	10^{-13}	13

H. RELATIONSHIP BETWEEN K_a AND K_b AND BETWEEN pK_a AND pK_b FOR WEAK ACIDS AND BASES

(a)   When a weak acid, HA, is dissolved in water, it ionizes as shown previously to form H_3O^+ and the corresponding conjugate base, A^-. A K_a expression can be written for the ionization.

$$HA + H_2O \rightleftharpoons H_3O^+ + A^-$$

$$K_a = \frac{[H_3O^+][A^-]}{[HA]}$$

If we *start* with the conjugate base, A^-, and dissolve it in water, it ionizes as a typical base; it accepts a proton from H_2O to form OH^- and the corresponding conjugate acid, HA. A K_b expression can be written for this ionization.

$$A^- + HOH \rightleftharpoons HA + OH^-$$

$$K_b = \frac{[HA][OH^-]}{[A^-]}$$

Solving the K_a and K_b expressions for $[H_3O^+]$ and $[OH^-]$:

$$[H_3O^+] = \frac{[HA]K_a}{[A^-]}$$

$$[OH^-] = \frac{[A^-]K_b}{[HA]}$$

$$[H_3O^+][OH^-] = K_w$$

Substituting:

$$\frac{[HA]K_a}{[A^-]} \times \frac{[A^-]K_b}{[HA]} = K_w$$

$$K_a \times K_b = K_w$$

Taking logarithms:

$$\log K_a + \log K_b = \log K_w$$

$$-\log K_a - \log K_b = -\log K_w$$

Just as $-\log [H^+]$ has been defined as pH, we can define $-\log K_a$ as pK_a, $-\log K_b$ as pK_b, and $-\log K_w$ as pK_w (which equals 14).

$$\boxed{\therefore \quad \mathbf{p}K_a + \mathbf{p}K_b = \mathbf{14}}$$

(b) We can derive the same two expressions by considering a weak organic base such as an amine, R-NH$_2$, and its conjugate acid, R-NH$_3^+$. As shown earlier, the base ionizes in water to form OH$^-$ and the conjugate acid, R-NH$_3^+$. If we *start* with the conjugate acid, it ionizes as any typical acid to yield H$_3$O$^+$ and the corresponding conjugate base. The K_b and K_a expressions can be written as usual.

$$R\text{-}NH_2 + HOH \rightleftharpoons R\text{-}NH_3^+ + OH^-$$

$$K_b = \frac{[R\text{-}NH_3^+][OH^-]}{[R\text{-}NH_2]}$$

$$R\text{-}NH_3^+ + H_2O \rightleftharpoons H_3O^+ + R\text{-}NH_2$$

$$K_a = \frac{[H_3O^+][R\text{-}NH_2]}{[R\text{-}NH_3^+]}$$

$$[H_3O^+] = \frac{[R\text{-}NH_3^+]K_a}{[R\text{-}NH_2]}$$

$$[OH^-] = \frac{[R\text{-}NH_2]K_b}{[R\text{-}NH_3^+]}$$

$$[H_3O^+][OH^-] = K_w$$

$$\frac{[R\text{-}NH_3^+]K_a}{[R\text{-}NH_2]} \times \frac{[R\text{-}NH_2]K_b}{[R\text{-}NH_3^+]} = K_w$$

$$\boxed{K_a \times K_b = K_w}$$

and

$$\boxed{\mathbf{p}K_a + \mathbf{p}K_b = \mathbf{14}}$$

For polyprotic acids, the K_a values are recorded and numbered in order of decreasing acid strength (K_{a_1}, K_{a_2}, etc.) The K_b values are numbered in order of decreasing base strength. However, remember that the conjugate base of the strongest acid group is the weakest basic group, and vice versa, and the K_a and K_b values are numbered accordingly as shown below.

$$H_2A \underset{K_{b_2}}{\overset{K_{a_1}}{\rightleftharpoons}} H^+ + HA^-$$

$$K_{a_2} \updownarrow K_{b_1}$$

$$H^+ + A^=$$

Thus, in using either of the above expressions, we must be sure to use the K_a and K_b (or pK_a and pK_b) of *the same ionization*. For the diprotic acid illustrated above, the correct expressions are as follows:

$$K_{a_1} \times K_{b_2} = K_w \qquad pK_{a_1} + pK_{b_2} = pK_w$$
$$K_{a_2} \times K_{b_1} = K_w \qquad pK_{a_2} + pK_{b_1} = pK_w$$

I. DERIVATION OF HENDERSON-HASSELBALCH EQUATION

A useful expression showing the relationship between (a) K_a of a weak acid, HA, and the pH of a solution of the weak acid and (b) K_b of a weak base and the pOH of a solution of the weak base can be derived.

(a)
$$K_a = \frac{[H^+][A^-]}{[HA]}$$

Cross-multiplying:

$$[H^+][A^-] = K_a[HA]$$

Rearranging terms:

$$[H^+] = K_a \frac{[HA]}{[A^-]}$$

Taking logarithms of both sides:

$$\log [H^+] = \log K_a + \log \frac{[HA]}{[A^-]}$$

Multiplying both sides by -1:

$$-\log [H^+] = -\log K_a - \log \frac{[HA]}{[A^-]}$$

$$pH = pK_a - \log \frac{[HA]}{[A^-]}$$

$$\boxed{pH = pK_a + \log \frac{[A^-]}{[HA]}}$$

(b) $$K_b = \frac{[M^+][OH^-]}{[MOH]}$$ $$K_b = \frac{[R\text{-}NH_3^+][OH^-]}{[R\text{-}NH_2]}$$

$$[M^+][OH^-] = K_b[MOH]$$ $$[R\text{-}NH_3^+][OH^-] = K_b[R\text{-}NH_2]$$

$$[OH^-] = K_b\frac{[MOH]}{[M^+]}$$ $$[OH^-] = K_b\frac{[R\text{-}NH_2]}{[R\text{-}NH_3^+]}$$

$$\log[OH^-] = \log K_b$$ $$\log[OH^-] = \log K_b$$

$$+ \log\frac{[MOH]}{[M^+]}$$ $$+ \log\frac{[R\text{-}NH_2]}{[R\text{-}NH_3^+]}$$

$$-\log[OH^-] = -\log K_b$$ $$-\log[OH^-] = -\log K_b$$

$$- \log\frac{[MOH]}{[M^+]}$$ $$- \log\frac{[R\text{-}NH_2]}{[R\text{-}NH_3^+]}$$

$$pOH = pK_b - \log\frac{[MOH]}{[M^+]}$$ $$pOH = pK_b - \log\frac{[R\text{-}NH_2]}{[R\text{-}NH_3^+]}$$

$$\boxed{pOH = pK_b + \log\frac{[M^+]}{[MOH]}}$$ $$\boxed{pOH = pK_b + \log\frac{[R\text{-}NH_3^+]}{[R\text{-}NH_2]}}$$

Note that when the concentrations of conjugate acid and conjugate base are equal, $pH = pK_a$ and $pOH = pK_b$. This same relationship can be seen from the K_a or K_b expressions; when $[A^-] = [HA]$, $[H^+] = K_a$ and when $[R\text{-}NH_2] = [R\text{-}NH_3^+]$, $[OH^-] = K_b$.

J. HYDROLYSIS OF SALTS OF WEAK ACIDS AND BASES

(a)   Salts of weak acids (the conjugate base anion of weak acids) react with water to produce the weak parent acid (conjugate acid) and OH^- ions.

$$A^- + HOH \rightleftharpoons HA + OH^-$$

We can see that the "hydrolysis" is nothing more than the ionization of the conjugate base as described in Part H. A new constant, the "hydrolysis constant," K_h, is frequently derived.

$$K_{eq} = \frac{[HA][OH^-]}{[A^-][HOH]}$$ $$K_{eq}[HOH] = K_h = \frac{[HA][OH^-]}{[A^-]}$$

The "hydrolysis constant," K_h, is identical to K_b. The K_h (or K_b) can be defined in terms of K_a for the conjugate acid and K_w.

$$[OH^-] = \frac{K_w}{[H^+]} \qquad K_h = \frac{[HA]K_w}{[A^-][H^+]} \qquad \frac{[HA]}{[A^-][H^+]} = \frac{1}{K_a}$$

$$\therefore \quad K_h = \frac{K_w}{K_a}$$

which is expected because

$$K_h = K_b = \frac{K_w}{K_a}$$

(b)   Similarly, salts of weak bases (the conjugate acid of weak bases) react with water to produce the weak parent base and H^+ ions.

$$R\text{-}NH_3^+ + HOH \rightleftharpoons R\text{-}NH_2 + H_3O^+$$

$$NH_4^+ + HOH \rightleftharpoons NH_3 + H_3O^+$$

or

$$NH_4^+ + HOH \rightleftharpoons \text{"}NH_4OH\text{"} + H^+$$

Again, we see that the "hydrolysis" is nothing more than the usual ionization of the conjugate acid. In this instance, the K_h that is derived is identical to K_a for the conjugate acid.

$$K_{eq} = \frac{[R\text{-}NH_2][H_3O^+]}{[R\text{-}NH_3^+][HOH]} \qquad K_{eq}[HOH] = K_h = \frac{[R\text{-}NH_2][H_3O^+]}{[R\text{-}NH_3^+]}$$

$$[H_3O^+] = \frac{K_w}{[OH^-]} \qquad K_h = \frac{[R\text{-}NH_2]K_w}{[R\text{-}NH_3^+][OH^-]} \qquad \frac{R\text{-}NH_2}{[R\text{-}NH_3^+][OH^-]} = \frac{1}{K_b}$$

$$\therefore \quad K_h = \frac{K_w}{K_b}$$

which is expected because

$$K_h = K_a = \frac{K_w}{K_b}$$

General Rule

For salts (conjugate bases) of weak acids:

$$K_h = K_b = \frac{K_w}{K_a}$$

For salts (conjugate acids) of weak bases:

$$K_h = K_a = \frac{K_w}{K_b}$$

The pH of a solution of a salt depends on the nature of both ions comprising the salt. Salts of weak acids and strong bases (for example, sodium acetate) produce a basic solution (because the acetate undergoes hydrolysis—ionizes as a base). Salts of weak bases and strong acids (for example, R-NH_3Cl or NH_4Cl) produce an acidic solution (because the R-NH_3^+ or NH_4^+ undergoes hydrolysis—ionizes as an acid). Both ions in salts of weak acids and weak bases (for example, NH_4CN) undergo hydrolysis. The pH of their solution then depends on the relative K_h values.

K. DISPROPORTIONATION REACTIONS

An intermediate ion of a polyprotic acid when dissolved in water undergoes both ionization as an acid (reaction 1) and ionization as a base or hydrolysis (reaction 2).

(1) $\quad HA^- + H_2O \rightleftharpoons A^= + H_3O^+ \qquad K_{eq_1} = K_{a_2}$

(2) $\quad HA^- + H_2O \rightleftharpoons H_2A + OH^- \qquad K_{eq_2} = K_{h_1} = K_{b_2} = \dfrac{K_w}{K_{a_1}}$

If reaction 1 proceeds further to the right (that is, has a greater K_{eq}) than reaction 2, the solution is acidic. If reaction 2 proceeds further to the right than reaction 1, the solution is basic. The acidity or alkalinity is not as great as we might expect judging from the relative values of K_{eq_1} and K_{eq_2}, however, because of a further compensating reaction that takes place. If reaction 1 goes further to the right than reaction 2, then some of the excess H_3O^+ produced reacts with unreacted HA^- according to reaction 3.

(3) $\qquad HA^- + H_3O^+ \rightleftharpoons H_2A + H_2O \qquad K_{eq_3} = \dfrac{1}{K_{a_1}}$

If, however, reaction 2 proceeds further than reaction 1, some of the excess OH^- produced reacts with HA^- according to reaction 4.

(4) $\quad HA^- + OH^- \rightleftharpoons A^= + H_2O \qquad K_{eq_4} = \dfrac{1}{K_{h_2}} = \dfrac{1}{K_{b_1}} = \dfrac{K_{a_2}}{K_w}$

Thus, the two major reactions taking place in the solution are either 1 plus 3 or 2 plus 4. The sum of either pair of reactions is identical and is called a "disproportionation" reaction (5). The equilibrium constant for a disproportionation reaction always is the ratio of the K_a for the next acid dissociation to the K_a for the preceding acid dissociation (K_{a_2}/K_{a_1}), as shown below.

(1) $HA^- + H_2O \rightleftharpoons A^= + H_3O^+$ $K_{eq_1} = K_{a_2}$

(3) $HA^- + H_3O^+ \rightleftharpoons H_2A + H_2O$ $K_{eq_3} = \dfrac{1}{K_{a_1}}$

Sum: (5) $2HA^- \rightleftharpoons A^= + H_2A$ $K_{eq_5} = K_{eq_1} \times K_{eq_3} = K_{dis}$

$$K_{dis} = \frac{K_{a_2}}{K_{a_1}}$$

(2) $HA^- + H_2O \rightleftharpoons H_2A + OH^-$ $K_{eq_2} = \dfrac{K_w}{K_{a_1}}$

(4) $HA^- + OH^- \rightleftharpoons A^= + H_2O$ $K_{eq_4} = \dfrac{K_{a_2}}{K_w}$

Sum: (5) $2HA^- \rightleftharpoons A^= + H_2A$ $K_{eq_5} = K_{eq_2} \times K_{eq_4} = K_{dis}$

$$K_{dis} = \frac{K_w}{K_{a_1}} \times \frac{K_{a_2}}{K_w}$$

$$K_{dis} = \frac{K_{a_2}}{K_{a_1}}$$

We can arrive at the same disproportionation reaction by assuming that 1 molecule of HA^- releases a proton and another molecule accepts the proton.

$HA^- \rightleftharpoons H^+ + A^=$ $K_{eq} = K_{a_2}$

$HA^- + H^+ \rightleftharpoons H_2A$ $K_{eq} = \dfrac{1}{K_{a_1}}$

Sum: (5) $2HA^- \rightleftharpoons H_2A + A^=$ $K_{eq} = \dfrac{K_{a_2}}{K_{a_1}} = K_{dis}$

The disproportionation reaction tends to equalize the concentrations of the two ionic forms on either side of the intermediate ion. When writing reaction 5, we assume that the component reactions proceed to the same extent (we cancel H_2O and OH^- or H_2O and H_3O^+). Actually, reaction 1 or reaction 2 proceeds slightly further than reaction 3 or 4; that is why the solution is acidic or basic. However, the actual *amounts* of OH^- or H_3O^+ produced (by reaction 1 or 2) and utilized (by reaction 3 or 4) are much greater than the *difference between* the amounts produced and utilized. Consequently, we can safely use reaction 5 as a basis for calculating the concentrations of all ionic forms (H_2A, HA^-, and $A^=$) in the solution.

The H^+ ion concentration and subsequently the pH of the solution may then be calculated from any K_{eq} expression containing the above components:

$$K_{a_1} = \frac{[H^+][HA^-]}{[H_2A]} \qquad K_{a_2} = \frac{[H^+][A^=]}{[HA^-]}$$

$$K_{h_1} = \frac{[H_2A][OH^-]}{[HA^-]} = \frac{K_w}{K_{a_1}} \qquad K_{h_2} = \frac{[HA^-][OH^-]}{[A^=]} = \frac{K_w}{K_{a_2}}$$

We can also derive equations from which the H^+ ion concentration and the pH may be determined directly from the K_{a_2} and K_{a_1} values.

$$K_{dis} = \frac{[H_2A][A^=]}{[HA^-]^2} = \frac{K_{a_2}}{K_{a_1}}$$

$$[H_2A] = [A^=]$$

$$\therefore \quad \frac{[H_2A]^2}{[HA^-]^2} = \frac{K_{a_2}}{K_{a_1}}$$

$$\frac{[H_2A]}{[HA^-]} = \frac{\sqrt{K_{a_2}}}{\sqrt{K_{a_1}}}$$

Substituting the $[H_2A]/[HA^-]$ ratio into the K_{a_1} expression:

$$K_{a_1} = \frac{[H^+][HA^-]}{[H_2A]} \qquad [H^+] = \frac{K_{a_1}[H_2A]}{[HA^-]} = \frac{K_{a_1}\sqrt{K_{a_2}}}{\sqrt{K_{a_1}}}$$

$$\frac{K_{a_1}}{\sqrt{K_{a_1}}} = \sqrt{K_{a_1}} \qquad \therefore \quad [H^+] = \sqrt{K_{a_1}}\sqrt{K_{a_2}}$$

$$\boxed{[H^+] = \sqrt{K_{a_1}K_{a_2}}}$$

$$\log [H^+] = \tfrac{1}{2}(\log K_{a_1} + \log K_{a_2})$$

$$-\log [H^+] = -\tfrac{1}{2}(\log K_{a_1} + \log K_{a_2})$$

$$-\log [H^+] = \frac{-\log K_{a_1} - \log K_{a_2}}{2}$$

$$\boxed{pH = \frac{pK_{a_1} + pK_{a_2}}{2}}$$

By the same reasoning, we can show that the pH at an intermediate equivalence point during the titration of a polyprotic acid (or amino acid) is the average of the pK_a values on either side of the equivalence point. For amino acids the pH is designated P_I (isoelectric point) if the intermediate ion in question is the one that carries no *net* charge. At the P_I the major ionic specie present is AA^0. However, because of the disproportionation reaction, small (and essentially equal) amounts of AA^{-1} and AA^{+1} are also present. The pH is designated pH_m if the major ionic specie present carries the maximum number of charges, regardless of sign.

L. PRIMARY FORMULAE FOR CALCULATING THE pH DURING A TITRATION

The Titration of a Strong Acid with a Strong Base

This titration is shown in Figure AI-1.

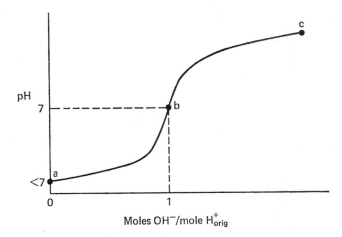

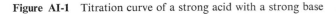

Figure AI-1   Titration curve of a strong acid with a strong base

a. $pH = -\log [H^+]_{orig}$ where $[H^+]_{orig} = [HA]$

a-b. $pH = -\log [H^+]_{remain}$

b. $pH = pOH = 7$

b-c. $pOH = -\log [OH^-]_{excess}$

Titration of a Strong Base with a Strong Acid

This titration is shown in Figure AI-2.

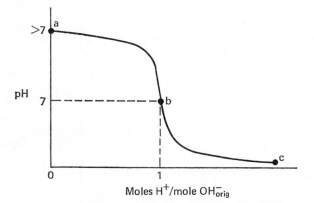

Figure AI-2   Titration curve of a strong base with a strong acid

a. $pOH = -\log [OH^-]_{orig}$ $[OH^-]_{orig} = [MOH]$
a-b. $pOH = -\log [OH^-]_{remain}$
b. $pOH = pH = 7$
b-c. $pH = -\log [H^+]_{excess}$

Titration of a Weak Monoprotic Acid with a Strong Base

This titration is shown in Figure AI-3.

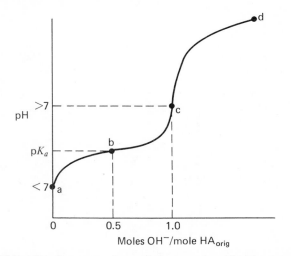

Figure AI-3   Titration curve of a weak monoprotic acid with a strong base

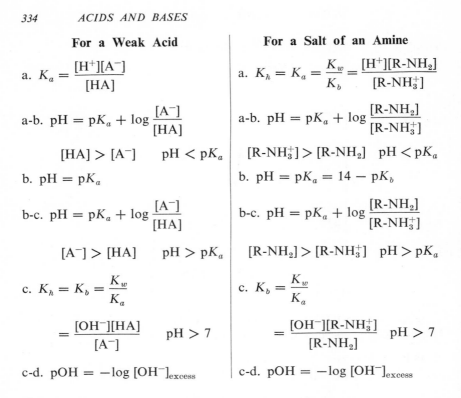

For a Weak Acid

a. $K_a = \dfrac{[\text{H}^+][\text{A}^-]}{[\text{HA}]}$

a-b. $\text{pH} = \text{p}K_a + \log \dfrac{[\text{A}^-]}{[\text{HA}]}$

$\quad [\text{HA}] > [\text{A}^-] \qquad \text{pH} < \text{p}K_a$

b. $\text{pH} = \text{p}K_a$

b-c. $\text{pH} = \text{p}K_a + \log \dfrac{[\text{A}^-]}{[\text{HA}]}$

$\quad [\text{A}^-] > [\text{HA}] \qquad \text{pH} > \text{p}K_a$

c. $K_h = K_b = \dfrac{K_w}{K_a}$

$\quad = \dfrac{[\text{OH}^-][\text{HA}]}{[\text{A}^-]} \qquad \text{pH} > 7$

c-d. $\text{pOH} = -\log [\text{OH}^-]_{\text{excess}}$

For a Salt of an Amine

a. $K_h = K_a = \dfrac{K_w}{K_b} = \dfrac{[\text{H}^+][\text{R-NH}_2]}{[\text{R-NH}_3^+]}$

a-b. $\text{pH} = \text{p}K_a + \log \dfrac{[\text{R-NH}_2]}{[\text{R-NH}_3^+]}$

$\quad [\text{R-NH}_3^+] > [\text{R-NH}_2] \quad \text{pH} < \text{p}K_a$

b. $\text{pH} = \text{p}K_a = 14 - \text{p}K_b$

b-c. $\text{pH} = \text{p}K_a + \log \dfrac{[\text{R-NH}_2]}{[\text{R-NH}_3^+]}$

$\quad [\text{R-NH}_2] > [\text{R-NH}_3^+] \quad \text{pH} > \text{p}K_a$

c. $K_b = \dfrac{K_w}{K_a}$

$\quad = \dfrac{[\text{OH}^-][\text{R-NH}_3^+]}{[\text{R-NH}_2]} \quad \text{pH} > 7$

c-d. $\text{pOH} = -\log [\text{OH}^-]_{\text{excess}}$

Titration of a Weak Monoprotic Base with a Strong Acid

This titration is illustrated in Figure AI-4.

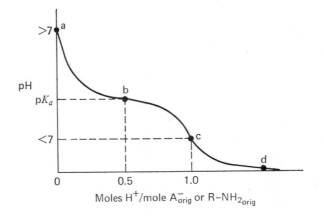

Figure AI-4   Titration curve of a weak monoprotic base with a strong acid

For a Salt of a Weak Acid

a. $K_h = K_b = \dfrac{K_w}{K_a} = \dfrac{[OH^-][HA]}{[A^-]}$

a-b. $pH = pK_a + \log \dfrac{[A^-]}{[HA]}$

$[A^-] > [HA] \qquad pH > pK_a$

b. $pH = pK_a$

b-c. $pH = pK_a + \log \dfrac{[A^-]}{[HA]}$

$[HA] > [A^-] \qquad pH > pK_a$

c. $K_a = \dfrac{[H^+][A^-]}{[HA]}$

c-d. $pH = -\log [H^+]_{excess}$

For an Amine

a. $K_b = \dfrac{K_w}{K_a} = \dfrac{[OH^-][R\text{-}NH_3^+]}{[R\text{-}NH_2]}$

a-b. $pH = pK_a + \log \dfrac{[R\text{-}NH_2]}{[R\text{-}NH_3^+]}$

$[R\text{-}NH_2] > [R\text{-}NH_3^+] \quad pH > pK_a$

b. $pH = pK_a = 14 - pK_b$

b-c. $pH = pK_a + \log \dfrac{[R\text{-}NH_2]}{[R\text{-}NH_3^+]}$

$[R\text{-}NH_3^+] > [R\text{-}NH_2] \quad pH < pK_a$

c. $K_h = K_a = \dfrac{K_w}{K_b} = \dfrac{[H^+][R\text{-}NH_2]}{[R\text{-}NH_3^+]}$

c-d. $pH = -\log [H^+]_{excess}$

Titration of a Weak Diprotic Acid with a Strong Base

This titration is shown in Figure AI-5.

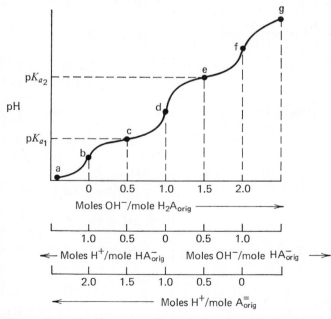

Figure AI-5 Titration curve of a weak diprotic acid with a strong base

For a Weak Acid

a-b. $pH = -\log [H^+]_{\text{excess}}$

b. $K_{a_1} = \dfrac{[H^+][HA^-]}{[H_2A]}$

b-c. $pH = pK_{a_1} + \log \dfrac{[HA^-]}{[H_2A]}$

$[H_2A] > [HA^-]$ $pH < pK_{a_1}$

c. $pH = pK_{a_1}$ $[H_2A] = [HA^-]$

c-d. $pH = pK_{a_1} + \log \dfrac{[HA^-]}{[H_2A]}$

$[HA^-] > [H_2A]$ $pH > pK_{a_1}$

d. $pH = \dfrac{pK_{a_1} + pK_{a_2}}{2}$

$[H^+] = \sqrt{K_{a_1}K_{a_2}}$

$K_{\text{dis}} = \dfrac{K_{a_2}}{K_{a_1}} = \dfrac{[H_2A][A^=]}{[HA^-]^2}$

d-e. $pH = pK_{a_2} + \log \dfrac{[A^=]}{[HA^-]}$

$[HA^-] > [A^=]$ $pH < pK_{a_2}$

e. $pH = pK_{a_2}$ $[HA^-] = [A^=]$

e-f. $pH = pK_{a_2} + \log \dfrac{[A^=]}{[HA^-]}$

$[A^=] > [HA^-]$ $pH > pK_{a_2}$

f. $K_{h_2} = K_{b_1} = \dfrac{K_w}{K_{a_2}} = \dfrac{[OH^-][HA^-]}{[A^=]}$

f-g. $pOH = -\log [OH^-]_{\text{excess}}$

For a Salt of an Amine

a-b. $pH = -\log [H^+]_{\text{excess}}$

b. $K_{h_1} = K_{a_1} = \dfrac{K_w}{K_{b_2}}$

$= \dfrac{[H^+][R\text{-}NH_3^+NH_2]}{[R\text{-}(NH_3^+)_2]}$

b-c. $pH = pK_{a_1} + \log \dfrac{[R\text{-}NH_3^+NH_2]}{[R\text{-}(NH_3^+)_2]}$

$[R\text{-}(NH_3^+)_2] > [R\text{-}NH_3^+NH_2]$

$pH < pK_{a_1}$

c. $pH = pK_{a_1}$

$[R\text{-}(NH_3^+)_2] = [R\text{-}NH_3^+NH_2]$

c-d. $pH = pK_{a_1} + \log \dfrac{[R\text{-}NH_3^+NH_2]}{[R\text{-}(NH_3^+)_2]}$

$[R\text{-}NH_3^+NH_2] > [R\text{-}(NH_3^+)_2]$

$pH > pK_{a_1}$

d. $pH = \dfrac{pK_{a_1} + pK_{a_2}}{2}$

$[H^+] = \sqrt{K_{a_1}K_{a_2}}$

$K_{\text{dis}} = \dfrac{K_{a_2}}{K_{a_1}}$

$= \dfrac{[R\text{-}(NH_3^+)_2][R\text{-}(NH_2)_2]}{[R\text{-}NH_3^+NH_2]^2}$

d-e. $pH = pK_{a_2} + \log \dfrac{[R\text{-}(NH_2)_2]}{[R\text{-}NH_3^+NH_2]}$

$[R\text{-}NH_3^+NH_2] > [R\text{-}(NH_2)_2]$

$pH < pK_{a_2}$

e. $pH = pK_{a_2}$

$[R\text{-}NH_3^+NH_2] = [R\text{-}(NH_2)_2]$

e-f. $pH = pK_{a_2} + \log \dfrac{[R\text{-}(NH_2)_2]}{[R\text{-}NH_3^+NH_2]}$

$[R\text{-}(NH_2)_2] > [R\text{-}NH_3^+NH_2]$

$pH > pK_{a_2}$

f. $K_{b_1} = \dfrac{K_w}{K_{a_2}} = \dfrac{[OH^-][R\text{-}NH_3^+NH_2]}{[R\text{-}(NH_2)_2]}$

f-g. $pOH = -\log [OH^-]_{\text{excess}}$

Titration of a Neutral Amino Acid with a Strong Acid and a Strong Base

This titration is shown in Figure AI-6.

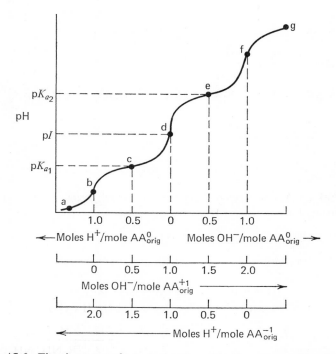

Figure AI-6 Titration curve of a neutral amino acid with a strong acid and strong base

Position	Predominant Ionic Form
b	$AA^{+1} = R\text{-}NH_3^+\text{-}COOH$
d	$AA^0 = R\text{-}NH_3^+\text{-}COO^-$
f	$AA^{-1} = R\text{-}NH_2\text{-}COO^-$

a-b. $pH = -\log [H^+]_{excess}$

b. $K_{a_1} = \dfrac{[H^+][AA^0]}{[AA^{+1}]}$

b-c. $pH = pK_{a_1} + \log \dfrac{[AA^0]}{[AA^{+1}]}$ $[AA^{+1}] > [AA^0]$ $pH < pK_{a_1}$

c. $pH = pK_{a_1}$ $[AA^{+1}] = [AA^0]$

c-d. $\text{pH} = pK_{a_1} + \log \dfrac{[\text{AA}^0]}{[\text{AA}^{+1}]}$ $[\text{AA}^0] > [\text{AA}^{+1}]$ $\text{pH} > pK_{a_1}$

d. $\text{pH} = pI = \dfrac{pK_{a_1} + pK_{a_2}}{2}$ $[\text{AA}^{+1}] = [\text{AA}^{-1}]$ $K_{\text{dis}} = \dfrac{[\text{AA}^{+1}][\text{AA}^{-1}]}{[\text{AA}^0]^2}$

d-e. $\text{pH} = pK_{a_2} + \log \dfrac{[\text{AA}^{-1}]}{[\text{AA}^0]}$ $[\text{AA}^0] > [\text{AA}^{-1}]$ $\text{pH} < pK_{a_2}$

e. $\text{pH} = pK_{a_2}$ $[\text{AA}^0] = [\text{AA}^{-1}]$

e-f. $\text{pH} = pK_{a_2} + \log \dfrac{[\text{AA}^{-1}]}{[\text{AA}^0]}$ $[\text{AA}^{-1}] > [\text{AA}^0]$ $\text{pH} > pK_{a_2}$

f. $K_{b_1} = \dfrac{[\text{OH}^-][\text{AA}^0]}{[\text{AA}^{-1}]} = \dfrac{K_w}{K_{a_2}}$

f-g. $\text{pOH} = -\log [\text{OH}^-]_{\text{excess}}$

Titration of an Acidic Amino Acid with a Strong Acid and a Strong Base

This titration is illustrated in Figure AI-7.

Position	Predominant Ionic Form
b	$\text{AA}^{+1} = \text{R-NH}_3^+\text{-[COOH]}_2$
d	$\text{AA}^0 = \text{R-NH}_3^+\text{-COOH-COO}^-$
f	$\text{AA}^{-1} = \text{R-NH}_3^+\text{-[COO}^-]_2$
h	$\text{AA}^{-2} = \text{RNH}_2\text{-[COO}^-]_2$

a-b. $\text{pH} = -\log [\text{H}^+]_{\text{excess}}$

b. $K_{a_1} = \dfrac{[\text{H}^+][\text{AA}^0]}{[\text{AA}^{+1}]}$

b-c. $\text{pH} = pK_{a_1} + \log \dfrac{[\text{AA}^0]}{[\text{AA}^{+1}]}$ $[\text{AA}^{+1}] > [\text{AA}^0]$ $\text{pH} < pK_{a_1}$

c. $\text{pH} = pK_{a_1}$ $[\text{AA}^{+1}] = [\text{AA}^0]$

c-d. $\text{pH} = pK_{a_1} + \log \dfrac{[\text{AA}^0]}{[\text{AA}^{+1}]}$ $[\text{AA}^0] > [\text{AA}^{+1}]$ $\text{pH} > pK_{a_1}$

d. $\text{pH} = pI = \dfrac{pK_{a_1} + pK_{a_2}}{2}$ $[\text{AA}^{+1}] = [\text{AA}^{-1}]$ $K_{\text{dis}} = \dfrac{[\text{AA}^{+1}][\text{AA}^{-1}]}{[\text{AA}^0]^2}$

d-e. $\text{pH} = pK_{a_2} + \log \dfrac{[\text{AA}^{-1}]}{[\text{AA}^0]}$ $[\text{AA}^0] > [\text{AA}^{-1}]$ $\text{pH} < pK_{a_2}$

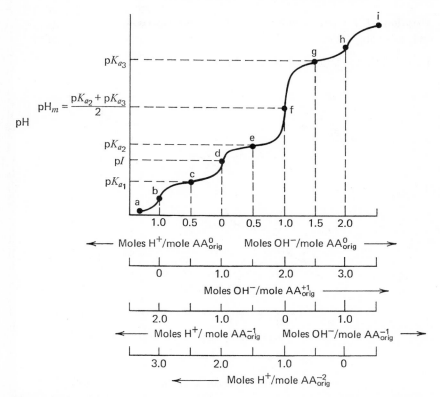

Figure AI-7 Titration curve of an acidic amino acid with a strong acid and strong base

e. $pH = pK_{a_2}$ $[AA^0] = [AA^{-1}]$

e-f. $pH = pK_{a_2} + \log \dfrac{[AA^{-1}]}{[AA^0]}$ $[AA^{-1}] > [AA^0]$ $pH > pK_{a_2}$

f. $pH = pH_m = \dfrac{pK_{a_3} + pK_{a_2}}{2}$ $[AA^0] = [AA^{-2}]$ $K_{dis} = \dfrac{[AA^{-2}][AA^0]}{[AA^{-1}]^2}$

f-g. $pH = pK_{a_3} + \log \dfrac{[AA^{-2}]}{[AA^{-1}]}$ $[AA^{-1}] > [AA^{-2}]$ $pH < pK_{a_3}$

g. $pH = pK_{a_3}$ $[AA^{-2}] = [AA^{-1}]$

g-h. $pH = pK_{a_3} + \log \dfrac{[AA^{-2}]}{[AA^{-1}]}$ $[AA^{-2}] > [AA^{-1}]$ $pH > pK_{a_3}$

h. $K_{b_1} = \dfrac{[OH^-][AA^{-1}]}{[AA^{-2}]} = \dfrac{K_w}{K_{a_3}}$

h-i. $pOH = -\log [OH^-]_{excess}$

Titration of a Basic Amino Acid with a Strong Acid and a Strong Base

This titration is shown in Figure AI-8.

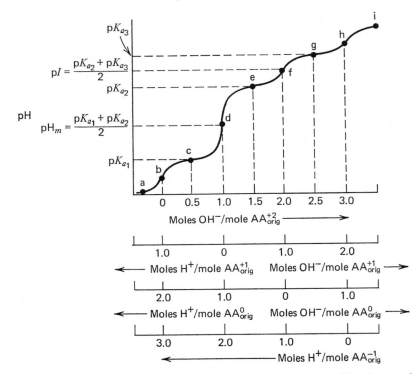

Figure AI-8 Titration curve of a basic amino acid with a strong acid and strong base

Position	Predominant Ionic Form
b	$AA^{+2} = R\text{-}[NH_3^+]_2\text{-COOH}$
d	$AA^{+1} = R\text{-}[NH_3^+]_2\text{-COO}^-$
f	$AA^0 = R\text{-}NH_3^+\text{-}NH_2\text{-COO}^-$
h	$AA^{-1} = R\text{-}[NH_2]_2\text{-COO}^-$

a-b. $pH = -\log [H^+]_{excess}$

b. $K_{a_1} = \dfrac{[H^+][AA^{+1}]}{[AA^{+2}]}$

b-c. $pH = pK_{a_1} + \log \dfrac{[AA^{+1}]}{[AA^{+2}]}$ $[AA^{+2}] > [AA^{+1}]$ $pH < pK_{a_1}$

c. $pH = pK_{a_1}$ $[AA^{+2}] = [AA^{+1}]$

c-d. $pH = pK_{a_1} + \log \dfrac{[AA^{+1}]}{[AA^{+2}]}$ $[AA^{+1}] > [AA^{+2}]$ $pH > pK_{a_1}$

d. $pH = pH_m = \dfrac{pK_{a_1} + pK_{a_2}}{2}$ $[AA^{+2}] = [AA^0]$

$$K_{dis} = \frac{K_{a_2}}{K_{a_1}} = \frac{[AA^{+2}][AA^0]}{[AA^{+1}]^2}$$

d-e. $pH = pK_{a_2} + \log \dfrac{[AA^0]}{[AA^{+1}]}$ $[AA^{+1}] > [AA^0]$ $pH < pK_{a_2}$

e. $pH = pK_{a_2}$ $[AA^{+1}] = [AA^0]$

e-f. $pH = pK_{a_2} + \log \dfrac{[AA^0]}{[AA^{+1}]}$ $[AA^0] > [AA^{+1}]$ $pH > pK_{a_2}$

f. $pH = pI = \dfrac{pK_{a_2} + pK_{a_3}}{2}$ $K_{dis} = \dfrac{K_{a_3}}{K_{a_2}} = \dfrac{[AA^{+1}][AA^{-1}]}{[AA^0]^2}$

f-g. $pH = pK_{a_3} + \log \dfrac{[AA^{-1}]}{[AA^0]}$ $[AA^0] > [AA^{-1}]$, $pH < pK_{a_3}$

g. $pH = pK_{a_3}$ $[AA^0] = [AA^{-1}]$

g-h. $pH = pK_{a_3} + \log \dfrac{[AA^{-1}]}{[AA^0]}$ $[AA^{-1}] > [AA^0]$ $pH > pK_{a_3}$

h. $K_{b_1} = \dfrac{[OH^-][AA^0]}{[AA^{-1}]} = \dfrac{K_w}{K_{a_3}}$

h-i. $pOH = -\log [OH^-]_{excess}$

TABLE AI-2. PROPERTIES OF COMMERCIAL CONCENTRATED SOLUTIONS OF ACIDS AND BASES

Compound	MW	Specific Gravity	Percent w/w	g/100 ml	Approximate N	Milliliters Required for 1 Liter of 1 N Solution
HCl	36.5	1.19	37	44	12.1	82.5
HNO$_3$	63.0	1.42	70	91	15.8	63.5
H$_2$SO$_4$	98.1	1.84	96	173	35.2	29
H$_3$PO$_4$	98.0	1.71	85	146	44.5	22.5
HCOOH	46.0	1.20	88	105.6	24	41.6
CH$_3$COOH	60.0	1.06	100	106	17.4	57.5
NH$_3$	17.0	0.91	28	22.8	14.8	67.5

TABLE AI-3. IONIZATION (DISSOCIATION) CONSTANTS AND pK_a VALUES OF SOME WEAK ACIDS

Conjugate Acid	MW	Conjugate Base	K_a	pK_a ($= -\log K_a$)
Acetic acid	60.05	Acetate^{-1}	1.75×10^{-5}	4.76
Ascorbic acid	176.12	Ascorbate^{-1}	8×10^{-5} (K_{a_1})	4.1 (pK_{a_1})
Ascorbate^{-1}		Ascorbate^{-2}	1.6×10^{-12} (K_{a_2})	11.79 (pK_{a_2})
Benzoic acid	122.12	Benzoate^{-1}	6.30×10^{-5}	4.20
Boric acid	61.84	Borate^{-1}	5.8×10^{-10}	9.24
n-Butyric acid	88.10	Butyrate^{-1}	1.5×10^{-5}	4.82
Carbonic acid	62.03	Bicarbonate^{-1}	4.31×10^{-7} (K_{a_1})	6.37 (pK_{a_1})
Bicarbonate^{-1}		Carbonate^{-2}	5.6×10^{-11} (K_{a_2})	10.25 (pK_{a_2})
Citric acid	192.12	Citrate^{-1}	8.7×10^{-4} (K_{a_1})	3.06 (pK_{a_1})
Citrate^{-1}		Citrate^{-2}	1.8×10^{-5} (K_{a_2})	4.74 (pK_{a_2})
Citrate^{-2}		Citrate^{-3}	4.0×10^{-6} (K_{a_3})	5.40 (pK_{a_3})
3,6-Endomethylene-1,2,3,6-tetrahydro-phthalic acid, "EMTA"	182.2		5×10^{-5} (K_{a_1})	4.3 (pK_{a_1})
			1×10^{-7} (K_{a_2})	7.0 (pK_{a_2})
Formic acid	46.03	Formate^{-1}	1.77×10^{-4}	3.75
Fumaric acid	116.07	Fumarate^{-1}	9.3×10^{-4} (K_{a_1})	3.03 (pK_{a_1})
Fumarate^{-1}		Fumarate^{-2}	3.4×10^{-5} (K_{a_2})	4.47 (pK_{a_2})
Glycerophosphoric acid	172.08	Glycerophosphate^{-1}	3.4×10^{-2} (K_{a_1})	1.47 (pK_{a_1})
Glycerophosphate^{-1}		Glycero-phosphate^{-2}	6.4×10^{-7} (K_{a_2})	6.19 (pK_{a_2})
Hippuric acid	179.17	Hippurate^{-1}	2.3×10^{-4}	3.64
Hydrocyanic acid	27.03	Cyanide^{-1}	4.9×10^{-10}	9.31

			K_a	pK_a
Hydrofluoric	20.01	Fluoride^{-1}	1×10^{-3}	3.00
Hydrogen sulfide	34.08	Hydrosulfide^{-1}	5.7×10^{-8} (K_{a_1})	7.24 (pK_{a_1})
Hydrosulfide^{-1}		Sulfide^{-2}	1.2×10^{-15} (K_{a_2})	14.92 (pK_{a_2})
Hydroquinone	110.11	Hydroquinone^{-1}	1.1×10^{-10}	9.96
N-2-hydroxy-ethylpiperazine-N'-2-ethane-sulfonic acid, "HEPES"	238.3		2.82×10^{-8}	7.55
Itaconic acid	130.10	Itaconate^{-1}	1.46×10^{-4} (K_{a_1})	3.84 (pK_{a_1})
Itaconate^{-1}		Itaconate^{-2}	2.8×10^{-6} (K_{a_2})	5.55 (pK_{a_2})
Lactic acid	90.08	Lactate^{-1}	1.39×10^{-4}	3.86
Maleic acid	116.07	Maleate^{-1}	1.0×10^{-2} (K_{a_1})	2.0 (pK_{a_1})
Maleate^{-1}		Maleate^{-2}	5.5×10^{-7} (K_{a_2})	6.26 (pK_{a_2})
Malic acid	134.09	Malate^{-1}	4×10^{-4} (K_{a_1})	3.40 (pK_{a_1})
Malate^{-1}		Malate^{-2}	9×10^{-6} (K_{a_2})	5.05 (pK_{a_2})
Malonic acid	104.06	Malonate^{-1}	1.4×10^{-3} (K_{a_1})	2.85 (pK_{a_1})
Malonate^{-1}		Malonate^{-2}	8.0×10^{-7} (K_{a_2})	6.10 (pK_{a_2})
2-(N-morpholino)-ethane sulfonic acid, "MES"	195.2		7.06×10^{-7}	6.15
Nitrous acid	47.02	Nitrite^{-1}	4×10^{-4}	3.40
Oxalic acid	126.07	Oxalate^{-1}	6.5×10^{-2} (K_{a_1})	1.19 (pK_{a_1})
Oxalate^{-1}		Oxalate^{-2}	6.1×10^{-5} (K_{a_2})	4.21 (pK_{a_2})
Phenol	94.11	Phenolate^{-1}	1.3×10^{-10}	9.89
Phosphoric acid (H_3PO_4)	98.00	$H_2PO_4^{-1}$	7.5×10^{-3} (K_{a_1})	2.12 (pK_{a_1})
$H_2PO_4^{-1}$		HPO_4^{-2}	6.2×10^{-8} (K_{a_2})	7.21 (pK_{a_2})
HPO_4^{-2}		PO_4^{-3}	4.8×10^{-13} (K_{a_3})	12.32 (pK_{a_3})

TABLE AI-3 (continued).

Conjugate Acid	MW	Conjugate Base	K_a	pK_a ($= -\log K_a$)
Propionic acid	74.08	Propionate^{-1}	1.34×10^{-5}	4.87
Pyrophosphoric acid ($H_4P_2O_7$)	177.99	$H_3P_2O_7^{-1}$	1.4×10^{-1} (K_{a_1})	0.85 (pK_{a_1})
		$H_2P_2O_7^{-2}$	1.1×10^{-2} (K_{a_2})	1.96 (pK_{a_2})
$H_3P_2O_7^{-1}$		$H_2P_2O_7^{-2}$	1.1×10^{-2} (K_{a_2})	1.96 (pK_{a_2})
$H_2P_2O_7^{-2}$		$HP_2O_7^{-3}$	2.1×10^{-7} (K_{a_3})	6.68 (pK_{a_3})
$HP_2O_7^{-3}$		$P_2O_7^{-4}$	4.06×10^{-10} (K_{a_4})	9.39 (pK_{a_4})
Selenious acid	128.98	Selenite^{-1}	3×10^{-3} (K_{a_1})	2.52 (pK_{a_1})
Selenite^{-1}		Selenite^{-2}	5×10^{-8} (K_{a_2})	7.30 (pK_{a_2})
Succinic acid	118.09	Succinate^{-1}	6.4×10^{-5} (K_{a_1})	4.19 (pK_{a_1})
Succinate^{-1}		Succinate^{-2}	2.7×10^{-6} (K_{a_2})	5.57 (pK_{a_2})
Sulfuric acid	98.08	Bisulfate^{-1}	$\sim 4 \times 10^{-1}$ (K_{a_1})	~ 0.40 (pK_{a_1})
Bisulfate^{-1}		Sulfate^{-2}	1.2×10^{-2} (K_{a_2})	1.92 (pK_{a_2})
Sulfurous acid	82.08	Bisulfite^{-1}	1.72×10^{-2} (K_{a_1})	1.76 (pK_{a_1})
Bisulfite^{-1}		Sulfite^{-2}	6.24×10^{-8} (K_{a_2})	7.20 (pK_{a_2})
Tartaric acid	168.10	Tartarate^{-1}	9.6×10^{-4} (K_{a_1})	3.02 (pK_{a_1})
Tartarate^{-1}		Tartarate^{-2}	2.9×10^{-4} (K_{a_2})	4.54 (pK_{a_2})
Telluric acid	229.66	Bitellurate^{-1}	6×10^{-7} (K_{a_1})	6.22 (pK_{a_1})
Bitellurate^{-1}		Tellurate^{-2}	2×10^{-8} (K_{a_2})	7.70 (pK_{a_2})
N-tris(hydroxy-methyl)methyl-2-amino ethane sulfonic acid, "TES"	229.3		3.16×10^{-8}	7.5
Uric acid	168.11	Urate^{-1}	1.3×10^{-4}	3.89

TABLE AI-4. IONIZATION CONSTANTS AND pK_b AND pK_a VALUES OF SOME WEAK BASES

Base	MW	K_b	pK_b	pK_a
Ammonia	17.03	1.8×10^{-5}	4.74	9.26
Aniline	93.12	3.82×10^{-10}	9.42	4.58
Benzidine	184.23	9.3×10^{-10} (K_{b_1})	9.03 (pK_{b_1})	4.97 (pK_{a_2})
		5.6×10^{-11} (K_{b_2})	10.25 (pK_{b_2})	3.75 (pK_{a_1})
N,N-bis (2-hydroxy-ethyl)glycine, "Bicine"	163.2	2.24×10^{-6}	5.65	8.35
Brucine	394.45	9×10^{-7} (K_{b_1})	6.05 (pK_{b_1})	7.95 (pK_{a_2})
		2×10^{-12} (K_{b_2})	11.7 (pK_{b_2})	2.30 (pK_{a_1})
n-Butylamine	73.14	4.1×10^{-4}	3.39	10.61
Caffeine	194.19	4.1×10^{-4}	3.39	10.61
Cocaine	303.35	2.6×10^{-6}	5.59	8.41
Codeine	299.36	9.0×10^{-7}	6.05	7.95
Colchicine	399.43	4.5×10^{-13}	12.35	1.65
Creatine	131.14	1.9×10^{-11}	10.72	3.28
Creatinine	113.12	3.7×10^{-11}	10.43	3.57
Diethylamine	73.14	9.6×10^{-4}	3.02	10.98
Dimethylamine	45.97	5.20×10^{-4}	3.28	10.72
Ethanolamine	61.08	2.77×10^{-5}	4.56	9.44
Ethylamine	45.08	5.6×10^{-4}	3.25	10.75
Guanine	151.13	8.4×10^{-12}	11.09	2.91
Hydrazine	32.05	3×10^{-6}	5.52	8.48
Hydroquinone	110.11	4.7×10^{-6}	5.33	8.67
Hydroxylamine	33.03	1.07×10^{-8}	7.97	6.03
Lead hydroxide	464.44	9.6×10^{-4}	3.02	10.98
Methylamine	31.06	4.38×10^{-4}	3.36	10.64
Morphine	285.33	7.4×10^{-7}	6.13	7.87
Nicotine	162.23	7×10^{-7} (K_{b_1})	6.15 (pK_{b_1})	7.85 (pK_{a_2})
		1.4×10^{-11} (K_{b_2})	10.85 (pK_{b_2})	3.15 (pK_{a_1})
Novocaine	236.31	7×10^{-6}	5.15	8.85
n-Propylamine	59.11	3.9×10^{-4}	3.41	10.59
Pyridine	79.10	1.71×10^{-9}	8.77	5.23
Quinine	324.41	1.1×10^{-6} (K_{b_1})	5.96 (pK_{b_1})	8.04 (pK_{a_2})
		1.35×10^{-10} (K_{b_2})	9.87 (pK_{b_2})	4.13 (pK_{a_1})
Silver hydroxide	124.88	1.1×10^{-4}	3.96	10.04
Strychnine	334.40	1×10^{-6} (K_{b_1})	6.0 (pK_{b_1})	8.0 (pK_{a_2})
		2×10^{-12} (K_{b_2})	11.7 (pK_{b_2})	2.3 (pK_{a_1})
Triethylamine	101.19	5.65×10^{-4}	3.24	10.76
Trimethylamine	59.11	5.45×10^{-5}	4.26	9.74
Trimethylene-diamine	74.13	2.8×10^{-4}	3.55	10.45
Tris(hydroxymethyl)-aminomethane, "TRIS"	121.14	1.26×10^{-6}	5.9	8.1
N-tris(hydroxymethyl)-methylglycine, "Tricine"	179.2	1.41×10^{-6}	5.85	8.15
Urea	60.06	1.5×10^{-14}	13.82	0.18
Zinc hydroxide	99.40	4.4×10^{-5} (K_{b_2})	4.36 (pK_{b_2})	9.64 (pK_{a_1})

$$K_b = \frac{[OH^-][base\text{-}NH^+]}{[base\text{-}N]} \qquad pK_b = -\log K_b$$

$$K_a = \frac{[H^+][base\text{-}N]}{[base\text{-}NH^+]} \qquad pK_a = -\log K_a$$

$$pK_a = 14 - pK_b$$

For bases with more than 1 nitrogen capable of accepting a proton:

$$K_{b_1} = \frac{[OH^-][base\text{-}N_3H^+]}{[base\text{-}N_3]}$$

$$K_{b_2} = \frac{[OH^-][base\text{-}N_3H_2^{++}]}{[base\text{-}N_3H^+]}$$

$$K_{b_3} = \frac{[OH^-][base\text{-}N_3H_3^{+++}]}{[base\text{-}N_3H_2^{++}]}$$

TABLE AI-5. *ACTIVITY COEFFICIENTS OF SOME IONS IN AQUEOUS SOLUTION*

	Total Ionic Concentration (M)		
Ion	0.001 M	0.01 M	0.1 M
H^+	0.975	0.933	0.86
OH^-	0.975	0.925	0.805
Acetate^{-1}	0.975	0.928	0.82
$H_2PO_4^{-1}$	0.975	0.928	0.744
HPO_4^{-2}	0.903	0.740	0.445
PO_4^{-3}	0.796	0.505	0.16
H_2citrate^{-1}	0.975	0.926	0.81
Hcitrate^{-2}	0.903	0.741	0.45
Citrate^{-3}	0.796	0.51	0.18
HCO_3^{-1}	0.975	0.928	0.82
CO_3^{-2}	0.903	0.742	0.445

TABLE AI-6. IONIZATION CONSTANTS, pK_a, pK_b, AND pI VALUES OF SOME COMMON AMINO ACIDS*

Compound	MW	Conjugate Acid	K_a	pK_a	Conjugate Base	K_b	pK_b	pI
α-Alanine	89.1	α-COOH	4.47×10^{-3}	2.35	α-COO⁻	2.24×10^{-12}	11.65	6.02
		α-NH₃⁺	2.04×10^{-10}	9.69	α-NH₂	4.90×10^{-5}	4.31	
β-Alanine	89.1	α-COOH	2.51×10^{-4}	3.60	α-COO⁻	3.98×10^{-11}	10.40	6.90
		β-NH₃⁺	6.46×10^{-11}	10.19	β-NH₂	1.55×10^{-4}	3.81	
Arginine	174.2	α-COOH	6.76×10^{-3}	2.17	α-COO⁻	1.48×10^{-12}	11.83	10.76
		α-NH₃⁺	9.12×10^{-10}	9.04	α-NH₂	1.10×10^{-5}	4.96	
		Guanidinium-NH₂⁺	3.31×10^{-13}	12.48	Guanidinium-NH	3.02×10^{-2}	1.52	
Asparagine	132.1	α-COOH	9.55×10^{-3}	2.02	α-COO⁻	1.05×10^{-12}	11.98	5.41
		α-NH₃⁺	1.58×10^{-9}	8.8	α-NH₂	6.31×10^{-6}	5.2	
Aspartic acid	133.1	α-COOH	8.13×10^{-3}	2.09	α-COO⁻	1.23×10^{-12}	11.91	2.87
		β-COOH	1.38×10^{-4}	3.86	β-COO⁻	7.25×10^{-11}	10.14	
		α-NH₃⁺	1.51×10^{-10}	9.82	α-NH₂	6.61×10^{-5}	4.18	
Citrulline	175.2	α-COOH	3.72×10^{-3}	2.43	α-COO⁻	2.69×10^{-12}	11.57	5.92
		α-NH₃⁺	3.89×10^{-10}	9.41	α-NH₂	2.57×10^{-5}	4.59	
Cysteine	121.2	α-COOH	1.95×10^{-2}	1.71	α-COO⁻	5.13×10^{-13}	12.29	5.02
		α-NH₃⁺	4.68×10^{-9}	8.33	α-NH₂	2.14×10^{-6}	5.67	
		β-SH	1.66×10^{-11}	10.78	β-S⁻	6.03×10^{-4}	3.22	
Cystine	240.3	α-COOH	2.24×10^{-2}	1.65	α-COO⁻	4.47×10^{-13}	12.35	5.06
		α-COOH	5.50×10^{-3}	2.26	α-COO⁻	1.82×10^{-12}	11.74	
		α-NH₃⁺	1.41×10^{-8}	7.85	α-NH₂	7.08×10^{-7}	6.15	
		α-NH₃⁺	1.41×10^{-10}	9.85	α-NH₂	7.08×10^{-5}	4.15	
Glutamic acid	147.1	α-COOH	6.46×10^{-3}	2.19	α-COO⁻	1.55×10^{-12}	11.81	3.22
		γ-COOH	5.62×10^{-5}	4.25	γ-COO⁻	1.78×10^{-10}	9.75	
		α-NH₃⁺	2.14×10^{-10}	9.67	α-NH₂	4.68×10^{-5}	4.33	

TABLE AI-6 (continued)

Compound	MW	Conjugate Acid	K_a	pK_a	Conjugate Base	K_b	pK_b	pI
Glutamine	146.1	α-COOH	6.76×10^{-3}	2.17	α-COO⁻	1.48×10^{-12}	11.83	5.65
		α-NH₃⁺	7.41×10^{-10}	9.13	α-NH₂	1.35×10^{-5}	4.87	
Glycine	75.1	α-COOH	4.57×10^{-3}	2.34	α-COO⁻	2.19×10^{-12}	11.66	5.97
		α-NH₃⁺	2.51×10^{-10}	9.6	α-NH₂	3.98×10^{-5}	4.4	
Histidine	155.2	α-COOH	1.51×10^{-2}	1.82	α-COO⁻	6.61×10^{-13}	12.18	7.58
		Imidazole-NH⁺	1.0×10^{-7}	6.0	Imidazole-N	1.0×10^{-9}	8.0	
		α-NH₃⁺	6.76×10^{-10}	9.17	α-NH₂	1.48×10^{-5}	4.83	
Homocysteine	135.2	α-COOH	6.03×10^{-3}	2.22	α-COO⁻	1.66×10^{-12}	11.78	5.54
		α-NH₃⁺	1.35×10^{-9}	8.87	α-NH₂	7.41×10^{-6}	5.13	
		γ-SH	1.38×10^{-11}	10.86	γ-S⁻	7.25×10^{-4}	3.14	
Homocystine	268.3	α-COOH	2.57×10^{-2}	1.59	α-COO⁻	3.89×10^{-13}	12.41	5.53
		α-COOH	2.88×10^{-3}	2.54	α-COO⁻	3.47×10^{-12}	11.46	
		α-NH₃⁺	3.02×10^{-9}	8.52	α-NH₂	3.31×10^{-6}	5.48	
		α-NH₃⁺	3.63×10^{-10}	9.44	α-NH₂	2.76×10^{-5}	4.56	
Hydroxylysine	162.2	α-COOH	7.41×10^{-3}	2.13	α-COO⁻	1.35×10^{-12}	11.87	9.15
		α-NH₃⁺	2.40×10^{-9}	8.62	α-NH₂	4.17×10^{-6}	5.38	
		ε-NH₃⁺	2.14×10^{-10}	9.67	ε-NH₂	4.68×10^{-5}	4.33	
Hydroxyproline	131.1	α-COOH	1.20×10^{-2}	1.92	α-COO⁻	8.32×10^{-13}	12.08	5.83
		α-NH₃⁺	1.86×10^{-10}	9.73	α-NH₂	5.37×10^{-5}	4.27	
Isoleucine	131.2	α-COOH	4.37×10^{-3}	2.36	α-COO⁻	2.29×10^{-12}	11.64	6.02
		α-NH₃⁺	2.09×10^{-10}	9.68	α-NH₂	4.78×10^{-5}	4.32	
Leucine	131.2	α-COOH	4.37×10^{-3}	2.36	α-COO⁻	2.29×10^{-12}	11.64	5.98
		α-NH₃⁺	2.51×10^{-10}	9.60	α-NH₂	3.98×10^{-5}	4.40	
Lysine	146.2	α-COOH	6.61×10^{-3}	2.18	α-COO⁻	1.51×10^{-12}	11.82	9.74
		α-NH₃⁺	1.12×10^{-9}	8.95	α-NH₂	8.91×10^{-6}	5.05	
		ε-NH₃⁺	2.95×10^{-11}	10.53	ε-NH₂	3.39×10^{-4}	3.47	

Name	MW	Acid group	K	pK	Base group	K	pK	
Methionine	149.2	α-COOH	5.25×10^{-3}	2.28	α-COO⁻	1.91×10^{-12}	11.72	5.75
		α-NH₃⁺	6.17×10^{-10}	9.21	α-NH₂	1.62×10^{-5}	4.79	
Ornithine	132.2	α-COOH	1.15×10^{-2}	1.94	α-COO⁻	8.71×10^{-13}	12.06	9.70
		α-NH₃⁺	2.24×10^{-9}	8.65	α-NH₂	4.47×10^{-6}	5.35	
		δ-NH₃⁺	1.74×10^{-11}	10.76	δ-NH₂	5.76×10^{-4}	3.24	
Phenylalanine	165.2	α-COOH	1.48×10^{-2}	1.83	α-COO⁻	6.76×10^{-13}	12.17	5.98
		α-NH₃⁺	7.41×10^{-10}	9.13	α-NH₂	1.35×10^{-5}	4.87	
Proline	115.1	α-COOH	1.02×10^{-2}	1.99	α-COO⁻	9.77×10^{-13}	12.01	6.10
		α-NH₃⁺	2.51×10^{-11}	10.60	α-NH₂	3.98×10^{-4}	3.40	
Serine	105.1	α-COOH	6.17×10^{-3}	2.21	α-COO⁻	1.62×10^{-12}	11.79	5.68
		α-NH₃⁺	7.08×10^{-10}	9.15	α-NH₂	1.41×10^{-5}	4.85	
Taurine	125.1	-SO₃H	3.16×10^{-2}	1.5	-SO₃⁻	3.16×10^{-13}	12.5	5.12
		α-NH₃⁺	1.82×10^{-9}	8.74	α-NH₂	5.50×10^{-6}	5.26	
Threonine	119.1	α-COOH	2.35×10^{-3}	2.63	α-COO⁻	4.27×10^{-12}	11.37	6.53
		α-NH₃⁺	3.72×10^{-11}	10.43	α-NH₂	2.69×10^{-4}	3.57	
Tryptophan	204.2	α-COOH	4.17×10^{-3}	2.38	α-COO⁻	2.40×10^{-12}	11.62	5.88
		α-NH₃⁺	4.07×10^{-10}	9.39	α-NH₂	2.46×10^{-5}	4.61	
Tyrosine	181.2	α-COOH	6.31×10^{-3}	2.20	α-COO⁻	1.59×10^{-11}	11.80	5.65
		α-NH₃⁺	7.76×10^{-10}	9.11	α-NH₂	1.29×10^{-5}	4.89	
		-OH	8.51×10^{-11}	10.07	-O⁻	1.18×10^{-4}	3.93	
Valine	117.1	α-COOH	4.79×10^{-3}	2.32	α-COO⁻	2.09×10^{-12}	11.68	5.97
		α-NH₃⁺	2.40×10^{-10}	9.62	α-NH₂	4.17×10^{-5}	4.38	
			K_{a_1}	pK_{a_1}		K_{b_3}	pK_{b_3}	
			K_{a_2}	pK_{a_2}		K_{b_2}	pK_{b_2}	
			K_{a_3}	pK_{a_3}		K_{b_1}	pK_{b_1}	

* K and pK values are numbered as shown on the right

Appendix: BIOCHEMICAL ENERGETICS

II

A. COUPLED REACTIONS AND ENERGY-RICH COMPOUNDS

Chemical reactions may be classified as: (a) "exergonic"—those that yield energy (that is, are capable of doing work), (b) "endergonic"—those that utilize (require) energy (work must be done to make them go), and (c) those that neither yield nor require energy. While it may not be immediately obvious why certain reactions are more exergonic than others, the student intuitively recognizes that biosynthetic processes, such as the assembling of large macromolecules from their constitutive subunits, require energy. Work must be done to build complex structures from simple building blocks. Living cells are exceedingly complex and delicate. Yet they not only maintain their integrity over long periods of time but also grow and multiply. In terms of energetics, this is accomplished by catalyzing certain exergonic reactions and trapping some of the energy released in "energy-rich" compounds. Biosynthetic (endergonic) reactions then are driven by this trapped energy. For example, suppose that the overall reaction sequence by which A is converted to B is exergonic, releasing 15 kcal of energy:

(1) $$A \rightarrow B + 15 \text{ kcal}$$

All of this energy would be wasted as heat if the reaction sequence proceeded as written. In a living cell, a portion of the total energy is trapped by coupling the reaction sequence to the endergonic synthesis of an "energy-rich" compound $X \sim Y$. If, for example, the synthesis of $X \sim Y$ requires 8 kcal, then the overall coupled reaction releases only 7 kcal.

(2)

$$A \longrightarrow B + 7 \text{ kcal}$$

$$X + Y \qquad X \sim Y$$

Now suppose that a complex molecule, C-D, is to be synthesized from its components $C + D$. The reaction is endergonic and will go only if 5 kcal of energy are supplied.

(3)
$$5 \text{ kcal} + C + D \rightarrow C\text{-}D$$

The 5 kcal of energy can be supplied by the "energy-rich" compound, $X \sim Y$. Because $X \sim Y$ has stored 8 kcal and only 5 kcal are required, 3 kcal are released by the overall coupled reaction.

(4)

$$C + D \longrightarrow C\text{-}D + 3 \text{ kcal}$$

$$X \sim Y \qquad X + Y$$

It is apparent that if $X \sim Y$ were degraded into its components without coupling its breakdown to an endergonic reaction, 8 kcal of energy would be wasted.

(5)
$$X \sim Y \rightarrow X + Y + 8 \text{ kcal}$$

These energy relationships are illustrated in Figure AII-1. Note that none of the reactions proceed with 100% efficiency. Of the original 15 kcal made available, only 8 kcal are conserved in $X \sim Y$; of the 8 kcal conserved, only 5 kcal are trapped in C-D.

The coupled-reaction concept adequately illustrates the principle of energy conservation and utilization in living cells. However, the actual mechanisms of energy coupling in living cells seldom involve the simultaneous catalysis of two reactions. Instead, the net effect is generally

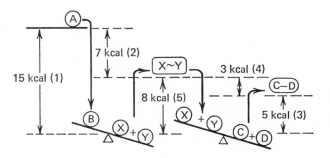

Figure AII-1 Energy relationships between coupled reactions

obtained by catalyzing two consecutive reactions involving a common intermediate.

As an actual example, consider the oxidation of glyceraldehyde-3-phosphate to 3-phosphoglyceric acid.

(6) 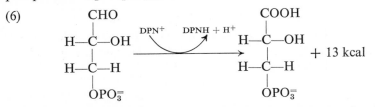 + 13 kcal

This oxidation-reduction reaction yields sufficient energy to drive the synthesis of ATP from ADP and P_i in a hypothetical coupled reaction.

(7) 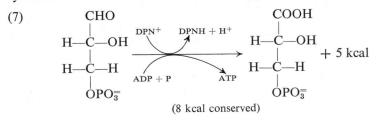 + 5 kcal

(8 kcal conserved)

In the living cell the two reactions are not actually coupled as shown. Instead, an energy-rich acyl phosphate is formed simultaneously with the oxidation.

(8) 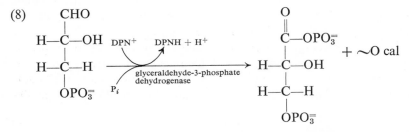 + ~0 cal

The energy is conserved by transferring the phosphate group to ADP in a subsequent reaction.

(9) 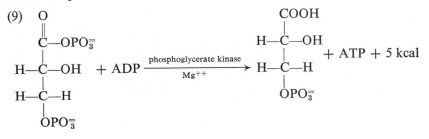 + ATP + 5 kcal

The sum of reactions 8 and 9 effectively equals the coupled reaction 7.

The formation of phosphoenolpyruvate (PEP) from 2-phospho-glyceric acid (2-PGA) illustrates a different principle.

(10)

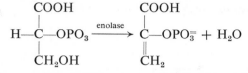

In this reaction an energy-rich compound is produced from an energy-poor precursor by a simple dehydration reaction. The student is tempted to ask "where did the energy come from?" The answer lies in the restricted definition of "energy-rich" as used by biochemists. In spite of the fact that 2-PGA is considered relatively "energy-poor," it yields about the same amount of energy as PEP if both are *burned* to CO_2, H_2O, and P_i. The dehydration reaction results in a rearrangement of electrons so that a much larger portion of the total potential energy becomes available upon *hydrolysis*. An "energy-rich" compound (to the biochemist) is one that releases a relatively large amount of energy (7–13 kcal per mole under standard-state conditions) upon hydrolysis.

Biosynthetic reactions also seldom involve the simultaneous coupling of two distinct reactions, although the net effect is the same. For example, consider the synthesis of glucose-6-phosphate from glucose and inorganic phosphate.

(11)   glucose + P_i + 3 kcal → G-6-P

The reaction is endergonic and requires 3 kcal/mole which can be supplied by the hydrolysis of ATP. The hypothetical coupled reaction is shown below.

(12)   glucose + P_i ⟶ G-6-P + HOH + 5 kcal
              ATP   ADP + P_i
             +HOH

The reaction actually catalyzed by hexokinase is the sum of the two individual reactions.

(13)   glucose + ATP $\xrightarrow[Mg^{++}]{hexokinase}$ G-6-P + ADP + 5 kcal

Not only does ATP supply the energy, it supplies the phosphate group as well. In fact, the energy in energy-rich compounds is seldom released by hydrolysis in living cells. Rather, the potential energy is used as "group-transfer potential,"—the potential energy is used to transfer a portion of the energy-rich molecule to an acceptor which, in effect, "activates" the acceptor. ATP is used in cells as: (a) a phosphate donor, (b) a pyro-phosphate donor, (c) an AMP donor, and (d) an adenosine donor.

Let us consider another reaction involving the synthesis of glutamine from glutamic acid and ammonia. The formation of this amide requires energy which can be supplied by ATP. The simple condensation reaction and the hypothetical coupled reaction are shown below.

(14)

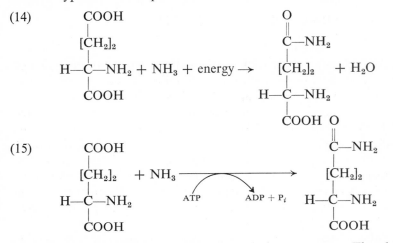

(15)

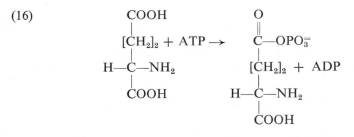

In the cell the overall reaction proceeds in two steps. The glutamic acid is first activated by ATP.

(16)

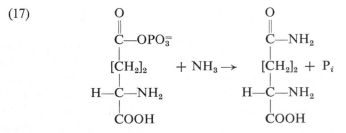

The activated glutamic acid (γ-glutamylphosphate) then condenses with NH_3.

(17)

$$
\begin{array}{c}
\text{O} \\
\| \\
\text{C--OPO}_3^= \\
| \\
\text{[CH}_2\text{]}_2 \\
| \\
\text{H--C--NH}_2 \\
| \\
\text{COOH}
\end{array}
+ NH_3 \rightarrow
\begin{array}{c}
\text{O} \\
\| \\
\text{C--NH}_2 \\
| \\
\text{[CH}_2\text{]}_2 \\
| \\
\text{H--C--NH}_2 \\
| \\
\text{COOH}
\end{array}
+ P_i
$$

The net effect is identical to the coupled reaction 15.

B. FREE ENERGY CHANGE (ΔF)

The energy released or utilized in a chemical reaction represents the *difference* between the energy contents of the products and the reactants. At constant temperature and pressure the energy difference is called the "free energy difference," ΔF, and is the total energy change occurring as a result of the reaction. By definition, ΔF is the free energy content of the products minus the free energy content of the reactants. Thus, for the exergonic reaction 1, we obtain the following:

(1) $$A \rightarrow B + 15 \text{ kcal}$$

$$\Delta F = F_B - F_A$$

In order for A to yield B plus energy, the free energy content of A must be greater than the free energy content of B.

$$\Delta F = (\text{some value}) - (\text{some larger value})$$
$$\Delta F = \text{a negative value}$$
$$\Delta F = -15 \text{ kcal}$$

Thus, because of the definition of ΔF, exergonic reactions have negative ΔF values while endergonic reactions have positive values. When a reaction is written in reverse, the sign of ΔF changes. For example, the synthesis of $X \sim Y$ *requires* 8 kcal per mole. Therefore, ΔF for the synthetic (endergonic) reaction is $+8$ kcal/mole.

$$X + Y \rightarrow X \sim Y \qquad \therefore \quad \Delta F = +8 \text{ kcal/mole}$$

The degradation of $X \sim Y$ *releases* 8 kcal/mole.

$$X \sim Y \rightarrow X + Y \qquad \Delta F = -8 \text{ kcal/mole}$$

C. RELATIONSHIP BETWEEN ΔF AND REACTANT AND PRODUCT CONCENTRATIONS

Consider the reaction $A \rightarrow B$. What is the relationship between the amount of energy released and the concentrations of A and B? By analogy, let us assume that the concentrations of A and B are liquid volumes separated into two arms of a U-tube as shown in Figure AII-2. Let us also place a waterwheel in the arm containing A. When the stopcock separating A and B is opened (i.e., when an enzyme catalyzing the A→B reaction is added), A is converted to B. As the level of A drops, the waterwheel turns (work is done and energy is released). A is converted to B until equilibrium is attained (indicated by the dotted line). Note that

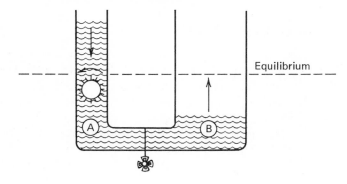

(a)

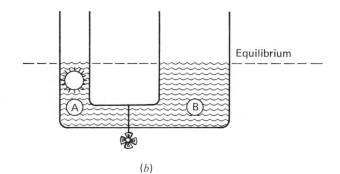

(b)

Figure AII-2 (a) [A]/[B] ratio at start of reaction is greater than ratio at equilibrium (b) [B]/[A] ratio at end of reaction equals the equilibrium ratio (K_{eq})

the *volumes* of A and B at equilibrium are not equal, indicating that the equilibrium lies far in favor of B. We can see from this analogy that the energy released depends on how far from equilibrium the original A/B ratio is. The greater the A/B ratio, the more work can be done in converting A to B. We can also see that if we start with an A/B ratio identical to the equilibrium A/B ratio (a B/A ratio equal to K_{eq}), no net transformation of A to B occurs; therefore, no work can be done (i.e., $\Delta F = 0$). At equilibrium the conversion of A to B or B to A *requires* energy. If we start with a B/A ratio greater than K_{eq} (the system is displaced from equilibrium in favor of B), then work must be done (energy is required) to convert still more A to B (i.e., ΔF is positive). However, the conversion of B to A now occurs with the production of energy (a minus ΔF value) until equilibrium (the minimum energy level of the system) is attained.

The U-tube analogy is not perfect because most reactions that occur in a living cell never attain equilibrium. Rather, the reactants and products are maintained (within narrow limits) at *steady-state* levels which may be quite different from the equilibrium levels. Nevertheless, the analogy illustrates how the energy released or utilized in a reaction depends on the displacement of the system from equilibrium. A mathematical statement of the ΔF of a reaction then must contain two terms—one that indicates the actual concentrations of the reactants and products and one that states the equilibrium concentrations. Such an expression is shown below:

$$\Delta F = -RT \ln K_{eq} + RT \ln \frac{[P_1]^a[P_2]^b \cdots}{[R_1]^c[R_2]^d \cdots}$$

$$= -2.3RT \log K_{eq} + 2.3RT \log \frac{[P_1]^a[P_2]^b \cdots}{[R_1]^c[R_2]^d \cdots}$$

(18) $$\boxed{\Delta F = -1363 \log K_{eq} + 1363 \log \frac{[P_1]^a[P_2]^b \cdots}{[R_1]^c[R_2]^d \cdots}} \quad \text{at } 25^\circ \text{ C}$$

where ΔF is the free energy change of the reaction when the products (P_1, P_2, etc.) and reactants (R_1, R_2, etc) are maintained at steady-state concentrations. The exponents a, b, c, and d are the coefficients of P_1, P_2, etc. in the balanced equation.

We can see that at equilibrium the [P]/[R] ratio equals K_{eq}; hence, $\Delta F = 0$.

To catalog and compare ΔF values for various reactions, chemists have agreed upon a "standard state" where all reactants and products are considered to be maintained at steady-state concentrations of $1\ M$. The standard state for gases is considered to be 1 atm partial pressure. Under this condition the [P]/[R] term (regardless of the exponents) is zero. The ΔF under standard-state conditions is designated ΔF°.

$$\Delta F = -1363 \log K_{eq} + 1363 \log 1$$
$$= -1363 \log K_{eq} + 0$$

(19) $$\boxed{\Delta F^\circ = -1363 \log K_{eq}}$$

Thus, the ΔF° value of a reaction is related to the K_{eq}. In fact, both the ΔF° and K_{eq} values impart the same information, namely, how far to the right a reaction will proceed. The correspondence between ΔF° and K_{eq} is shown in Table AII-1.

TABLE AII-1. CORRESPONDENCE BETWEEN K_{eq} AND $\Delta F°$ AT 25° C

K_{eq}	$\log K_{eq}$	$\Delta F°$
0.0001	−4	+5452 cal
0.001	−3	+4089 cal
0.01	−2	+2726 cal
0.1	−1	+1363 cal
1	0	0
10	1	−1363 cal
100	2	−2726 cal
1000	3	−4089 cal
10,000	4	−5452 cal

The ΔF under any particular set of concentration conditions can be calculated from reaction 18:

(18) $$\Delta F = -1363 \log K_{eq} + 1363 \log \frac{[P_1]^a[P_2]^b \cdots}{[R_1]^c[R_2]^d \cdots}$$

or

(20) $$\Delta F = \Delta F° + 1363 \log \frac{[P_1]^a[P_2]^b \cdots}{[R_1]^c[R_2]^d \cdots}$$

When the H^+ ion appears as a reactant or product, its standard state concentration is also taken as 1 M (i.e., pH = 0). However, most enzymes are denatured at pH 0 and, consequently, measurements of K_{eq} (to calculate $\Delta F°$) are impossible. Because of this, biochemists have adopted a modified standard state in which all reactants and products *except* H^+ are considered to be 1 M. The H^+ ion concentration is taken to be some physiological value (e.g., 10^{-7} M). The relationship between $\Delta F°$ and the modified standard state free energy change, designated $\Delta F'$, can be easily calculated. For example, consider a reaction that yields an H^+ ion as a product.

$$R \rightarrow P + H^+$$

$$\Delta F' = \Delta F° + 1363 \log \frac{[P][H^+]}{[R]}$$

$$= \Delta F° + 1363 \log [H^+] \quad \text{when [P] and [R] = 1 } M$$

$$= \Delta F° - 1363 \log \frac{1}{[H^+]}$$

(21) $$\Delta F' = \Delta F° - 1363 \text{ pH}$$

For a reaction involving the H^+ ion as a reactant the equation becomes:

(22)
$$\boxed{\Delta F' = \Delta F^\circ + 1363 \text{ pH}}$$

The ΔF of a reaction for any nonstandard concentration condition (at pH 7) can be calculated from the K'_{eq} or $\Delta F'$ values.

$$\Delta F = -1363 \log K'_{eq} + 1363 \log \frac{[P_1]^a [P_2]^b \cdots}{[R_1]^c [R_2]^d \cdots}$$

or (23)
$$\boxed{\Delta F = \Delta F' + 1363 \log \frac{[P_1]^a [P_2]^b \cdots}{[R_1]^c [R_2]^d \cdots}}$$

where K'_{eq} is the equilibrium constant for the reaction at pH 7.

The $\Delta F'$ values for the hydrolysis of some "energy-rich" and "energy-poor" compounds are listed in Table AII-2.

D. OXIDATION-REDUCTION REACTIONS

Many reactions that occur in living cells are oxidation-reduction reactions. Table AII-3 lists several compounds of biological importance and shows their relative tendencies to gain electrons at pH 7 under "standard conditions." The numerical values of E'_0 reflect the reduction potentials relative to the $H^+ + 1e^- \rightarrow \frac{1}{2} H_2$ half-reaction which is taken as -0.42 volt at pH 7. The value for the hydrogen half-reaction at pH 7 was calculated from the arbitrarily assigned value (E_0) of 0.00 volt under true standard-state conditions (1 M H^+ and 1 atm H_2). For those few half-reactions of biological importance that do not involve H^+ as a reactant, the E_0 and E'_0 values are essentially identical.

Because no substance can gain electrons without another substance losing electrons, a complete oxidation-reduction must be composed of two half-reactions. When any two of the half-reactions in Table AII-3 are coupled, the one with the greater tendency to gain electrons (the one with the *more positive* reduction potential) goes as written (as a reduction). Consequently, the other half-reaction (the one with the lesser tendency to gain electrons as shown by the *less positive* reduction potential) is driven backwards (as an oxidation). The reduced forms of those substances with highly negative reduction potentials are good reducing agents. The oxidized form of those substances with highly positive reduction potentials are good oxidizing agents, as shown on page 362.

TABLE AII-2. ΔF' VALUES FOR THE HYDROLYSIS OF SOME "ENERGY-RICH" AND "ENERGY-POOR" COMPOUNDS

General Type	Example	Hydrolysis Products	$\Delta F'$ (pH 6–8) kcal/mole	
Phosphosulfate anhydride	Adenosine phospho-sulfate (APS)	$AMP + SO_4^=$	-18	
Pyrophosphate	Inorganic P-P_i	$2\ P_i$	$-3?\quad -7?*$	
	ATP	$ADP + P_i$	-7.7 to -8	
	ATP	$AMP + P$-P_i	$-11?\quad -8.6?$	
	ADP	$AMP + P_i$	-6.4	
Acyl phosphate	Acetyl phosphate	Acetic acid $+ P_i$	-10	"Energy-rich"
	1,3-diphosphoglyceric acid	3-PGA $+ P_i$	-12	
Thioester	Acetyl CoA	Acetic acid $+$ CoA-SH	-8.2	
Enolic phosphate	PEP	Ketopyruvic acid $+ P_i$	-12.8	
Guanidinium phosphate	Creatine phosphate	Creatine $+ P_i$	-10.5	
Hemiacetal phosphate	α-D-glucose-1-phosphate	α- and β-D-glucose $+ P_i$	-5	
Simple phosphate ester	AMP	Adenosine $+ P_i$	-2	
	Glucose-6-phosphate	Glucose $+ P_i$	-3	"Energy-poor"
	α-glycerophosphate	Glycerol $+ P_i$	-2.5	

* Recent evidence suggests that the ordering of water molecules around the P_i produced upon hydrolysis may contribute a large negative ΔS value to the $\Delta F = \Delta H - T\Delta S$ expression. Inorganic pyrophosphate, therefore, may be relatively "energy poor." If the $\Delta F'$ of hydrolysis of PP_i is only -3 kcal/mole, then it is likely that the $\Delta F'$ for the reaction $ATP + H_2O \rightarrow AMP + PP_i$ is higher than generally believed (ca. -11 kcal/mole).

TABLE AII-3. STANDARD REDUCTION POTENTIALS OF SOME OXIDATION REDUCTION HALF-REACTIONS*

Reaction	Half-Reaction (Written as a Reduction)	E_0' at pH 7.0 (volts)
1	$\frac{1}{2}O_2 + 2H^+ + 2e^- \rightarrow H_2O$	0.816
2	$Fe^{+3} + 1e^- \rightarrow Fe^{+2}$	0.771
3	$\frac{1}{2}O_2 + H_2O + 2e^- \rightarrow H_2O_2$	0.30
4	cytochrome-a-$Fe^{+3} + 1e^- \rightarrow$ cytochrome-a-Fe^{+2}	0.29
5	cytochrome-c-$Fe^{+3} + 1e^- \rightarrow$ cytochrome-c-Fe^{+2}	0.25
6	2,6-dichlorophenolindophenol$_{(ox)} + 2H^+ + 2e^- \rightarrow$ 2,6-DCPP$_{(red)}$	0.22
7	crotonyl-CoA $+ 2H^+ + 2e^- \rightarrow$ butyryl-CoA	0.19
8	methemoglobin-$Fe^{+3} + 1e^- \rightarrow$ hemoglobin-Fe^{+2}	0.139
9	ubiquinone $+ 2H^+ + 2e^- \rightarrow$ ubiquinone-H_2	0.10
10	dehydroascorbate $+ 2H^+ + 2e^- \rightarrow$ ascorbate	0.06
11	metmyoglobin-$Fe^{+3} + 1e^- \rightarrow$ myoglobin-Fe^{+2}	0.046
12	fumarate $+ 2H^+ + 2e^- \rightarrow$ succinate	0.030
13	FAD $+ 2H^+ + 2e^- \rightarrow FADH_2$	−0.06
14	oxalacetate $+ 2H^+ + 2e^- \rightarrow$ malate	−0.102
15	pyruvate $+ NH_3 + 2H^+ + 2e^- \rightarrow$ alanine	−0.13
16	α-ketoglutarate $+ NH_3 + 2H^+ + 2e^- \rightarrow$ glutamate $+ H_2O$	−0.14
17	acetaldehyde $+ 2H^+ + 2e^- \rightarrow$ ethanol	−0.163
18	pyruvate $+ 2H^+ + 2e^- \rightarrow$ lactate	−0.190
19	riboflavin $+ 2H^+ + 2e^- \rightarrow$ riboflavin-H_2	−0.200
20	1,3-diphosphoglyceric acid $+ 2H^+ + 2e^- \rightarrow$ GAP $+ P_i$	−0.29
21	acetoacetate $+ 2H^+ + 2e^- \rightarrow \beta$-hydroxybutyrate	−0.290
22a	$DPN^+(NAD^+) + 2H^+ + 2e^- \rightarrow$ DPNH(NADH) $+ H^+$	−0.320
b	$TPN^+(NADP^+) + 2H^+ + 2e^- \rightarrow$ TPNH(NADPH) $+ H^+$	−0.320
23	pyruvate $+ CO_2 + 2H^+ + 2e^- \rightarrow$ malate	−0.33
24	uric acid $+ 2H^+ + 2e^- \rightarrow$ xanthine	−0.36
25	acetyl-CoA $+ 2H^+ + 2e^- \rightarrow$ acetaldehyde $+$ CoA	−0.41
26	$CO_2 + 2H^+ + 2e^- \rightarrow$ formate	−0.420
27	$H^+ + 1e^- \rightarrow \frac{1}{2}H_2$	−0.420
28	ferredoxin-$Fe^{+3} + 1e^- \rightarrow$ ferredoxin-Fe^{+2}	−0.432
29	acetate $+ 2H^+ + 2e^- \rightarrow$ acetaldehyde	−0.60
30	acetate $+ CO_2 + 2H^+ + 2e^- \rightarrow$ pyruvate	−0.70

* Standard conditions: Unit activity of all components except H^+ which is maintained at $10^{-7} M$. Gases are at 1 atm pressure.

Oxidized Form			Reduced Form		Relative E_0'
A	$+2H^+ + 2e^- \rightarrow$		AH_2		$+3$
B	$+2H^+ + 2e^- \rightarrow$		BH_2		$+2$
C	$+2H^+ + 2e^- \rightarrow$		CH_2		$+1$
D	$+2H^+ + 2e^- \rightarrow$		DH_2		0
E	$+2H^+ + 2e^- \rightarrow$		EH_2		-1
F	$+2H^+ + 2e^- \rightarrow$		FH_2		-2
G	$+2H^+ + 2e^- \rightarrow$		GH_2		-3

Increasing strength as an oxidizing agent (A → G, upward arrow on left)

Increasing strength as a reducing agent (downward arrow on right)

For example, EH_2 is a much better reducing agent than DH_2, CH_2, BH_2, or AH_2 but not as good a reducing agent as FH_2 of GH_2. C is a much better oxidizing agent than D, E, F, or G, but not as good an oxidizing agent as B or A. In other words, the better oxidizing agent is that substance which has the greater tendency to become reduced; the better reducing agent is that substance which has the greater tendency to become oxidized.

The relative tendency of the *overall* oxidation-reduction reaction to go may be calculated from the difference between the reduction potentials of the component half-reactions:

$\Delta E_0' = [E_0'$ of the half-reaction containing the oxidizing agent]
$\quad\quad - [E_0'$ of the half-reaction containing the reducing agent]

For example, consider the overall reaction resulting from the coupling of half-reactions 12 and 22a (Table AII-3). Half-reaction 12 goes as a reduction because it has the higher reduction potential. Half-reaction 22a then goes as an oxidation.

$$\text{Fumarate} + 2H^+ + 2e^- \rightarrow \text{Succinate}$$
$$\text{DPNH} + H^+ \quad\quad \rightarrow \text{DPN}^+ + 2H^+ + 2e^-$$

Overall: $\text{Fumarate} + \text{DPNH} + H^+ \rightarrow \text{DPN}^+ + \text{Succinate}$

The fumarate is reduced (it gains electrons); therefore, it is the "oxidizing agent." The DPNH is oxidized (it loses electrons); therefore, it is the "reducing agent."

$$\Delta E_0' = (0.03) - (-0.32)$$
$$\Delta E_0' = 0.03 + 0.32$$
$$\Delta E_0' = +0.35 \text{ volt}$$

The $\Delta F'$ of the reaction may be calculated from the following relationship:

$$\boxed{\Delta F' = -n\mathscr{F}\,\Delta E_0'}$$

where n is the number of electrons transferred per mole and $\mathscr{F}$ is Faraday's constant (23,063), a factor converting volts/equivalent to calories/equivalent. Note that $\Delta E_0'$ must be positive in order to obtain a negative ΔF, that is, a positive $\Delta E_0'$ indicates a "spontaneous" reaction. The K_{eq}' of the reaction is related to the $\Delta F'$ (and hence to $\Delta E_0'$) in the following ways:

$$\Delta F' = -RT \ln K_{eq}'$$

or

$$\Delta F' = -2.3RT \log K_{eq}'$$

$$\boxed{\Delta F' = -1363 \log K_{eq}' \text{ at } 25° \text{ C } (T = 298° \text{ K})}$$

$$\Delta F' = -n\mathscr{F} \Delta E_0' = -2.3RT \log K_{eq}'$$

$$\Delta E_0' = \frac{-2.3RT}{-n\mathscr{F}} \log K_{eq}'$$

$$\boxed{\Delta E_0' = \frac{0.06}{n} \log K_{eq}'} \qquad \text{at } 30° \text{ C}$$

or

$$\boxed{\Delta E_0' = \frac{0.059}{n} \log K_{eq}'} \qquad \text{at } 25° \text{ C}$$

The reduction potential of a half-reaction in which the oxidized and reduced forms of the substance are present at nonstandard concentrations may be calculated from the following expression, called the Nernst Equation.

$$E = E_0' + \frac{RT}{n\mathscr{F}} \ln \frac{[\text{oxidized form}]}{[\text{reduced form}]}$$

or

$$E = E_0' + \frac{2.3RT}{n\mathscr{F}} \log \frac{[\text{oxidized form}]}{[\text{reduced form}]}$$

At 30° C, the $2.3RT/\mathscr{F}$ term is 0.06 volt.

$$\boxed{E = E_0' + \frac{0.06}{n} \log \frac{[\text{oxidized form}]}{[\text{reduced form}]}}$$

E. pH ELECTRODES

Consider the two half-reactions shown below.

$$E_0$$

(1) $X + ne^- \rightarrow X^{-n}$ Zv

(2) $Y + ne^- + nH^+ \rightarrow YH_2$ Mv

Reaction 1 represents any half-reaction whose E_0 value is known (for example, the normal calomel electrode: $Hg_2Cl_2 + 2e^- \rightarrow 2Hg^0 + 2Cl^-$, or the silver/silver chloride electrode: $AgCl + 1e^- \rightarrow Ag^0 + Cl^-$). Reaction 2 represents any other half-reaction that is H^+ ion dependent (for example, $H^+ + 1e^- \rightarrow \frac{1}{2} H_2$). If the E_0 value of reaction 1 and the concentrations of X and X^{-n} are known, then the E value for half-reaction 1 is known. If the ΔE value of the overall oxidation-reduction reaction can be measured and the E_0 value for reaction 2 and the concentrations of Y and YH_2 are known, then the only remaining unknown value (the H^+ ion concentration) can be calculated. In this manner, oxidation-reduction reactions can be used to measure the pH of a solution.

In practice, commercial "pH meters" employing a "glass electrode" and a reference electrode are used to make pH measurements. The E value (potential) of the glass electrode does not result from an oxidation-reduction reaction but rather from the transfer of H^+ ions through a thin glass membrane. A setup for measuring pH using a glass electrode is shown in Figure AII-3.

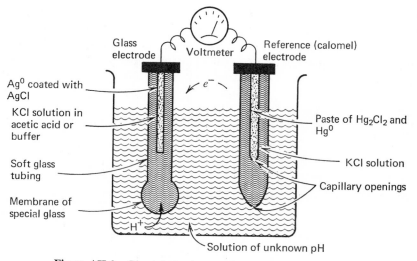

Figure AII-3   Glass electrode assembly for pH measurements

The glass electrode assembly is composed of a thin-walled bulb of a special H^+ ion-permeable glass blown on the end of a piece of soft glass tubing. The inner space contains a KCl solution in acetic acid or buffer and some kind of standard electrode such as the $Ag^0/AgCl$ electrode. The concentration of Ag^+ ions in the solution is fixed by the K_{SP} of AgCl and the concentration of Cl^- ions provided by the KCl. The H^+ ion concentration is fixed by the buffer or weak acetic acid. The $Ag^0/AgCl$ electrode has a certain potential relative to the reference electrode. Superimposed on this potential is an additional potential resulting from the difference in H^+ ion concentration on either side of the glass membrane. The potential of the glass electrode is given by the usual relation: $E = E_0 + 0.059$ log $[H^+]$. The E_0 value of the glass electrode not only depends on the potential of the $Ag^0/AgCl$ half-cell, but also depends on the pH of the internal buffer and the particular characteristics of the glass used to construct the membrane. The ΔE_0 of the overall cell depends on these factors and the E_0 value of the reference cell. Because the glass electrode has no fixed E_0 value, it does not provide an absolute measure of the H^+ ion concentration (such as the $\frac{1}{2}H_2/H^+$ electrode would). The glass electrode must be calibrated periodically against buffers of known pH.

We can visualize the glass electrode as operating in the following way: If the electrode is placed into a solution with a pH lower than that of the internal buffer, H^+ ions migrate from the external solution into the thin glass membrane. This, in turn, causes positive ions to be displaced from the inside surface of the glass membrane to the internal solution. (These ions may be Na^+ or K^+ or Ca^{++} from the glass.) The momentary excess of positive charges in the glass electrode compartment causes electrons to flow from the reference electrode compartment ($2Hg^0 + 2Cl^- \rightarrow Hg_2Cl_2 + 2e^-/2AgCl + 2e^- \rightarrow 2Ag^0 + 2Cl^-$). If the pH of the external solution were higher than that in the glass electrode compartment, then H^+ ions would migrate out of the glass electrode compartment, leaving it with a momentary negative charge. Electrons would then flow from the glass electrode to the reference electrode ($2Ag^0 + 2Cl^- \rightarrow 2AgCl + 2e^-/Hg_2Cl_2 + 2e^- \rightarrow 2Hg^0 + 2Cl^-$). The voltmeter measures the potential of the electrons which is proportional to the H^+ ion gradient established across the glass membrane.

Appendix: ENZYME KINETICS

III

A. EFFECT OF ENZYMES ON ACTIVATION ENERGY

Although a reaction may have a large negative ΔF, it does not necessarily proceed at a measurable velocity. For example, the oxidation of glucose to CO_2 and H_2O has a $\Delta F'$ of -686 kcal/mole. Yet we all know that crystalline glucose (or a sterile solution of glucose) in the presence of air does not spontaneously oxidize. The rate of a reaction depends on the number of molecules that attain a certain minimum "activation energy" per unit time. The activation energy is a barrier that must be overcome. The situation is analogous to a ball lying on a balcony several floors above the ground. The ball could do work if it were to roll off the balcony and fall to the ground. However, if the balcony is bounded by a wall, work has to be done first to lift the ball over the wall before it could drop. Employing the U-tube analogy shown in Figure AIII-1, we can see that the transformation of A to B could do work, but energy must be supplied first to lift A to a sufficient activated state. Activation energy for a chemical reaction could be supplied by adding heat to the system (i.e., by raising the temperature). Living cells, however, can only exist within narrow temperature limits and, furthermore, have no mechanism for specifically adding heat to certain reactions and not to others. Instead of *supplying* the activation energy for specific chemical reactions, cells operate by *lowering* the activation energy. *Enzymes* are protein catalysts that increase the rates of specific chemical reactions by lowering the activation energy required.

B. DERIVATION OF MICHAELIS-MENTEN EQUATION

The overall sequence of events during an enzyme-catalyzed reaction, $S \overset{E}{\rightleftharpoons} P$ may be depicted as shown below:

$$E + S \rightleftharpoons ES \rightleftharpoons EX \rightleftharpoons EY \rightleftharpoons EZ \rightleftharpoons \text{E-P} \rightleftharpoons P + E$$

The enzyme [E] first combines with the substrate [S] to form an enzyme-substrate complex [ES]. On the surface of the enzyme the substrate may go through one or more transitional forms (X, Y, Z) and finally be converted to the product [P]. The product then dissociates, allowing the free enzyme [E] to begin again.

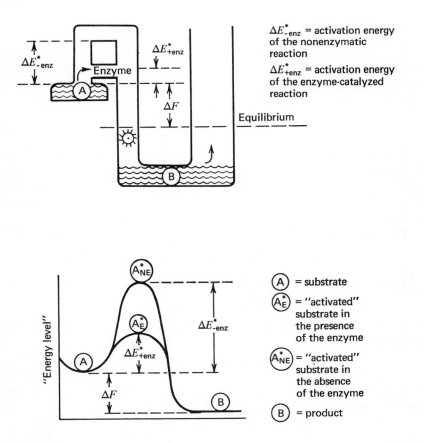

ΔE^*_{-enz} = activation energy of the nonenzymatic reaction

ΔE^*_{+enz} = activation energy of the enzyme-catalyzed reaction

Equilibrium

(A) = substrate

(A^*_E) = "activated" substrate in the presence of the enzyme

(A^*_{NE}) = "activated" substrate in the absence of the enzyme

(B) = product

Figure AIII-1 Energy relationships of a reaction in the presence and in the absence of an enzyme

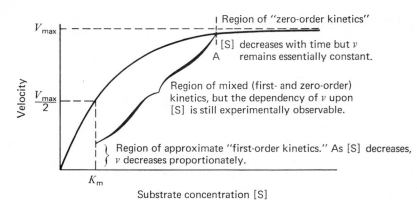

Figure AIII-2   Plot of initial velocity versus substrate concentration

For simplicity, assume that the sequence of events in the overall enzyme-catalyzed reaction involves only: (a) the reversible combination of the enzyme and substrate to form an ES complex and then (b) the reversible decomposition of the ES complex to form free enzyme and product.

$$E + S \underset{k_2}{\overset{k_1}{\rightleftharpoons}} ES \underset{k_4}{\overset{k_3}{\rightleftharpoons}} P + E$$

Assume also that a steady-state situation exists (after the first few milliseconds of the reaction) where the concentration of the ES complex does not change. Therefore the rate of formation of ES must equal its rate of decomposition. The rates of formation and decomposition can be written in terms of the appropriate rate constants, as shown below.

ES can be formed by two processes: $E + S \rightarrow ES$ and $E + P \rightarrow ES$.

$$\therefore \quad k_1[E][S] + k_4[P][E] = \text{rate of formation of ES}$$

ES can decompose by two processes: $ES \rightarrow E + S$ and $ES \rightarrow P + E$.

$$\therefore \quad k_2[ES] + k_3[ES] = \text{rate of decomposition of ES}$$

Because the rate of formation of ES = rate of decomposition, we can write the following

$$k_1[E][S] + k_4[P][E] = k_2[ES] + k_3[ES]$$

The [E] and [ES] terms may be factored out and the expression simplified by rearrangement.

$$[E](k_1[S] + k_4[P]) = [ES](k_2 + k_3)$$

$$\frac{[ES]}{[E]} = \frac{k_1[S] + k_4[P]}{k_2 + k_3}$$

The above expression may be rewritten in two parts.

$$\frac{[ES]}{[E]} = \frac{k_1[S]}{k_2 + k_3} + \frac{k_4[P]}{k_2 + k_3}$$

The concentration of P during the first few moments of the reaction is very low, consequently its influence on the *initial* velocity of the reaction is negligible. If we consider only the early part of the enzyme-catalyzed reaction, the rate of formation of ES from E + P may be neglected. As a result, the term containing k_4 and [P] may be safely neglected.

$$\frac{[ES]}{[E]} = \frac{k_1[S]}{k_2 + k_3}$$

Inverting:

$$\frac{[E]}{[ES]} = \frac{k_2 + k_3}{k_1[S]}$$

All the remaining rate constants may be combined into a single constant, K_m.

$$\frac{k_2 + k_3}{k_1} = K_m$$

$$\therefore \quad \frac{[E]}{[ES]} = \frac{K_m}{[S]}$$

Let $[E]_T$ = total enzyme concentration. Then:

$$[E]_T = [E] + [ES]$$
$$[E] = [E]_T - [ES]$$

Dividing by [ES]:

$$\frac{[E]}{[ES]} = \frac{[E]_T - [ES]}{[ES]}$$

$$\frac{[E]_T - [ES]}{[ES]} = \frac{[E]_T}{[ES]} - \frac{[ES]}{[ES]} = \frac{[E]_T}{[ES]} - 1$$

$$\therefore \quad \frac{[E]_T}{[ES]} - 1 = \frac{K_m}{[S]}$$

$$\frac{[E]_T}{[ES]} = \frac{K_m}{[S]} + 1$$

The $[E]_T$ and [ES] cannot be easily determined. However, both can be expressed in terms of initial velocities of the reaction (v) at any given substrate concentration [S] and the *maximum* initial velocity of the reaction (V_{max}) at saturating concentrations of S.

The *maximum* initial velocity (V_{max}) is proportional to the *total* enzyme present (and is observed when *all* the enzyme is in the ES complex).

The velocity at any given substrate concentration is proportional to the amount of enzyme present as the ES complex.

$$V_{max} \propto [E]_T$$

$$v \propto [ES]$$

Therefore, we can substitute the ratio of V_{max}/v for the ratio of $[E]_T/[ES]$.

$$\frac{V_{max}}{v} = \frac{K_m}{[S]} + 1$$

Multiplying both sides of the equation by [S]:

$$\frac{V_{max}[S]}{v} = \frac{K_m[S]}{[S]} + [S]$$

Cancelling [S] terms and combining terms:

$$\frac{V_{max}[S]}{v} = \frac{K_m + [S]}{1}$$

Inverting:

$$\frac{v}{V_{max}[S]} = \frac{1}{K_m + [S]}$$

$$\boxed{\frac{v}{V_{max}} = \frac{[S]}{K_m + [S]}}$$

Note that K_m is *not* simply the dissociation constant for the ES complex because it contains the k_3 term. The true dissociation constant, K_D, for the dissociation reaction is k_2/k_1.

$$[ES] \underset{k_1}{\overset{k_2}{\rightleftharpoons}} [E] + [S]$$

The Michaelis-Menten equation is valid throughout the entire range of [S]. However, when [S] is very large compared to K_m, the denominator of the expression is dominated by the [S] term. Under this condition, v is essentially constant and equal to V_{max}; that is, "zero-order kinetics" are observed. When [S] is very small compared to K_m, the denominator is dominated by the K_m term. Under this condition, v is directly proportional to [S]; that is, "first-order kinetics" are observed. When $[S] = K_m$, $v = V_{max}/2$.

$$v = \frac{V_{max}[S]}{K_m + [S]} \xrightarrow[K_M \langle\langle [S]]{[S] \rangle\rangle K_M} \frac{V_{max}[S]}{[S]} = V_{max} \quad \text{(a constant)}$$

$$v = \frac{V_{max}[S]}{K_m + [S]} \xrightarrow[K_M \rangle\rangle [S]]{[S] \langle\langle K_M} \frac{V_{max}[S]}{K_m}$$

$$\frac{V_{max}}{K_m} = \text{first-order rate constant} = k$$

$$\therefore \quad v = k[S]$$

For all practical purposes, we may assume that $v = V_{max}$ (i.e., "zero-order kinetics" are observed) when $[S] \geq 100 \, K_m$ and that $v = k[S]$ (i.e., "first-order kinetics" are observed) at $[S] \leq 0.01 \, K_m$.

C. COMPETITIVE INHIBITORS

Inhibitors are substances that decrease the velocity of an enzyme-catalyzed reaction. A "competitive" inhibitor is a compound that combines with the enzyme in such a way as to prevent the enzyme from combining with the substrate. Frequently, competitive inhibitors are either nonmetabolizable analogs or derivatives of the true substrate, or they are alternate substrates of the enzyme. Because such inhibitors resemble the "true" (or preferred or alternate) substrate structurally, they compete with that substrate for the active site. If the active site is already occupied by a substrate molecule, the inhibitor cannot combine with the enzyme, and vice-versa. Thus, the inhibitor and the substrate can only combine with free enzyme.

A classical example of a competitive inhibitor is malonic acid. Malonic acid inhibits the flavin-containing enzyme succinic dehydrogenase which catalyzes the oxidation of succinic acid to fumaric acid, as shown below.

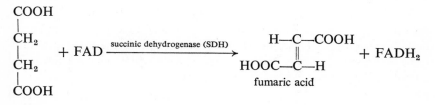

COOH
|
CH₂
| + FAD ——— succinic dehydrogenase (SDH) ———→
CH₂
|
COOH

succinic acid

Malonic acid resembles succinic acid sufficiently to combine with the enzyme.

malonic acid (MA)

However, because malonic acid has only one methylene group, it is obvious that no oxidation-reduction can take place. Only association of the enzyme and inhibitor, and dissociation of the EI complex, can occur.

$$\text{FAD-SDH} + \text{MA} \underset{k_{2i}}{\overset{k_{1i}}{\rightleftharpoons}} \text{FAD-SDH-MA}$$

$$\quad\;\;\text{(E)} \qquad\quad \text{(I)} \qquad\quad\;\; \text{(EI)}$$

A dissociation constant, K_i, can be written:

$$K_D = \frac{k_2}{k_1} = \frac{[\text{E}][\text{I}]}{[\text{EI}]} = K_i$$

Another example is the inhibition of the hexokinase-catalyzed reaction between glucose and ATP by fructose or mannose. Glucose, fructose, and mannose are all substrates of hexokinase and can be converted to product (hexose-6-phosphate). All three hexoses combine with the enzyme at the same active site. Consequently, the utilization of any one of the hexoses is inhibited in the presence of either of the other two.

There are many compounds that yield competitive inhibition kinetics, yet they bear no structural relationship to the substrate. The inhibitor is generally an end-product or near end-product of a metabolic pathway; the enzyme is one that catalyzes an early reaction (or a branch-point reaction) in the pathway. The phenomenon is called "feedback inhibition." The inhibitor ("effector" or "regulator") combines with the enzyme at a position other than the active (substrate) site. The combination of the inhibitor with the enzyme causes a change in the "conformation" (tertiary or quaternary structure) of the enzyme that results in a decreased affinity for the substrate.

A Michaelis-Menten expression relating v, V_{max}, [S], and K_m in the presence of a competitive inhibitor can be derived in a manner similar to that shown earlier. This time we must recognize that the enzyme at any time is present in three forms: free enzyme, [E]; enzyme-substrate complex, [ES]; and enzyme-inhibitor complex, [EI].

$$[\text{E}]_\text{T} = [\text{E}] + [\text{ES}] + [\text{EI}]$$

$$[\text{E}] = [\text{E}]_\text{T} - [\text{ES}] - [\text{EI}]$$

Dividing by [ES]:

$$\frac{[E]}{[ES]} = \frac{[E]_T - [ES] - [EI]}{[ES]}$$

As shown earlier:

$$\frac{[E]}{[ES]} = \frac{K_m}{[S]}$$

$$\therefore \quad \frac{K_m}{[S]} = \frac{[E]_T}{[ES]} - \frac{[ES]}{[ES]} - \frac{[EI]}{[ES]}$$

Also as shown earlier:

$$[EI] = \frac{[E][I]}{K_i}$$

and

$$\frac{1}{[ES]} = \frac{K_m}{[E][S]}$$

$$\therefore \quad \frac{K_m}{[S]} = \frac{[E]_T}{[ES]} - 1 - \frac{[E][I]}{K_i} \times \frac{K_m}{[E][S]}$$

$$\frac{[E]_T}{[ES]} = \frac{K_m}{[S]} + 1 + \frac{K_m[I][E]}{[S]K_i[E]}$$

Substituting V_{max}/v for $[E]_T/[ES]$:

$$\frac{V_{max}}{v} = \frac{K_m}{[S]} + \frac{K_m[I]}{[S]K_i} + 1$$

Multiplying both sides of the equation by [S]:

$$\frac{V_{max}[S]}{v} = \frac{K_m[S]}{[S]} + \frac{K_m[I][S]}{K_i[S]} + [S]$$

Cancelling [S] terms and factoring out K_m:

$$\frac{V_{max}[S]}{v} = K_m\left(1 + \frac{[I]}{K_i}\right) + [S]$$

Inverting:

$$\frac{v}{V_{max}[S]} = \frac{1}{K_m\left(1 + \frac{[I]}{K_i}\right) + [S]}$$

Cross-multiplying [S]:

$$\frac{v}{V_{max}} = \frac{[S]}{K_m\left(1 + \dfrac{[I]}{K_i}\right) + [S]}$$

or

$$\frac{v}{V_{max}} = \frac{[S]}{K_{m_{app}} + [S]}$$

where:

$$K_{m_{app}} = K_m\left[1 + \frac{[I]}{K_i}\right]$$

[I] = inhibitor concentration

K_i = the dissociation constant
for the EI complex

In the above equation, K_i represents the dissociation constant for the EI complex. The affinity constant would be $1/K_i$. Thus, the lower the value of K_i, the greater is the affinity of the inhibitor for the enzyme.

We can see from the above equation that the velocity and the degree of inhibition of an enzyme-catalyzed reaction in the presence of a competitive inhibitor depend on [S], [I], and the relative affinities of the substrate and inhibitor for the enzyme (indicated by K_m and K_i, respectively). In the presence of a competitive inhibitor, V_{max} remains unchanged. However, a greater concentration of substrate is required to achieve a given velocity.

D. NONCOMPETITIVE INHIBITORS

A "noncompetitive" inhibitor can combine with either the free enzyme or with the enzyme-substrate complex. The inhibitor generally combines with the enzyme somewhere other than at the active site. Consequently, the degree of inhibition is independent of the substrate concentration. The reversible formation of the enzyme-inhibitor complex is given by the equation shown below.

$$E_{avail} + I \rightleftharpoons E_{avail} I$$

The enzyme species that are available to combine with the noncompetitive inhibitor include E and ES.

$$\underset{[E]_{avail}}{([E] + [ES])} + I \rightleftharpoons \underset{[E_{avail}I]}{[(EI] + [ESI])}$$

For simplicity, the EI and ESI complexes ($E_{avail}I$) will be referred to as EI.

A dissociation constant can be written for the EI complex:

$$K_i = \frac{[E]_{avail}[I]}{[EI]} = \frac{([E] + [ES])[I]}{[EI]}$$

The total enzyme is present in three forms:

$$[E]_T = [E] + [ES] + [EI]$$

$$[E] + [ES] = [E]_T - [EI]$$

Substituting $[E]_T - [EI]$ for $[E] + [ES]$:

$$K_i = \frac{([E]_T - [EI])[I]}{[EI]}$$

The velocity at any substrate concentration in the absence of an inhibitor is proportional to the total enzyme concentration.

$$v \propto [E]_T$$

The velocity at any substrate concentration in the presence of a noncompetitive inhibitor is proportional to the enzyme concentration that is not tied up with inhibitor.

$$v_i \propto [E]_T - [EI]$$

$$\therefore \quad \frac{v}{v_i} = \frac{[E]_T}{[E]_T - [EI]}$$

The [EI] can be expressed in terms of $[E]_T$, [I], and K_i.

$$K_i = \frac{([E]_T - [EI])[I]}{[EI]}$$

$$K_i[EI] = ([E]_T - [EI])[I]$$

$$K_i[EI] = [E]_T[I] - [EI][I]$$

$$K_i[EI] + [EI][I] = [E]_T[I]$$

$$[EI](K_i + [I]) = [E]_T[I]$$

$$[EI] = \frac{[E]_T[I]}{K_i + [I]}$$

Substituting the above value for [EI] into the v/v_i expression:

$$\frac{v}{v_i} = \frac{[E]_T}{[E]_T - \dfrac{[E_T][I]}{K_i + [I]}}$$

Dividing numerator and denominator of the right-hand expression by $[E]_T$:

$$\frac{v}{v_i} = \frac{[E]_T/[E]_T}{\dfrac{[E]_T}{[E]_T} - \dfrac{[E]_T[I]}{[E]_T(K_i + [I])}} = \frac{1}{1 - \dfrac{[I]}{K_i + [I]}}$$

Substituting $\left(K_i + [I]\right)/\left(K_i + [I]\right)$ for 1 in the denominator of the right-hand expression to combine terms:

$$\frac{v}{v_i} = \frac{1}{\dfrac{K_i + [I]}{K_i + [I]} - \dfrac{[I]}{K_i + [I]}} = \frac{1}{\dfrac{K_i + [I] - [I]}{K_i + [I]}} = \frac{K_i + [I]}{K_i + [I] - [I]}$$

$$\frac{v}{v_i} = \frac{K_i + [I]}{K_i}$$

Inverting:

$$\boxed{\frac{v_i}{v} = \frac{K_i}{K_i + [I]}}$$

In the presence of a noncompetitive inhibitor, the velocity still depends on [S] until saturation is attained. However, the degree of inhibition at any [I] is independent of [S] and depends only on [I] and K_i. The K_m is unchanged; V_{max}, however, is decreased.

E. METHODS OF PLOTTING ENZYME KINETICS DATA

There are several ways of determining graphically the effects of varying substrate and/or inhibitor concentrations on the velocity of an enzyme-catalyzed reaction. The curves shown in Figures AIII-3 through AIII-8 are based on various forms of the Michaelis-Menten equation; the curves are based on the assumption that the enzyme obeys hyperbolic saturation kinetics. The Michaelis-Menten equation states

$$\boxed{\frac{v}{V_{max}} = \frac{[S]}{K_m + [S]}}$$

where:

v = velocity at any substrate concentration

V_{max} = maximum attainable velocity

[S] = substrate concentration

K_m = Michaelis-Menten constant = substrate concentration required for half-maximum velocity

For example, Figure AIII-4 (Lineweaver-Burk plot) is based on the following linear transformation of the Michaelis-Menten equation:
Inverting:

$$\frac{V_{max}}{v} = \frac{K_m + [S]}{[S]}$$

Cross-multiplying:

$$\frac{1}{v} = \frac{K_m + [S]}{V_{max}[S]}$$

Separating terms:

$$\frac{1}{v} = \frac{K_m}{V_{max}[S]} + \frac{[S]}{V_{max}[S]} \qquad \text{or} \qquad \frac{1}{v} = \frac{K_m}{V_{max}} \frac{1}{[S]} + \frac{1}{V_{max}}$$

This form is the equation for a straight line ($y = mx + b$). Thus, if we plot y versus x where $y = 1/v$ and $x = 1/[S]$, then m (the slope) = K_m/V_{max} and b (the intercept on the y axis) = $1/V_{max}$. We can also see that when $y = 0$ (i.e., $1/v = 0$), $mx = -b$ (i.e., $K_m/V_{max} \times 1/[S] = -1/V_{max}$), and $x = -b/m$ (i.e., $1/[S] = -1/V_{max} \times V_{max}/K_m$ or $1/[S] = -1/K_m$). Thus, the intercept on the $1/[S]$ axis is $-1/K_m$.

Figure AIII-5 (Woolf plot) is based on the further manipulation of the Lineweaver-Burk equation.

$$\frac{1}{v} = \frac{K_m}{V_{max}} \frac{1}{[S]} + \frac{1}{V_{max}}$$

Multiplying by [S]:

$$\frac{[S]}{v} = \frac{[S]K_m}{V_{max}} \frac{1}{[S]} + \frac{[S]}{V_{max}}$$

$$\frac{[S]}{v} = \frac{K_m}{V_{max}} + \frac{[S]}{V_{max}} = \frac{K_m}{V_{max}} + \frac{1}{V_{max}}[S]$$

or

$$\boxed{\frac{[S]}{v} = \frac{1}{V_{max}}[S] + \frac{K_m}{V_{max}}}$$

The above equation has the general form for a straight line, $y = mx + b$, where

$$y = \frac{[S]}{v}, \quad m = \frac{1}{V_{max}}, \quad x = [S], \quad b = \frac{K_m}{V_{max}}$$

When

$$\frac{[S]}{v} = 0, \quad \frac{1}{V_{max}}[S] = -\frac{K_m}{V_{max}}, \quad \frac{[S]}{V_{max}} = \frac{-K_m}{V_{max}}, \quad [S] = -K_m,$$

i.e., the intercept on the [S] axis gives $-K_m$.

Similarly, when $[S] = 0$, the intercept on the $[S]/v$ axis gives K_m/V_{max}.

The linear equation for Figure AIII-6 (Hofstee plot) is obtained as follows:

$$v = \frac{V_{max}[S]}{K_m + [S]}$$

Dividing numerator and denominator by [S]:

$$v = \frac{V_{max}[S]/[S]}{(K_m + [S])/[S]} = \frac{V_{max}}{(K_m + [S])/[S]} = \frac{V_{max}}{(K_m/[S]) + 1}$$

Cross-multiplying:

$$v\left[\frac{K_m}{[S]} + 1\right] = V_{max}$$

$$\frac{vK_m}{[S]} + v = V_{max}$$

or

$$\frac{v}{[S]}K_m + v = V_{max}$$

$$v = V_{max} - K_m\frac{v}{[S]}$$

or

$$\boxed{v = -K_m\frac{v}{[S]} + V_{max}}$$

In this form, $y = v$, $m = -K_m$, $x = v/[S]$, and $b = V_{max}$. We can see that when $v/[S] = 0$, $v = V_{max}$. Similarly, when $v = 0$, $v/[S] = V_{max}/K_m$.

Plot of Initial Velocity, v, versus Substrate Concentration, [S]

This plot is shown in Figure AIII-3.

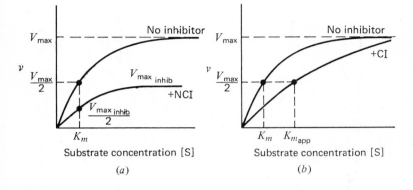

Figure AIII-3 Plots of initial velocity versus substrate concentration in the absence and in the presence of (a) a noncompetitive inhibitor (NCI) and (b) a competitive inhibitor (CI)

Plot of Reciprocal of Velocity, $1/v$, versus Reciprocal of Substrate Concentration

This plot, known as the Lineweaver-Burk plot, is shown in Figure AIII-4.

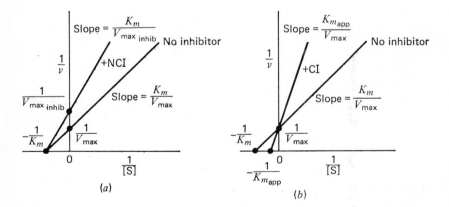

Figure AIII-4 Reciprocal (Lineweaver-Burk) plot of $1/v$ versus $1/[S]$ in the absence and in the presence of (a) a noncompetitive inhibitor (NCI) and (b) a competitive inhibitor (CI)

Plot of [S]/v versus [S]

This plot, known as the Woolf plot, is shown in Figure AIII-5.

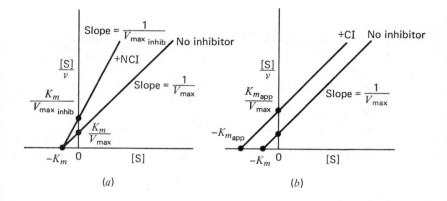

(a)   (b)

Figure AIII-5   Plot of [S]/v versus [S] (Woolf plot) in the absence and in the presence of (a) a noncompetitive inhibitor (NCI) and (b) a competitive inhibitor (CI)

Plot of v versus v/[S]

This plot, the Hofstee plot, is illustrated in Figure AIII-6.

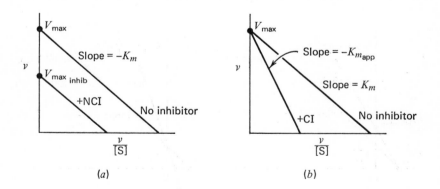

(a)   (b)

Figure AIII-6   Plot of v versus v/[S] (Hofstee plot) in the absence and in the presence of (a) a noncompetitive inhibitor (NCI) and (b) a competitive inhibitor (CI)

Plot of v versus $-\log$ [S]

This plot is shown in Figure AIII-7.

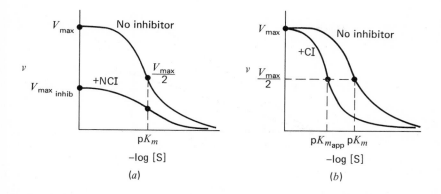

Figure AIII-7   Plot of v versus $-\log$ [S] in the absence and in the presence of (a) a noncompetitive inhibitor (NCI) and (b) a competitive inhibitor (CI)

Plot of $1/v$ versus [I]

Under some conditions it may be desirable to vary the inhibitor concentration and to maintain a constant substrate concentration. If the initial velocities are determined for two or more substrate concentrations (and varying inhibitor concentrations), the nature of the inhibitor (competitive or noncompetitive) may be determined from the curves shown in Figure AIII-8; and K_i can be directly determined.

Plot of v/v_i versus [I]

Another way of plotting data from experiments where the inhibitor concentration is varied is shown in Figure AIII-9. Note that for noncompetitive inhibition the slope of the curve is identical for all substrate concentrations employed. This is to be expected because, as we have seen, the degree of inhibition is independent of substrate concentration for a noncompetitive inhibitor at any given inhibitor concentration. If the degree of inhibition is independent of [S], the relative velocities expressed as v_i/v or v/v_i are also independent of [S]. For competitive inhibition the degree of inhibition decreases with increasing [S] at any given [I]. In other words, the relative velocities expressed as v_i/v increase and the relative velocities expressed as v/v_i decrease with increasing [S].

Sigmoidal Kinetics

Many enzymes do not exhibit classical Michaelis-Menten saturation (hyperbolic) kinetics. Instead, plots of velocity versus substrate concentration for these enzymes are sigmoidal. A sigmoidal response generally

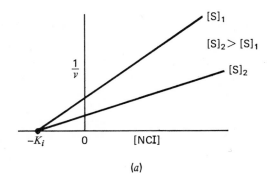

(a)

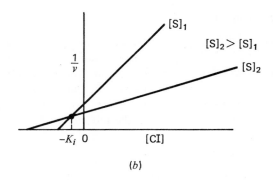

(b)

Figure AIII-8 The Dixon plot: (a) plot of $1/v$ versus concentration of noncompetitive inhibitor [NCI]   (b) plot of $1/v$ versus concentration of competitive inhibitor [CI]

indicates multiple and cooperative binding sites for the substrate (possibly on different subunits). The binding of a substrate to a site presumably results in a conformational change in the protein, so the affinity of the vacant site(s) for the substrate is increased. Thus, the substrate, in effect, is also an activator. The catalytic and effector sites may be different. A sigmoidal response allows a much more sensitive control of reaction rate by the substrate. As shown in Figures AIII-10 and AIII-11, the ratio of substrate concentrations required for 90% and 10% of V_{max} is 81 for classical Michaelis-Menten kinetics but less for sigmoidal kinetics.

Sigmoidal responses are described by the Hill equation which was originally derived for the oxygen-hemoglobin system.

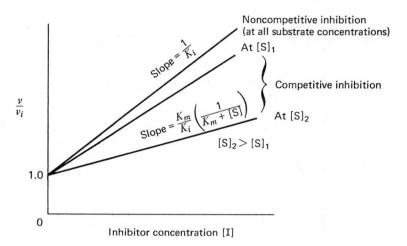

Figure AIII-9 Plot of relative velocity (v/v_i) versus inhibitor concentration [I]

The Hill equation can be rearranged into a straight line ($y = mx + b$) form as follows:

$$\frac{v}{V_{\max}} = \frac{[S]^n}{K' + [S]^n}$$

$$V_{\max}[S]^n = vK' + v[S]^n$$

$$V_{\max}[S]^n - v[S]^n = vK'$$

$$[S]^n(V_{\max} - v) = vK'$$

$$\frac{[S]^n(V_{\max} - v)}{v} = K'$$

$$\log [S]^n \frac{V_{\max} - v}{v} = \log K'$$

$$n \log [S] + \log \frac{V_{\max} - v}{v} = \log K'$$

$$\log \frac{V_{\max} - v}{v} = \log K' - n \log [S]$$

Multiplying both sides by -1:

$$-\log \frac{V_{\max} - v}{v} = n \log [S] - \log K'$$

or

$$\boxed{\log \frac{v}{V_{\max} - v} = n \log [S] - \log K'}$$

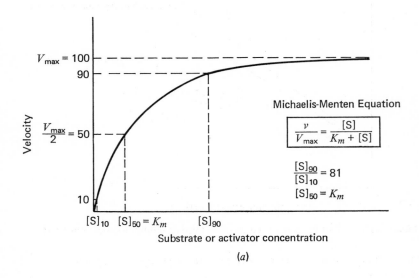

(a)

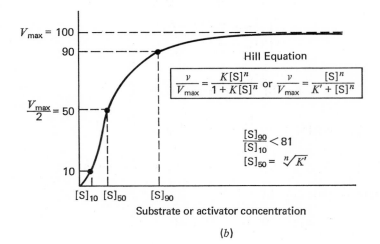

(b)

Figure AIII-10 (a) Hyperbolic saturation curve (b) Sigmoidal saturation curve

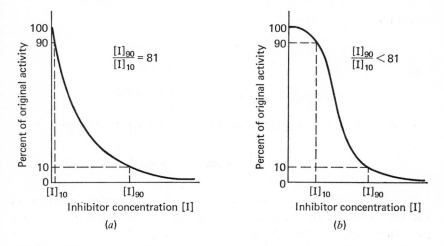

Figure AIII-11 (a) Hyperbolic inhibition curve (b) Sigmoidal inhibition curve

Thus, a plot of $\log v/(V_{\max} - v)$ versus $\log [S]$ is a straight line as shown in Figure AIII-12. The slope is n. When $v = \frac{1}{2}V_{\max}$, $v/(V_{\max} - v) = 1$ and $\log v/(V_{\max} - v) = 0$. The corresponding position on the $\log [S]$ axis gives $\log [S]_{50}$ (i.e., the logarithm of the substrate concentration that yields 50% of $V_{\max}$). We can see that for sigmoidal kinetics ($n \neq 1$), K' is *not* equivalent to the substrate concentration that yields 50% of $V_{\max}$

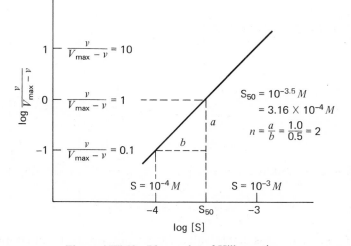

Figure AIII-12 Linear plot of Hill equation

but rather $[S]_{50}^n$. When

$$\log \frac{v}{V_{max} - v} = 0, \qquad [S] = [S]_{50}$$

and

$$n \log [S]_{50} = \log K' \qquad \therefore \qquad \boxed{K' = [S]_{50}^n}$$

For inhibitors that give a sigmoidal response, the linear expression is

$$\boxed{\log \frac{v}{V_{max} - v} = \log K_I' - n \log [I]}$$

The Hill equation says that the velocity of the reaction increases as the nth power of the substrate concentration when the substrate concentrations are very low compared to the K' value. We can see why this is so by considering the model shown below. Here we assume that the enzyme is a tetramer with one active site on each subunit. For simplicity, we assume first that the binding constants for all ES_n forms are the same:

$$\frac{[ES_1]}{[E][S]} = \frac{[ES_2]}{[ES_1][S]} = \frac{[ES_3]}{[ES_2][S]} = \frac{[ES_4]}{[ES_3][S]} = K_B$$

Furthermore, we assume that the only product-producing form is ES_4 and that each of the active sites on the tetramer can change S to P with equal facility.

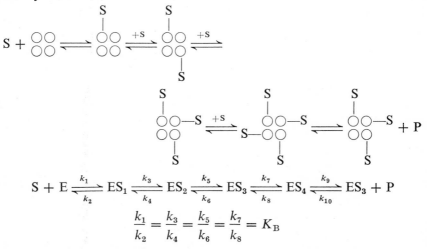

$$S + E \underset{k_2}{\overset{k_1}{\rightleftharpoons}} ES_1 \underset{k_4}{\overset{k_3}{\rightleftharpoons}} ES_2 \underset{k_6}{\overset{k_5}{\rightleftharpoons}} ES_3 \underset{k_8}{\overset{k_7}{\rightleftharpoons}} ES_4 \underset{k_{10}}{\overset{k_9}{\rightleftharpoons}} ES_3 + P$$

$$\frac{k_1}{k_2} = \frac{k_3}{k_4} = \frac{k_5}{k_6} = \frac{k_7}{k_8} = K_B$$

Let [E] be some reasonable value for an *in vitro* assay situation, e.g., 10^{-10} M. Let K_B be some reasonable value for a binding constant, e.g., 10^4. And let [S] be some usual "very low" substrate concentration for an *in vitro* assay, e.g., 10^{-6} M.

The velocity of the reaction is proportional to the concentration of the active species (the product-producing form, ES_4). Let us calculate the concentration of ES_4 at 10^{-6} M substrate.

$$\frac{[ES_1]}{[E][S]} = 10^4 \qquad [ES_1] = (10^4)(10^{-10})(10^{-6}) \qquad \boxed{[ES_1] = 10^{-12}\ M}$$

$$\frac{[ES_2]}{[ES_1][S]} = 10^4 \qquad [ES_2] = (10^4)(10^{-12})(10^{-6}) \qquad \boxed{[ES_2] = 10^{-14}\ M}$$

$$\frac{[ES_3]}{[ES_2][S]} = 10^4 \qquad [ES_3] = (10^4)(10^{-14})(10^{-6}) \qquad \boxed{[ES_3] = 10^{-16}\ M}$$

Similarly,

$$\boxed{[ES_4] = 10^{-18}\ M}$$

We could evaluate $[ES_4]$ in the following way:

$$E + S \rightleftharpoons ES_1 \qquad K_{B_1} = 10^4$$
$$ES_1 + S \rightleftharpoons ES_2 \qquad K_{B_2} = 10^4$$
$$ES_2 + S \rightleftharpoons ES_3 \qquad K_{B_3} = 10^4$$
$$ES_3 + S \rightleftharpoons ES_4 \qquad K_{B_4} = 10^4$$

Sum: $\quad E + 4S \rightleftharpoons ES_4 \qquad K_{B_{overall}} = K_{B_1} \times K_{B_2} \times K_{B_3} \times K_{B_4}$

$$K_{B_{overall}} = (10^4)^4 = 10^{16}$$

$$\frac{[ES_4]}{[E][S]^4} = K_{B_{overall}} = 10^{16}$$

$$[ES_4] = (10^{16})(10^{-10})(10^{-6})^4 = (10^6)(10^{-24})$$

$$\boxed{[ES_4] = 10^{-18}\ M}$$

Now let us calculate $[ES_4]$ when the substrate concentration is doubled.

$$\frac{[ES_1]}{[E][S]} = 10^4 \qquad [ES_1] = (10^4)(10^{-10})(2 \times 10^{-6})$$

$$\boxed{[ES_1] = 2 \times 10^{-12}\ M}$$

$$\frac{[ES_2]}{[ES_1][S]} = 10^4 \qquad [ES_2] = (10^4)(2 \times 10^{-12})(2 \times 10^{-6})$$

$$\boxed{[ES_2] = 4 \times 10^{-14}\ M}$$

$$\frac{[ES_3]}{[ES_2][S]} = 10^4 \qquad [ES_3] = (10^4)(4 \times 10^{-14})(2 \times 10^{-6})$$

$$\boxed{[ES_3] = 8 \times 10^{-16} \, M}$$

Similarly,

$$\boxed{[ES_4] = 16 \times 10^{-18} \, M}$$

If we continue to evaluate $[ES_4]$ at $[S] = 3 \times 10^{-6} \, M$, $4 \times 10^{-6} \, M$, etc., we shall obtain the results shown in Table AIII-1. Note that the results were calculated assuming concentrations of each ES_{n-1} that were not corrected for the amount that disappeared to form ES_n. This is a reasonable assumption only at very low substrate concentrations.

TABLE AIII-1.

[S]	[E]	$[ES_1]$	$[ES_2]$	$[ES_3]$	$[ES_4]$
1×10^{-6}	$\sim 0.99 \times 10^{-10}$	1×10^{-12}	1×10^{-14}	1×10^{-16}	1×10^{-18}
2×10^{-6}	$\sim 0.98 \times 10^{-10}$	2×10^{-12}	4×10^{-14}	8×10^{-16}	16×10^{-18}
3×10^{-6}	$\sim 0.97 \times 10^{-10}$	3×10^{-12}	9×10^{-14}	27×10^{-16}	81×10^{-18}
4×10^{-6}	$\sim 0.96 \times 10^{-10}$	4×10^{-12}	16×10^{-14}	64×10^{-16}	256×10^{-18}
5×10^{-6}	$\sim 0.95 \times 10^{-10}$	5×10^{-12}	25×10^{-14}	125×10^{-16}	625×10^{-18}

We can see that for any given increment in [S], the concentration of the product-producing form ES_4 and, hence, the reaction velocity increase by a factor of $([S]_2/[S]_1)^4$. In general, if $n > 1$, the v versus [S] curve increases exponentially at very low substrate concentrations. The exponential increase does not keep up indefinitely, however. The enzyme eventually becomes saturated. We can also see that the concentration of any ES_n increases by a factor less than $([S]_2/[S]_1)^n$ as the correction in the denominator term for the amount of ES_{n-1} converted to ES_n becomes significant. As a result the curve begins to flatten out as [S] increases, producing an overall sigmoidal shape.

If the binding constants are not identical for all $ES_{n-1} + S \rightleftharpoons ES_n$ reactions, then the amount of any ES_n form at any given [S] is different from the values shown in the table. However, v_2/v_1 still equals $([S]_2/[S]_1)^n$ at a very low [S]. An increase in the binding constants as we go from ES_1 to ES_2 to ES_3, etc., indicates "cooperative binding"—the binding of 1 molecule of substrate increases the affinity of the enzyme for another molecule of substrate.

If the binding constants are the same for all ES_n but both ES_{n-1} and ES_n are product producing, then the experimentally observed value of n is $n - 1$ because significantly more ES_{n-1} than ES_n is present. However, if the binding constant for ES_n is significantly higher than that for ES_{n-1} (such that both ES_n and ES_{n-1} are present at about the same concentrations), then the experimentally observed value of n is intermediate between the true n and $n - 1$ if both ES_n and ES_{n-1} are product producing.

In the model presented above, we assumed that all the substrate binding sites were catalytically active. If only two of the binding sites are catalytically active, n still equals 4 if only the ES_4 form is product producing.

The Hill equation can be derived in a manner similar to that described earlier for the Michaelis-Menten equation. However, to complete the derivation, it is necessary to assume: (a) only one product-producing form and (b) cooperative binding to the extent that the predominant enzyme-substrate complex present is the product-producing form. Then $v \propto [ES_n]$, $V_{max} \propto [E]_T$, and $[E] = [E]_T - [ES_n]$. The fact that fractional n values are observed experimentally indicates that these conditions are not really met. Because the number of substrate binding sites must be a whole number, the experimentally determined n value is frequently called an "interaction coefficient"; it is simply a measure of the sigmoidicity of the v versus [S] curve.

Reaction Kinetics

The curves and equations in the preceding sections show how the initial velocity of an enzyme-catalyzed reaction varies with substrate and inhibitor concentrations. The curves and equations below show how the velocity and the concentration of substrate and product vary with time.

1. *Zero-Order Kinetics*

If we examine the *velocity versus substrate concentration* curve, we see that at an initial [S] that is very large compared to K_m, the velocity is *essentially* independent of [S] and (*for all practical purposes*) constant. Therefore "zero-order kinetics" are observed. We can also see that as the reaction proceeds, substrate is converted to product but the velocity remains *essentially* constant until point A is reached. This is illustrated in Figure AIII-13 where:

$$v = \text{slope} = -\frac{\Delta[S]}{\Delta t} = \frac{\Delta[P]}{\Delta t} = k$$

$t_{1/2} = $ time required to convert 50% of S to P

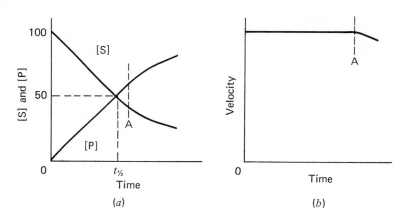

Figure AIII-13 (*a*) Substrate [S] and product [P] concentrations as a function of time for a reaction obeying zero-order kinetics   (*b*) Initial velocity versus time for a reaction obeying zero-order kinetics

(It should be noted that even at very high [S], v is still somewhat dependent on [S]; thus, the curve is asymptotic not horizontal. However, experimentally, the dependency of v upon [S] can not be observed.)

2. First-Order Kinetics

At substrate concentrations below "saturation" (point A) the dependency of v upon [S] can be experimentally observed. Below point A the velocity constantly changes as substrate is utilized. Throughout the entire concentration range of [S], we are really dealing with a mixed order of kinetics. However, from zero [S] to about [S] = 0.01 K_m, we may assume that the velocity is *directly* proportional to [S], i.e., that pure "first-order kinetics" are observed. If we assume first-order kinetics, we can then easily calculate and plot [S], [P], and v as a function of time as shown in Figure AIII-14, where:

$$v = \text{slope} = -\frac{\Delta[S]}{\Delta t} = \frac{\Delta[P]}{\Delta t} \neq k$$

The $v \propto$ [S] and [S] keeps changing.

$$v = -\frac{dS}{dt} = k[S]$$

Rearrranging:

$$-\frac{dS}{S} = k\,dt$$

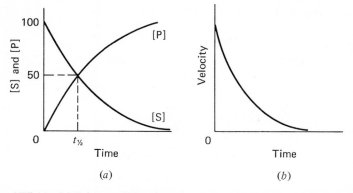

Figure AIII-14 (*a*) Substrate [S] and product [P] concentrations as a function of time for a reaction obeying first-order kinetics   (*b*) Initial velocity versus time for a reaction obeying first-order kinetics

Integrating between any two different substrate concentrations, S_1 and S_2, and the corresponding times, t_1 and t_2:

$$-\int_{S_1}^{S_2} \frac{dS}{S} = k\int_{t_1}^{t_2} dt$$

$$-\ln \frac{S_2}{S_1} = k(t_2 - t_1)$$

$$\ln \frac{S_1}{S_2} = k(t_2 - t_1)$$

$$2.3 \log \frac{S_1}{S_2} = k(t_2 - t_1)$$

If $S_1 =$ the initial substrate concentration, S_0, and $t_1 =$ zero time, then the above equation may be written as follows.

$$\boxed{2.3 \log \frac{S_0}{S_t} = kt}$$

where

$$t = \text{elapsed time}$$

$$S_t = \text{substrate concentration at time } t$$

In exponential form, the equation may be written:

$$\boxed{S_t = S_0 e^{-kt}}$$

The above equation may be rearranged into a linear ($y = mx + b$) form.

$$2.3 \log \frac{S_0}{S_t} = kt$$

$$\log \frac{S_0}{S_t} = \frac{kt}{2.3}$$

$$\log S_0 - \log S_t = \frac{kt}{2.3}$$

$$-\log S_t = \frac{kt}{2.3} - \log S_0$$

$$\boxed{+\log S_t = -\frac{k}{2.3} t + \log S_0}$$

Thus, a plot of log [S] versus t yields a straight line where the y intercept (log [S] intercept) is log S_0 and the slope is $-k/2.3$. (see Figure AIII-15). When $[S] = \frac{1}{2}S_0$, $t =$ the "half-life," $t_{1/2}$.

$$2.3 \log \frac{1}{0.5} = kt_{1/2}$$

$$2.3 \log 2 = kt_{1/2}$$

$$2.3(0.301) = kt_{1/2}$$

$$0.693 = kt_{1/2}$$

$$\boxed{\frac{0.693}{k} = t_{1/2}}$$

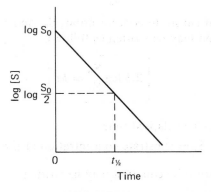

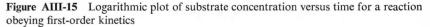

Figure AIII-15 Logarithmic plot of substrate concentration versus time for a reaction obeying first-order kinetics

F. INTEGRATED FORM OF MICHAELIS-MENTEN EQUATION

At any substrate concentration below "saturation" (point A), the velocity is dependent on the substrate concentration. If the *initial velocity* at several substrate concentrations can be determined, then V_{max} and K_m can be calculated or determined graphically. However, under some conditions it may be difficult to make accurate measurements of the initial velocity, although it is still possible to determine the substrate (or product) concentration during the reaction. If the decreasing velocity results only from the decreasing saturation of the enzyme, the K_m and V_{max} may be determined by using the integrated form of the Michaelis-Menten equation.

$$v = \frac{V_{max}[S]}{K_m + [S]}$$

Let

$$v = -\frac{dS}{dt}$$

[S] = substrate concentration present at any time during the reaction

$$-\frac{dS}{dt} = \frac{V_{max}[S]}{K_m + [S]}$$

Multiplying both sides of the equation by -1:

$$\frac{dS}{dt} = -\frac{V_{max}[S]}{K_m + [S]}$$

Rearranging:

$$\frac{1}{V_{max}\, dt} = -\frac{[S]}{(K_m + [S])\, dS}$$

Inverting:

$$V_{max}\, dt = -\frac{K_m + [S]}{[S]}\, dS$$

Integrating between any two times (e.g., zero-time, t_0, and any other time, t) and the corresponding two substrate concentrations (S_0 and S_t):

$$V_{max} \int_{t_0}^{t} dt = -\int_{S_0}^{S_t} \frac{K_m + [S]}{[S]}\, dS$$

Separating the terms in the right-hand expression:

$$V_{\max} \int_{t_0}^{t} dt = -\int_{S_0}^{S_t} K_m \frac{dS}{S} - \int_{S_0}^{S_t} \frac{S}{S} \, dS$$

$$V_{\max} \int_{t_0}^{t} dt = -K_m \int_{S_0}^{S_t} \frac{dS}{S} - \int_{S_0}^{S_t} dS$$

$$V_{\max} t = -K_m \ln \frac{S_t}{S_0} - (S_t - S_0)$$

or

$$V_{\max} t = +K_m \ln \frac{S_0}{S_t} + (S_0 - S_t)$$

$$\boxed{V_{\max} t = 2.3 \, K_m \log \frac{S_0}{S_t} + (S_0 - S_t)}$$

where

$(S_0 - S_t)$ = concentration of substrate utilized by time t

= concentration of product produced by time t

We can see that when S_0 is very small compared to K_m, $(S_0 - S_t)$ is very small compared to the log term. Under this condition, the equation reduces to the usual first-order equation. Similarly, when S_0 is very large compared to K_m, the S_0/S_t ratio approaches unity, the log term approaches zero, and the equation reduces to the usual zero-order equation. The above equation may be divided throughout by t and then rearranged as follows.

$$\frac{V_{\max} t}{t} = \frac{2.3 K_m}{t} \log \frac{S_0}{S_t} + \frac{(S_0 - S_t)}{t}$$

$$V_{\max} - \frac{(S_0 - S_t)}{t} = \frac{2.3 K_m}{t} \log \frac{S_0}{S_t}$$

$$\frac{V_{\max}}{K_m} - \frac{(S_0 - S_t)}{K_m t} = \frac{2.3}{t} \log \frac{S_0}{S_t}$$

or

$$\boxed{\frac{2.3}{t} \log \frac{S_0}{S_t} = -\frac{1}{K_m} \frac{(S_0 - S_t)}{t} + \frac{V_{\max}}{K_m}}$$

This is the equation for a straight line ($y = mx + b$) where

$$y = \frac{2.3}{t} \log \frac{S_0}{S_t}$$

$$x = \frac{(S_0 - S_t)}{t}$$

$$m(\text{the slope of the line}) = -\frac{1}{K_m}$$

$$b(\text{the intercept on the } y \text{ axis}) = \frac{V_{\max}}{K_m}$$

When the y value $= 0$, the x value $= V_{\max}$ as shown below.

$$0 = -\frac{1}{K_m} \frac{(S_0 - S_t)}{t} + \frac{V_{\max}}{K_m}$$

$$-\frac{V_{\max}}{K_m} = -\frac{1}{K_m} \frac{(S_0 - S_t)}{t}$$

$$\frac{-K_m V_{\max}}{-K_m} = \frac{(S_0 - S_t)}{t}$$

$$V_{\max} = \frac{(S_0 - S_t)}{t} = \frac{[P]}{t}$$

Thus, K_m and $V_{\max}$ may be determined by measuring the concentration of substrate utilized (or product produced) several times during the reaction and then plotting the appropriate values as shown in Figure AIII-16.

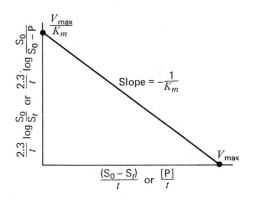

Figure AIII-16   Graphical method of determining K_m and $V_{\max}$ from [S] and [P] data obtained during the course of an enzyme-catalyzed reaction

G. ENZYME UNITS

Most enzyme-catalyzed reactions are conducted with cell-free extracts or partially purified preparations in which the actual amount or concentration of the enzyme present is unknown. Consequently, the amount of enzyme present is often expressed in terms of "units," and the specific activity of the preparation is expressed in terms of units/mg protein. A unit is usually defined as that amount of enzyme that yields 1 μmole of product per minute under optimum assay conditions (e.g., saturating substrate, activator, and coenzyme concentrations, otimum pH and ionic strength, etc.).

Appendix: SPECTROPHOTOMETRY

IV

A. ABSORPTION OF ELECTROMAGNETIC ENERGY

Spectrophotometry (the measurement of light absorption, transmission, emission, or scattering) is one of the most valuable analytical techniques available to biochemists. Unknown compounds may be identified by their characteristic absorption spectra in the ultraviolet, visible, or infrared. Concentrations of known compounds in solutions may be determined by measuring the light absorption at one or more wavelengths. Enzyme-catalyzed reactions frequently can be followed by measuring spectrophotometrically the appearance of a product or disappearance of a substrate.

The physical phenomena underlying light absorption in the various regions of the electromagnetic spectrum are shown in Table AIV-1.

TABLE AIV-1.

Region	X-rays	Ultraviolet	Visible	Infrared	Microwave
Wavelength	0.1–100 mμ	100–400 mμ	400–800 mμ	800 mμ–50 μ	50 μ–30 cm
Effect on molecule	Subvalence electrons excited to higher energy levels	Valence electrons excited to higher energy levels		Molecular vibration	Molecular rotation

B. SPECTROPHOTOMETERS

The essential components of a spectrophotometer (see Figure AIV-1) include: (a) a "light" source, (b) a collimator or focusing device which transmits an intense straight beam of light, (c) a monochromator (prism or grating) to divide the light beam into its component wavelengths, (d) a device for selecting the desired wavelength, (e) a cell compartment in which the sample (in a test tube or cuvette) is placed, (f) a photoelectric detector, and (g) an electrical meter to record the output of the detector.

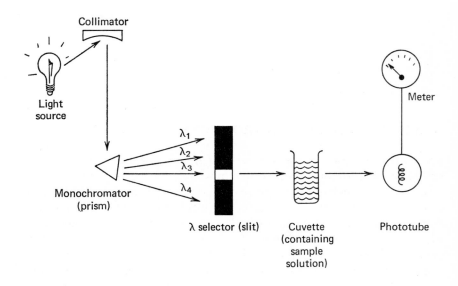

Figure AIV-1 Essential parts of a spectrophotometer

In place of (c) and (d), some simple colorimeters use filters which transmit only wavelength bands.

C. DERIVATION OF LAMBERT-BEER LAW

The fraction of the incident light that is absorbed by a solution depends on the thickness of the sample, the concentration of the absorbing compound in the solution, and the chemical nature of the absorbing compound. Light absorption follows an exponential rather than a linear law. The relationship between concentration, thickness of the light path, and the light absorbed by a particular substance are expressed mathematically as shown below.

$$-\frac{dI}{I} \propto c \, dl \quad \text{and} \quad -\frac{dI}{I} \propto l \, dc$$

or

$$\frac{dI}{I} = -kc \, dl \quad \text{and} \quad \frac{dI}{I} = -kl \, dc$$

where

$-dI =$ the small decrease in light transmission caused by increasing the thickness a small increment, dl, at constant concentration, or the small decrease in light transmission caused by increasing the concentration a small increment dc, at constant thickness

$\dfrac{dI}{I} =$ the fraction of the incident light absorbed

$k =$ proportionality constant, specific for the particular compound under consideration.

The above differential relationships may be integrated between any two thicknesses (e.g., 0 and l) or any two concentrations (e.g., 0 and c).

$$\int_{I_0}^{I} \frac{dI}{I} = -kc \int_{0}^{l} dl \qquad \int_{I_0}^{I} \frac{dI}{I} = -kl \int_{0}^{c} dc$$

$$\ln \frac{I}{I_0} = -kcl \qquad\qquad \ln \frac{I}{I_0} = -klc$$

$$\ln \frac{I_0}{I} = kcl \qquad\qquad \ln \frac{I_0}{I} = klc$$

or

$$2.3 \log \frac{I_0}{I} = kcl \qquad 2.3 \log \frac{I_0}{I} = klc$$

$$\log \frac{I_0}{I} = \frac{k}{2.3} cl \qquad \log \frac{I_0}{I} = \frac{k}{2.3} lc$$

$$\boxed{\log \frac{I_0}{I} = acl} \qquad \boxed{\log \frac{I_0}{I} = alc}$$

where $a =$ the "absorbancy index" or "extinction coefficient" for the particular absorbing compound.

If the concentration is expressed in molarity, a becomes the "molar extinction coefficient" a_m, or E. If the concentration is given in g/liter, a becomes the "specific extinction coefficient," a_s; $a_m = a_s \times$ MW. If the concentration is expressed in % w/v, then the extinction coefficient is given the symbol $E_{1\%}$. In most biochemical calculations, molarities and molar extinction coefficients are employed. Furthermore, the sample thickness is almost always 1 cm. Extinction coefficients vary with varying wavelengths. Thus, the symbol $a_{m_{340}}$ refers to the molar extinction coefficient at 340 mμ.

The $\log I_0/I$ term is called "absorbancy," A, or more commonly, "optical density," O.D.

$$\text{O.D.} = a_m c l$$

The optical density of a 1 M solution of a given substance at a given wavelength (λ) in a 1 cm cuvette would be equal numerically to a_{m_λ}. Note that optical density is a linear function of concentration.

The exponential nature of the Lambert-Beer law is illustrated below. Consider a beam of light passing through a 1 cm cuvette containing 1 mg/liter of a light-absorbing compound. Suppose 80% of the incident light is transmitted (20% of the incident light is absorbed).

$$I_0 = 1.00 \longrightarrow \boxed{c = 1 \text{ mg/liter}} \longrightarrow I_1 = 0.8$$

$$|\leftarrow l = 1 \text{ cm} \rightarrow|$$

Now let us place a second identical cuvette in the light path directly behind the first cuvette. What is the intensity of the light transmitted through both cuvettes? The Lambert-Beer law is *not* a linear relationship. Thus I_2 is *not* 0.6—each centimeter of path length *does not* absorb a constant *amount* of light. Rather, each centimeter absorbs a constant *fraction* of the incident light. Consequently, the second cuvette (the second centimeter) absorbs 20% of the *incident* light. However, the incident light on the second cuvette is I_1 or 0.8. Thus, the second cuvette absorbs 20% of 0.8 which is 0.16. It transmits 80% of 0.8, which is 64% of the original incident light.

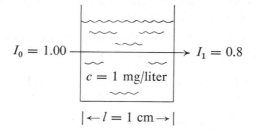

$$I_0 = 1.00 \longrightarrow \boxed{c = 1 \text{ mg/liter}} \longrightarrow I_1 = 0.80 \longrightarrow \boxed{c = 1 \text{ mg/liter}} \longrightarrow I_2 = 0.64$$

$$|\leftarrow l = 1 \text{ cm} \rightarrow| \qquad |\leftarrow l = 1 \text{ cm} \rightarrow|$$

Similarly, if we place three 1 cm cuvettes in series, each one absorbs 20% of the incident light and transmits 80%.

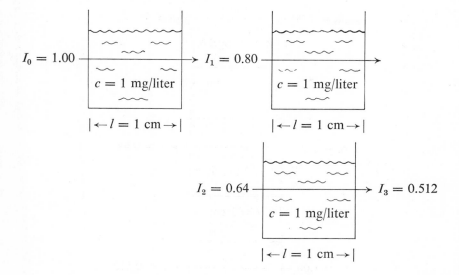

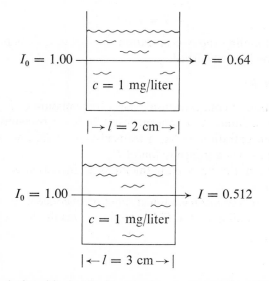

We would obtain exactly the same results if we used single cuvettes with 2 cm and 3 cm thicknesses.

The same relationship holds for varying concentrations in a cuvette of constant thickness.

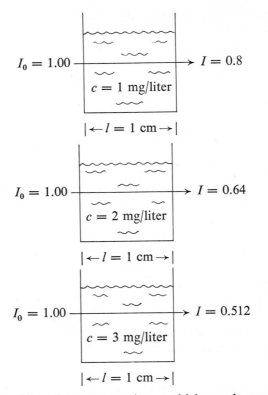

Note that doubling the concentration or thickness does not result in a doubling of the light absorbed.

General Principles

1. If a solution of concentration c has a transmission of I (as a decimal fraction), then a solution of concentration $2c$ has a transmission of I^2, a solution of concentration $3c$ has a transmission of I^3, and a solution of concentration nc has a transmission of I^n.

2. Similarly, if the light path thickness is increased n-fold, the new transmission is I^n.

3. The O.D. of a solution is a linear function of concentration. Doubling c results in a doubling of the O.D., tripling c results in a tripling of the O.D., etc.

Compound	λ_{max}(mμ)	Molar extinction coefficient* $(a_m \times 10^{-3})$
Adenine	260.5	13.3
Adenosine	259.5	14.9
AMP, ADP, ATP	259	15.4
Cytidine	271	9.1
Cytosine	267	6.1
CMP, CDP, CTP	271	9.1
DPN$^+$, TPN$^+$	259	18
DPNH, TPNH	339	6.2
	259	15
Flavin adenine dinucleotide (FAD)	450	11.3
	375	9.3
	260	37
Guanine	275.5	8.1
	246	10.7
Guanosine	252.5	13.6
Nicotinamide	260	4.6
Phenylalanine (in 0.1 N HCl)	257.5	0.19
Phenylalanine (in 0.1 N NaOH)	258	0.206
Pyridoxal phosphate	388	4.9
	330	2.5
Riboflavin	450	12.2
	375	10.6
	260	27.7
Riboflavin phosphate (FMN)	450	12.2
	375	10.4
	260	27.1
Thiamine hydrochloride	267	9.0
	235	11.5
Thymidine	267	9.7
	207.5	9.6
Thymine	264	7.9
Tryptophan (in 0.1 N HCl)	278	5.6
	218	33.5
Tryptophan (in 0.1 N NaOH)	280.5	5.43
	221.5	34.6
Tyrosine (in 0.1 N HCl)	274.5	1.34
	223	8.2
Tyrosine (in 0.1 N NaOH)	293.5	2.33
	240	11.1
Uracil	259.5	8.2
Uridine	262	10.1
UMP, UDP, UTP	261	8.1

* Extinction coefficients are given for a 1 cm light path.

Appendix: ISOTOPES IN BIOCHEMISTRY

V

A. ISOTOPES

"Isotopes" are atoms that contain the same number of protons (have the same atomic number) but different numbers of neutrons (have different atomic weights). Most naturally occurring elements exist as mixtures of isotopes. For example, magnesium exists as Mg^{24}, Mg^{25}, and Mg^{26}, which account for about 78.6%, 10.11%, and 11.29%, respectively, of the total magnesium in nature. Because of the mass distribution, the weighted average atomic weight of magnesium is 24.31. The chemical properties of an element are determined by its atomic number, not its atomic weight. Consequently, all three isotopes of magnesium react almost identically. "Isotope effects" caused by the slight mass differences are generally negligible in biochemical studies.

B. MODES OF RADIOACTIVE DECAY

Many of the naturally occurring and man-made isotopes are unstable. The nuclei of these isotopes decay to more stable forms by one or more of the processes shown in Table AV-1. Such isotopes are called "radioactive isotopes." Because of the decay process, many isotopes have long since disappeared from nature. For example, Mg^{23} and Mg^{27} no longer constitute any measurable proportion of the magnesium in nature, although they can be produced by appropriate nuclear reactions for use in research.

TABLE AV-1.

Decay Process	Nuclear Transformation	Net Equation
Beta particle emission	$_0n^1 \rightarrow {_{+1}}p^1 + {_{-1}}\beta^0$	$_{AN}I^{AW} \rightarrow {_{AN+1}}I^{AW} + {_{-1}}\beta^0$
Positron particle emission	$_{+1}p^1 \rightarrow {_0}n^1 + {_{+1}}\beta^0$	$_{AN}I^{AW} \rightarrow {_{AN-1}}I^{AW} + {_{+1}}\beta^0$
Alpha particle emission	Loss of $_{+2}He^4$ (α)	$_{AN}I^{AW} \rightarrow {_{AN-2}}I^{AW-4} + {_{+2}}He^4$
Electron capture (EC)	$_{+1}p^1 + {_{-1}}e^0 \rightarrow {_0}n^1$	$_{AN}I^{AW} + {_{-1}}e^0 \rightarrow {_{AN-1}}I^{AW} + \gamma$
Particle emission followed by isomeric transition of still unstable nucleus		$_{AN}^{*}I^{AW} \rightarrow {_{AN}}I^{AW} + \gamma$

Most radioisotopes used in biochemical studies are beta and/or gamma emitters.

C. EQUATIONS OF RADIOACTIVE DECAY

The decay of radioactive isotopes is a simple exponential (first-order) process.

$$-\frac{dN}{dt} = \lambda N$$

where:

$-\dfrac{dN}{dt} =$ the number of atoms decaying per unit time (i.e., the count rate)

$N =$ the total number of radioactive atoms present at any given time

$\lambda =$ a decay constant, different for each isotope

The negative sign indicates that the number of radioactive atoms *decreases* with time. It is obvious from the above relationship that a sample containing twice as many radioactive atoms shows twice as many disintegrations per unit time, etc.

Although λ is a proportionality constant, we can see its physical significance by rearranging the above equation.

$$-\frac{dN}{dt} = \lambda N$$

$$\lambda = -\frac{dN}{dtN}$$

$$\lambda = \frac{-dN/N}{dt}$$

In other words, λ is the fraction of the radioactive atoms that decays per unit time.

The differential decay equation may be integrated to obtain a far more useful relationship.

$$-\frac{dN}{dt} = \lambda N, \qquad \frac{dN}{dt} = -\lambda N$$

$$\frac{dN}{N} = -\lambda dt, \qquad \int \frac{dN}{N} = -\lambda \int dt$$

Integrating between the limits of N_0 (the original number of radioactive atoms) and N (the number of radioactive atoms at any other time) and between the limits of zero time and any other time:

$$\int_{N_0}^{N} \frac{dN}{N} = -\lambda \int_{t=0}^{t} dt$$

$$\ln \frac{N}{N_0} = -\lambda t$$

$$\boxed{\ln \frac{N_0}{N} = \lambda t}$$

or

$$\boxed{2.303 \log \frac{N_0}{N} = \lambda t}$$

or, in exponential form

$$\boxed{N = N_0 e^{-\lambda t}}$$

Note that the relationship is identical to those derived earlier for other common first-order processes (e.g., see the sections on enzyme kinetics and spectrophotometry).

D. HALF-LIFE

The "half-life" ($t_{1/2}$) of a radioactive isotope is the time required for half of the original number of atoms to decay. The relationship between $t_{1/2}$

and λ is shown below.

$$\ln \frac{N_0}{N} = \lambda t \qquad \text{or} \qquad 2.303 \log \frac{N_0}{N} = \lambda t$$

$$\ln \frac{1}{0.5} = \lambda t_{1/2} \qquad \bigg| \qquad 2.303 \log \frac{1}{0.5} = \lambda t_{1/2}$$

$$\ln 2 = \lambda t_{1/2} \qquad \bigg| \qquad 2.303 \log 2 = \lambda t_{1/2}$$

$$0.693 = \lambda t_{1/2} \qquad \bigg| \qquad (2.303)(0.301) = \lambda t_{1/2}$$

$$0.693 = \lambda t_{1/2}$$

$$\boxed{\lambda = \frac{0.693}{t_{1/2}} \qquad \text{or} \qquad t_{1/2} = \frac{0.693}{\lambda}}$$

The amount of radioactivity remaining in a sample can easily be determined by constructing a semilog plot as shown in Figure AV-1. A single point (50% at one half-life) is sufficient to plot the curve.

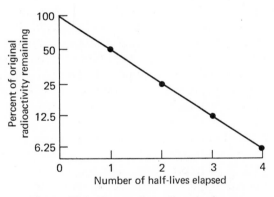

Figure AV-1   Decay of a radioactive isotope

E. THE CURIE

The curie is a standard unit of radioactive decay. It was originally defined as the rate at which 1 g of radium[226] decays. Because of the relatively long half-life of Ra^{226}, the isotope served as a convenient standard. The curie is now defined as the quantity of any radioactive substance in which the decay rate is 3.700×10^{10} disintegrations per second (2.22×10^{12} DPM). Because the efficiency of most radiation detection devices is less than 100%, a given number of curies almost always yields a lower than theoretical count rate. Hence, there is the distinction between DPM and CPM. For example, a sample containing 1 mc of radioactive material has a decay rate of 2.22×10^6 DPM. If only 30% of the disintegrations are detected, the observed count rate is 6.66×10^5 CPM.

F. SPECIFIC ACTIVITY

Specific activity (S.A.) refers to the amount of radioactivity per unit amount of substance. Specific activity may be given in terms of curies per gram, (c/g), millicuries per mg (mc/mg), millicuries per millimole (mc/mmole), disintegrations per minute per millimole (DPM/mmole), counts per minute per micromole (CPM/μmole), or in any other convenient way. Once the specific activity of a compound is known, any given count rate can be equated to the amount of the compound in a sample.

G. RADIOACTIVE TRACER AND DILUTION ANALYSES

Determination of Unknown Amounts of Unlabeled Compounds

Radioactive tracers may be used to determine the amount of a single substance in a mixture. The tracer technique is especially useful where quantitative recovery of the substance in question is difficult or impossible. Basically, the technique involves adding a known amount of the radioactive compound to the mixture and then (after thorough equilibration) reisolating a small amount of the compound. The amount of compound recovered is unimportant, provided it is sufficient to weigh and to count. From the specific activity of the reisolated compound as well as a knowledge of the total number of counts originally added, the amount of nonradioactive compound in the mixture may be calculated as shown below. Let:

$A_0 =$ CPM of tracer added in a weight very small compared to the weight of the nonradioactive compound in the mixture

$M_u =$ unknown amount of the nonradioactive compound in the mixture

$\therefore \quad \dfrac{A_0}{M_u} =$ S.A.$_u =$ specific activity of the compound in the mixture after addition of the tracer

$M_r =$ amount of pure compound reisolated from the mixture after equilibration with tracer

$A_r =$ CPM in the reisolated sample

$\dfrac{A_r}{M_r} =$ S.A.$_r =$ specific activity of the reisolated compound

It is obvious that the ratio of radioactivity to mass—the specific activity of the compound—is the same in the mixture as it is after isolation.

$$\text{S.A.}_u = \text{S.A.}_r \qquad\qquad \frac{A_o}{M_u} = \frac{A_r}{M_r}$$

$$\therefore \quad \boxed{M_u = M_r \frac{A_o}{A_r} \quad \text{or} \quad M_u = \frac{A_o}{\text{S.A.}_r}}$$

The above calculation is based on the assumption that the amount of the added tracer is negligible compared to the amount of the compound in the mixture. This condition is easily met in practice because many compounds of biological interest are available in radioactive form with very high specific activities.

If the amount of the radioactive tracer is significant compared to the amount of the nonradioactive compound in the mixture, this equation must be corrected. Let:

$$A_o = \text{CPM of tracer added}$$
$$M_o = \text{amount of tracer added}$$

$$\therefore \quad \frac{A_o}{M_o} = \text{specific activity of tracer added}$$

$$= \text{S.A.}_o$$

$$M_u = \text{unknown amount of nonradioactive compound in the mixture}$$

$$\therefore \quad \frac{A_o}{M_u + M_o} = \text{specific activity of compound in the mixture after adding tracer}$$

$$= \text{S.A.}_u$$

$$M_r = \text{amount of pure compound reisolated from the mixture after equilibration of radioactive and nonradioactive compounds}$$

$$A_r = \text{CPM in the reisolated sample}$$

$$\therefore \quad \frac{A_r}{M_r} = \text{specific activity of reisolated sample}$$

$$= \text{S.A.}_r$$

As before, the specific activity of the compound in the reisolated sample is the same as it was in the mixture.

$$S.A._u = S.A._r$$

$$\frac{A_o}{M_u + M_o} = \frac{A_r}{M_r}$$

$$A_r M_u + A_r M_o = A_o M_r$$

$$A_r M_u = A_o M_r - A_r M_o$$

$$M_u = \frac{A_o M_r - A_r M_o}{A_r} = \frac{A_o M_r}{A_r} - \frac{A_r M_o}{A_r}$$

$$\boxed{M_u = M_r \frac{A_o}{A_r} - M_o \qquad \text{or} \qquad M_u = \frac{A_o}{S.A._r} - M_o}$$

We can see that as M_o becomes small compared to M_u, the equation reduces to that derived earlier. Another useful form of the above relationship is derived below. Dividing by M_o:

$$\frac{M_u}{M_o} = \frac{M_r A_o}{A_r M_o} - \frac{M_o}{M_o}$$

$$\frac{M_u}{M_o} = \frac{M_r A_o}{A_r M_o} - 1$$

$$\boxed{M_u = \left[\frac{S.A._o}{S.A._r} - 1\right] M_o}$$

If we concern ourselves only with specific activities, then several points are obvious. First, the specific activity of the reisolated compound is less than the specific activity of the original tracer. Second, the degree to which the specific activity decreases (i.e., the dilution factor) is directly related to the amount of unlabeled compound in the mixture. In other words, the dilution of specific activity is identical to the dilution of radioactive tracer.

$$\text{dilution} = \frac{S.A._r}{S.A._o} = \frac{M_o}{M_o + M_u}$$

$$S.A._r M_o + S.A._r M_u = S.A._o M_o$$

$$S.A._r M_u = S.A._o M_o - S.A._r M_o$$

$$M_u = \frac{\text{S.A.}_o M_o - \text{S.A.}_r M_o}{\text{S.A.}_r}$$

$$M_u = \left[\frac{\text{S.A.}_o - \text{S.A.}_r}{\text{S.A.}_r}\right] M_o$$

$$M_u = \left[\frac{\text{S.A.}_o}{\text{S.A.}_r} - \frac{\text{S.A.}_r}{\text{S.A.}_r}\right] M_o$$

$$\boxed{M_u = \left[\frac{\text{S.A.}_o}{\text{S.A.}_r} - 1\right] M_o}$$

Determination of Unknown Amounts of Radioactive Compounds

Dilution analysis can also be used to determine how much of a radioactive compound is present in a mixture, provided the specific activity of the compound is known. For example, if an organism is grown on labeled $C^{14}O_2$ or uniformly labeled glucose-C^{14} as the sole carbon source, then all carbon compounds in the organism have the same specific activity (CPM/μmole of C) as the original radioactive carbon source. Let:

M_u = unknown amount of a radioactive compound in a mixture

A_u = CPM in the above unknown weight

$\therefore \dfrac{A_u}{M_u}$ = specific activity of the radioactive compound

$\phantom{\therefore \dfrac{A_u}{M_u}}$ = S.A.$_u$

M_o = amount of nonradioactive compound added to the mixture

$\therefore \dfrac{A_u}{M_u + M_o}$ = specific activity of the compound in the mixture after adding the nonradioactive tracer

$\phantom{\therefore \dfrac{A_u}{M_u + M_o}}$ = S.A.$_o$

M_r = amount of compound reisolated from the mixture after equilibration of radioactive and nonradioactive compound

A_r = CPM in recovered sample

$\therefore \dfrac{A_r}{M_r}$ = specific activity of recovered compound

$\phantom{\therefore \dfrac{A_r}{M_r}}$ = S.A.$_r$

$\text{S.A.}_r = \text{S.A.}_o$

$\text{S.A.}_r = \dfrac{A_u}{M_u + M_o}$

If the amount of the added nonradioactive compound ("carrier") is very large compared to the amount of the radioactive compound in the mixture, the equation may be simplified:

$$\text{S.A.}_r = \frac{A_u}{M_o} \qquad A_u = \text{S.A.}_r M_o$$

By definition:

$$\text{S.A.}_u = \frac{A_u}{M_u} \quad \therefore \quad M_u = \frac{A_u}{\text{S.A.}_u}$$

Substituting:

$$\boxed{M_u = \frac{\text{S.A.}_r}{\text{S.A.}_u} M_o}$$

If the amount of radioactive compound is significant compared to the amount of carrier added, then suitable corrections must be made.

$$\text{S.A.}_r = \frac{A_u}{M_u + M_o}$$

$$\text{S.A.}_r M_u + \text{S.A.}_r M_o = A_u$$

$$A_u = \text{S.A.}_u M_u$$

$$\text{S.A.}_r M_u + \text{S.A.}_r M_o = \text{S.A.}_u M_u$$

$$\text{S.A.}_r M_u - \text{S.A.}_u M_u = -\text{S.A.}_r M_o$$

$$\text{S.A.}_u M_u - \text{S.A.}_r M_u = \text{S.A.}_r M_o$$

$$M_u[\text{S.A.}_u - \text{S.A.}_r] = \text{S.A.}_r M_o$$

$$\boxed{M_u = \frac{\text{S.A.}_r}{\text{S.A.}_u - \text{S.A.}_r} M_o}$$

Radioactive Derivative Analysis

An unknown amount of an unlabeled compound in a mixture may be determined, even though the labeled compound is unavailable, if a suitable radioactive derivative can be prepared and isolated. By reacting the mixture with a suitable radioactive reagent of known specific activity, the compound in question is converted to a derivative of the same specific activity. The amount of the derivative in the mixture can be quantitated by the calculations outlined above.

The number of moles of radioactive derivative in the mixture (M_u) before adding the carrier derivative (M_o) equals the number of moles of compound in question in the mixture.

TABLE AV-2. RADIOISOTOPES COMMONLY USED IN BIOLOGICAL RE-SEARCH

Isotope	Half-life	Decay Energy (MeV) Beta	Decay Energy (MeV) Gamma
Calcium-45	163 days	0.254	
Carbon-14	5700 yr	0.154	
Cesium-137	33 yr	0.52	0.032
		1.18	0.662
Chlorine-36	4.4×10^5 yr	0.714	
Chromium-51	27.8 days		0.267
			0.32
Cobalt-60	5.3 yr	0.31	1.17
			1.33
Gold-198	2.69 days	0.290	0.411
		0.97	0.676
		1.38	1.087
Hydrogen-3	12.3 yr	0.0179	
Iodine-131	8.1 days	0.250	0.080
		0.31	0.284
		0.608	0.364
			0.638
Iron-55	2.9 yr		K-capture: 0.231
Iron-59	45.1 days	0.29	0.19
		0.46	1.10
		1.56	1.29
Lead-210	25 yr	0.018	0.047
		0.029	
Manganese-54	314 days	1.0	0.84
Molybdenum-99	66 hr	<0.2	0.04
		0.445	0.367
		1.23	0.740
			0.780
Nickel-63	85 yr	0.063	
Phosphorus-32	14.3 days	1.718	
Potassium-42	12.4 hr	1.98	1.51
		3.58	
Rubidium-86	18.7 days	1.82	1.08
		0.72	
Selenium-75	128 days		0.025, 0.066, 0.081, 0.097, 0.121, 0.136, 0.199, 0.265, 0.280, 0.305, 0.402
Sodium-22	2.6 yr	0.58	0.510
		1.8	1.28
Sodium-24	15.06 hr	1.390	1.38
			2.758

TABLE AV-2 (*Continued*)

Isotope	Half-life	Decay Energy (MeV)	
		Beta	Gamma
Strontium-90	28 yr	0.54	
Sulfur-35	87.1 days	0.167	
Technetium-99	2.1×10^5 yr	0.293	
Zinc-65	245 days		0.201
			1.11
Zirconium-95	65 days	0.84	0.72
		0.371	

H. STABLE ISOTOPES

Isotopes that are not radioactive are called "stable isotopes." If a particular isotope contains a greater number of neutrons than the most common form of the element, the isotope is frequently referred to as a "heavy isotope." Stable isotopes commonly used in biochemical research include: H^2 (deuterium), C^{13}, N^{15}, O^{18}, O^{17}, S^{34}, Ca^{44}, Fe^{54}, Fe^{57}, Zn^{66}, and Zn^{68}.

I. ATOM PERCENT EXCESS

The degree to which a radioactive compound is labeled is given in terms of its specific activity (CPM/μmole, mc/mmole, etc.). The degree of labeling of a compound with a stable isotope is expressed in terms of "atom percent excess." This term represents the proportion of stable isotope *over and above that normally present in nature*. For example, if P_n is the normal abundance (%) of a given isotope in nature and P_e is the abundance (%) in a labeled (enriched) compound, then $P_e - P_n$ is the atom percent excess (APE).

Stable isotopes may be used for carrier dilution analyses in much the same manner as radioactive tracers. The calculations are based on the dilution of the atom percent excess instead of specific radioactivities. The basis of the calculations is the same as for radioisotopes; the degree to which the atom percent excess of the stable isotope is diluted is directly related to the amount of unlabeled compound equilibrated with the labeled compound before reisolation.

$$\text{dilution} = \frac{(\text{APE}_r)}{(\text{APE}_o)} = \frac{M_o}{M_o + M_u}$$

where:

M_o = amount of labeled compound added

M_u = amount of unknown compound in the mixture

$$(\text{APE}_o)M_o = (\text{APE}_r)M_o + (\text{APE}_r)M_u$$

$$(\text{APE}_r)M_u = (\text{APE}_o)M_o - (\text{APE}_r)M_o$$

$$M_u = \frac{(\text{APE}_o)M_o - (\text{APE}_r)M_o}{(\text{APE}_r)}$$

$$M_u = \frac{[(\text{APE}_o) - (\text{APE}_r)]}{(\text{APE}_r)}M_o$$

$$M_u = \left[\frac{(\text{APE}_0)}{(\text{APE}_r)} - \frac{(\text{APE}_r)}{(\text{APE}_r)}\right]M_o$$

$$\boxed{M_u = \left[\frac{(\textbf{APE}_o)}{(\textbf{APE}_r)} - 1\right]M_o}$$

To avoid complications resulting from differences in the molecular weights of the tracer compound and the unlabeled compound, the amounts of the two, M_u and M_o, in the above formula should be expressed as moles rather than as weights.

Appendix: OPTICAL ROTATION

VI

TABLE AVI-1. SPECIFIC ROTATION OF SOME CARBO-
HYDRATES AND DERIVATIVES

Compound	Specific Rotation $[\alpha]_D^{T=20-25° C}$
β-D-Arabinose	$-175^* \rightarrow -103^*$
α-L-Arabinose	$+55.4 \rightarrow +105$
β-L-Arabinose	$+190.6 \rightarrow +104.5$
β-D-Fructose	$-133.5 \rightarrow -92$
D-Galactonic acid	$-11.2 \rightarrow +57.6$
α-D-Galactosamine	$+121 \rightarrow +80$
α-D-Galactose	$+150.7 \rightarrow +80.2$
β-D-Galactose	$+52.8 \rightarrow +80.2$
β-D-Galacturonic acid	$+27 \rightarrow +55.6$
D-Gluconic acid	$-6.7 \rightarrow +11.9$
α-D-Glucosamine	$+100 \rightarrow +47.5$
α-D-Glucose	$+112 \rightarrow +52.7$
β-D-Glucose	$+18.7 \rightarrow +52.7$
α-L-Glucose	$-95.5 \rightarrow -51.4$
β-D-Glucuronic acid	$+11.7 \rightarrow +36.3$
D-Glyceraldehyde	$+13.5$
α-D-Mannose	$+29.3 \rightarrow +14.5$
β-D-Mannose	$-16.3 \rightarrow +14.5$
β-D-Mannuronic acid	$-47.9 \rightarrow -23.9$
α-L-Rhamnose	$-8.6 \rightarrow +8.2$
α-D-Xylose	$+9.36 \rightarrow +18.8$

* The first figure given indicates the $[\alpha]_D^T$ of the original form;
the second figure given indicates the $[\alpha]_D^T$ of the equilibrium
mixture of α and β forms after mutarotation. The aldonic acids
equilibrate with the lactone.

Appendix: PRACTICE PROBLEM ANSWERS

VII

CHAPTER I. STRONG ACIDS AND BASES

1. $17.4\ M$

2.

	(a)	(b)	(c)	(d)
pH	2	4	2.3	9.57
pOH	12	10	11.7	4.43
H^+ ions/liter	6.023×10^{21}	6.023×10^{19}	3.0×10^{21}	1.63×10^{14}
OH^- ions/liter	6.023×10^{11}	6.023×10^{13}	1.2×10^{12}	2.23×10^{19}

	(e)	(f)	(g)	(h)	(i)
pH	6.8	11.46	0	-1	4.52
pOH	7.2	2.54	14	15	9.48
H^+ ions/liter	9.05×10^{16}	2.1×10^{12}	6.023×10^{23}	6.023×10^{24}	1.81×10^{14}
OH^- ions/liter	4.01×10^{16}	1.75×10^{21}	6.023×10^{9}	6.023×10^{8}	1.99×10^{19}

3.

	(a)	(b)	(c)	(d)
$[H^+]$	$1.86 \times 10^{-3}\ M$	$5.14 \times 10^{-6}\ M$	$1.66 \times 10^{-7}\ M$	$2.24 \times 10^{-9}\ M$
$[OH^-]$	$5.4 \times 10^{-12}\ M$	$1.95 \times 10^{-9}\ M$	$6.02 \times 10^{-8}\ M$	$4.46 \times 10^{-6}\ M$
H^+ ions/liter	1.12×10^{21}	3.1×10^{18}	1.0×10^{17}	1.35×10^{15}
OH^- ions/liter	3.26×10^{12}	1.18×10^{15}	3.6×10^{16}	2.7×10^{18}

	(e)	(f)	(g)
$[H^+]$	$3.02 \times 10^{-10}\ M$	$3.9 \times 10^{-12}\ M$	$1\ M$
$[OH^-]$	$3.31 \times 10^{-5}\ M$	$2.57 \times 10^{-3}\ M$	$1 \times 10^{-14}\ M$
H^+ ions/liter	1.82×10^{14}	2.3×10^{12}	6.023×10^{23}
OH^- ions/liter	1.99×10^{19}	1.55×10^{21}	6.023×10^{9}

4. $[H^+] = 0.4\ M$, $[OH^-] = 2.5 \times 10^{-14}\ M$, pH $= 0.398$, pOH $= 13.602$.

5. pOH $= 3.42$, pH $= 10.58$

6. $a_{H^+} = 0.071$, $\gamma_{H^+} = 0.71$

7. pH $= 12.86$, pOH $= 1.14$

8. (a) 20.8 ml, (b) 33.3 ml, (c) 68 ml, (d) 180 ml, (e) 4.4×10^{-3} ml

9. (a) 12.2 M, (b) 8.2 ml con HCl/500 ml of solution, (c) 14.3 ml con HCl/350 ml of solution, (d) 666 ml con HCl/liter of solution, (e) 1.64×10^{-3} ml con HCl/liter of solution

10. (a) 400 g NaOH/5 liters of solution, (b) 0.25 g NaOH/2 liters of solution, (c) 356 g NaOH/500 ml of solution

11. 202 ml

12. 1.06 g

13. 24.5×10^4 ml (24.5 liters)

14.

Milliliters Acid Added	pH
0	12.69
50	12.57
100	12.43
150	12.30
200	12.15
250	11.95
300	11.70
350	11.18
375 (equivalence point)	7.0
400	2.85

CHAPTER II. WEAK ACIDS AND BASES

1. (a) $K_a = 1.27 \times 10^{-4}$, (b) pH $= 2.28$, (c) 1210 ml, (d) 1.75×10^{21} ions

2. (a) $[H^+] = 5 \times 10^{-5} M$, (b) $1.85 \times 10^{-2}\%$, (c) $K_a = 9.25 \times 10^{-9}$

3. (a) $[OH^-] = 1.49 \times 10^{-12} M$, (b) 4.46%

4. (a) 100 K_a, (b) 12 K_a, (c) 0.123 K_a

5. (a) pH $= 11.7$, (b) 2.36%

6. (a) pH $= 2.64$, (b) $2.28 \times 10^{-3} M$, (c) $7 \times 10^{-7} M$, (d) primary ionization $= 1.14\%$, secondary ionization $= 3.5 \times 10^{-4}\%$

7. (a) $pK_a = 3.21$ and $pK_b = 10.79$, (b) $pK_a = 4.54$ and $pK_b = 9.46$, (c) $pK_a = 4.47$ and $pK_b = 9.53$, (d) $pK_a = 5.14$ and $pK_b = 8.86$

8. (a) $pK_b = 4.68$ and $pK_a = 9.32$, (b) $pK_b = 5.51$ and $pK_a = 8.49$, (c) $pK_b = 4.11$ and $pK_a = 9.89$, (d) $pK_b = 3.04$ and $pK_a = 10.96$

9. pH $= 11.4$

10. (a) 1.46, (b) 4.67, (c) 9.76, (d) 12.74, (e) 9.03, (f) 4.98, (g) 11.59, (h) 9.35, (i) 5.07, (j) 11.08, (k) 13.30, (l) 5.07

11. 1080 ml

12. 487.5 ml

13. $[H^+] = 5 \times 10^{-13} M$, pH $= 12.3$

14. $[H^+] = 1.66 \times 10^{-9} M$, pH $= 8.78$

15. $[H^+] = 7.25 \times 10^{-5} M$, pH $= 4.14$

16. $[H^+] = 1.6 \times 10^{-13} M$, pH $= 12.8$

17. $\dfrac{[H^+]([PrNH_2] + [OH^-])}{([PrNH_3^+] - [OH^-])} \neq K_a$ at constant $[H^+]$

18.

Milliliters Base Added	pH
0	2.88
25	4.28
50	4.76
75	5.24
100 (equivalence point)	8.88
125	12.19

19.

Milliliters Base Added	pH
0	2.01
56.25 (first pK)	4.19
112.5 (first equivalence point)	4.88
168.75 (second pK)	5.57
225.0 (second equivalence point)	9.87
300	12.52

20. $[NH_3] = 0.103 M$
$[NH_4Cl] = 0.047 M$

21. (a) pH $= 12.32$, (b) pH $= 12.16$

22. pH $= 8.28$

23. pH $= 10.15$

24. pH $= 9.56$

25. 28.56 g sodium formate $+$ 80 ml 1 M formic acid/2 liters of solution

26. (a) 400 ml 2 M H_3PO_4 $+$ 1070 ml 1 N KOH/2 liters of solution, (b) 1000 ml 0.8 M H_3PO_4 $+$ 42.8 g NaOH/2 liters of solution, (c) 54 ml 14.8 M H_3PO_4 $+$ 1070 ml 1 M KOH/2 liters of solution, (d) 533 ml KH_2PO_4 $+$ 267 ml $Na_2 HPO_4$/2 liters of solution, (e) 72.5 g KH_2PO_4 $+$ 46.5 g K_2HPO_4/2 liters of solution, (f) 139.1 g K_2HPO_4 $+$ 355 ml 1.5 M HCl/2 liters of solution, (g) 666.7 ml K_2HPO_4 $+$ 133.0 ml 2 M H_2SO_4/2 liters of solution, (h) 108.8 g KH_2PO_4 $+$ 133.5 ml 2 M KOH/2 liters of solution, (i) 533.3 ml 1.5 M KH_2PO_4 $+$ 267 ml 1 M NaOH/2 liters of solution, (j) 131.2 g Na_3PO_4 $+$ 1330 ml 1 M HCl/2 liters of solution

27. 21 ml glacial acetic acid and 62.2 g potassium acetate/5 liters of solution

28. (a) pH $= 6.83$, (b) pH $= 1.3$, (c) $HPO_4^= + H^+ \rightleftharpoons H_2PO_4^-$

29. $\frac{1}{2} = $ tris^0/tris$^+$, (b) $\frac{1}{1} = $ tris^0/tris$^+$, (c) pH $= 8.1$, (d) pH $= 12.52$, (e) $—NH_3^+ \rightleftharpoons$ $—NH_2 + H^+$ replacing a large portion of the H^+ utilized and converting some of the tris$^+$ to tris^0.

30. $[H^+] = 0.0606 M$, $[OH^-] = 0.129 M$

CHAPTER III. AMINO ACIDS AND PEPTIDES

1. (a) pH $= 1.56$, (b) pH $= 6.02$, (c) pH $= 11.49$

2. (a) 18%, (b) 2.57%

3. 2.0%

4. (a) pH = 6.96, (b) pH = 1.82, (c) pH = 1.82

5. (a) 2.25 liters (2250 ml), (b) 200 ml, (c) 600 ml, (d) 1200 ml

6. (a) 250 ml, (b) 468.8 ml, (c) 1225 ml, (d) 468.8 ml

8. pH = 2.64

11. (a) $pI = 3.22$, $pH_m = 6.96$; (b) $pI = 5.68$, $pH_m = 5.68$; (c) $pI = 9.70$, $pH_m = 5.30$; (d) $pI = 2.27$, $pH_m = 6.01$; (e) $pI = 9.51$, $pH_m = 5.65$ and 10.30; (f) $pI = 10.53$, $pH_m = 5.64$; (g) $pI = 5.34$, $pH_m = 5.34$ and 10.78

13. (a)   —   (c)   —   (e)   —
            +   —   0
            +   —   0
            +   +   +
            +   +   +
     (b)   —   (d)   —
            +   +
            +   +
            +   +
            +   +

14. (a)

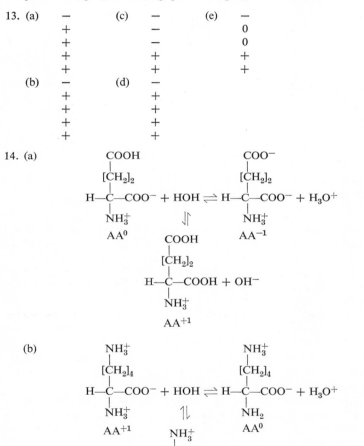

(c)

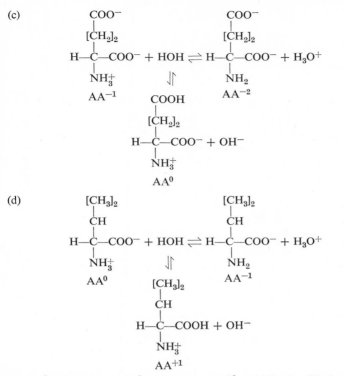

(d)

15. (a) $AA^0 = 0.183\ M$, $AA^{-1} = 0.017\ M$, $AA^{+1} = 0.017\ M$; (b) $AA^{+1} = 0.2\ M$, $AA^0 =$ negligible, $AA^{+2} =$ negligible; (c) $AA^{-1} = 0.2\ M$, $AA^{-2} =$ negligible, $AA^0 =$ negligible; (d) $AA^0 = 0.2\ M$, $AA^{-1} =$ negligible, $AA^{+1} =$ negligible

16. $AA^0 = 0.248\ M$, $AA^{+1} =$ negligible, $AA^{-1} = 0.002\ M$; (b) $AA^{+1} =$ negligible, $AA^0 =$ negligible, $AA^{-1} = 0.249\ M$, $AA^{-2} = 0.001\ M$; (c) $AA^{+2} =$ negligible, $AA^{+1} = 0.242\ M$, $AA^0 = 0.008\ M$, $AA^{-1} =$ negligible

17. (a) $AA^{+1} =$ negligible, $AA^0 = 0.045\ M$, $AA^{-1} = 0.255\ M$, $AA^{-2} =$ negligible; (b) $AA^{+1} =$ negligible, $AA^0 = 0.3\ M$, $AA^{-1} =$ negligible; (c) $AA^{+2} =$ negligible, $AA^{+1} = 0.3\ M$, $AA^0 =$ negligible, $AA^{-1} =$ negligible

18. (a) $AA^{+1} =$ negligible, $AA^0 = 0.022\ M$, $AA^{-1} = 0.178\ M$; (b) $AA^{+1} =$ negligible, $AA^0 =$ negligible, $AA^{-1} = 0.026\ M$, $AA^{-2} = 0.174\ M$; (c) $AA^{+2} =$ negligible, $AA^{+1} =$ negligible, $AA^0 = 0.102\ M$, $AA^{-1} = 0.098\ M$

19. (a) 31 g histidine + 48 ml 1 M HCl/liter of solution, (b) 22.5 g glycine + 91 ml 1 M HCl/liter of solution, (c) 21.9 g lysine + 71 ml 1 M HCl/liter of solution, (d) 22.3 g alanine + 61 ml 1 M KOH/liter of solution, (e) 22.1 g glutamate + 22.7 ml 1 M KOH/liter of solution

CHAPTER IV. BIOCHEMICAL ENERGETICS

1. (a) $5 \times 10^{-4}\ M$, (b) -5452 cal/mole, (c) 0
2. $-11,936$ cal/mole

3. $K'_{eq} = 1.35 \times 10^3$, $\Delta F' = -4274$ cal/mole

4. $K'_{eq} = 64.2$, $\Delta F' = -2464$ cal/mole

5. $-12,490$ cal/mole

6. (a) -3440 cal/mole, (b) $+1800$ cal/mole, (c) -2800 cal/mole, (d) ~ 0

7. > 0.218 M

8. 18 moles ATP/mole ethanol (assuming $\Delta F'_{ATP} = -7.7$ kcal/mole ATP), 17 moles ATP/mole ethanol (assuming $\Delta F'_{ATP} = -8.0$ kcal/mole ATP)

9. (a) [glucose-6-P] = 0.95 M, [glucose-1-P] = 0.05 M, [glucose-6-P]/[glucose-1-P] = 19; (b) [glucose-6-P] = 0.095 M, [glucose-1-P] = 0.005 M, [glucose-6-P]/[glucose-1-P] = 19; (c) [glucose-6-P] = 9.5×10^{-3} M, [glucose-1-P] = 5×10^{-4} M, [glucose-6-P]/[glucose-1-P] = 19; (d) [glucose-6-P] = 9.5×10^{-4} M, [glucose-1-P] = 5×10^{-5} M, [glucose-6-P]/[glucose-1-P] = 19; (e) [glucose-6-P] = 9.5×10^{-5} M, [glucose-1-P] = 5×10^{-6} M, [glucose-6-P]/[glucose-1-P] = 19

10. (a) isocitrate = 0.846 M, glyoxylate = 0.154 M, succinate = 0.154 M

$$\frac{isocitrate}{glyoxylate} = 5.49 \qquad \frac{glyoxylate}{succinate} = 1.0$$

(b) isocitrate = 0.0591 M, glyoxylate = 0.0409 M, succinate = 0.0409 M

$$\frac{isocitrate}{glyoxylate} = 1.45 \qquad \frac{glyoxylate}{succinate} = 1.0$$

(c) isocitrate = 0.0022 M, glyoxylate = 0.0078 M, succinate = 0.0078 M

$$\frac{isocitrate}{glyoxylate} = 0.355 \qquad \frac{glyoxylate}{succinate} = 1.0$$

(d) isocitrate = 4.0×10^{-5} M, glyoxylate = 9.6×10^{-4} M, succinate = 9.6×10^{-4} M

$$\frac{isocitrate}{glyoxylate} = 0.042 \qquad \frac{glyoxylate}{succinate} = 1.0$$

(e) isocitrate = ~ 0 M, glyoxylate = $\sim 1 \times 10^{-4}$ M, succinate = $\sim 1 \times 10^{-4}$ M

$$\frac{isocitrate}{glyoxylate} = \sim 0 \qquad \frac{glyoxylate}{succinate} = 1.0$$

11. (a) pyruvate + β-hydroxybutyrate $\rightarrow$ lactate + acetoacetate; (b) pyruvate is reduced to lactate, β-hydroxybutyrate is oxidized to acetoacetate, pyruvate is the oxidizing agent, β-hydroxybutyrate is the reducing agent; (c) $\Delta E'_0 = +0.100$ v, $\Delta F' = -4612.6$ cal/mole, $K'_{eq} = 2420$

12. (a) ubiquinone + succinate $\rightarrow$ ubiquinone-H_2 + fumarate; (b) $\Delta E = +0.070$ v, $\Delta F' = -3228.8$ cal/mole, $K'_{eq} \doteq 234$

13. (a) At $30°$ C, $E = -0.150$ v, (b) at $30°$ C, $E = -0.081$ v, (c) at $30°$ C, $E = E'_0 = -0.06$ v, (d) at $30°$ C, $E = -0.046$ v, (e) at $30°$ C, $E = -0.018$ v, (f) at $30°$ C, $E = +0.018$ v

14. (a)–(g) OAA + DPNH + H^+ $\rightarrow$ malate + DPN^+

15. $FADH_2/FAD > 10^3$

16. (a) 0 ATP, (b) 1 ATP, (c) 2 ATP

CHAPTER V. ENZYME KINETICS

1. (a) $1 \times 10^{-5} M$, (b) Verify by plotting v versus [S], showing a hyperbolic curve, and by plotting $1/v$ versus $1/[S]$ showing a straight line. You can also show that the original v versus [S] data yield the same K_m value regardless which values are substituted into the Michaelis-Menten equation, (c) 12 min^{-1}

2. (a) 0.107 μmole/liter-min, (b) 26.6 μmoles/liter-min, (c) 37.7 μmoles/liter-min, (d) 114.2 μmoles/liter-min, (e) 128 μmoles/liter-min

3. Problem 1: 600 μmoles/liter-min, 595 μmoles/liter-min, 500 μmoles/liter-min, 100 μmoles/liter-min, 54.5 μmoles/liter-min; Problem 2: 0.535 μmole/liter-min, 133.0 μmoles/liter-min, 188.5 μmoles/liter-min, 571.0 μmoles/liter-min, 640 μmoles/liter-min

4. (a) 5.4 μmoles/liter, (b) 8.55 μmoles/liter, (c) 16.2 μmoles/liter, (d) 28.62 μmoles/liter, (e) $2.7 \times 10^{-6}\%$, $4.28 \times 10^{-6}\%$, $8.1 \times 10^{-6}\%$, $14.31 \times 10^{-6}\%$

5. (a) $2.1 \times 10^{-6} M$ at 5 minutes, (b) $3.9 \times 10^{-6} M$ at 10 minutes

6. (a) 2.3%, (b) 2.3%, (c) 1.44×10^{-6} moles/liter-min, (d) 144.3 min, (e) 288.5 min

7. (a) 358, (b) 16, (c) 9

8. (a) 4.7 μmoles/liter-min, (b) 44.6 μmoles/liter-min, (c) 60.7 μmoles/liter-min

9. (a) 17.6 μmoles/liter-min, 87%; (b) $v_i = 3.7$ μmoles/liter-min, 91.8%; (c) $v_i = 210.2$ μmoles/liter-min, 1.3%

10. (a) $3.7 \times 10^{-4} M$, (b) $2.93 \times 10^{-2} M$

11. $6.66 \times 10^{-5} M$

12. (a) 20.7 μmoles/liter-min, (b) 89.3%

13. $K' = 2.5 \times 10^{-9} M^2$, $S_{50} = 5 \times 10^{-5} M$, $n = 2$, $V_{max} = 100$ μmoles/liter-min

14. (a) 1.6×10^{-3} μmoles/min, (b) 16 μmoles/liter-min, (c) 3.34×10^{-3} μmoles/mg protein-min, (d) 0.08 units/ml, (e) 0.0033 units/mg protein

15. (a) 89.5%, (b) 2.98-fold

CHAPTER VI. SPECTROPHOTOMETRY

1.

	(a)	(b)	(c)	(d)
260 mμ I	0.925	0.051	0.0049	0.176
260 mμ O.D.	0.034	1.29	2.31	0.754
340 mμ I	0.967	0.300	1.00	0.903
340 mμ O.D.	0.014	0.523	0	0.044

2. (a)

	(a)	(b)	(c)
260 mμ I	0.962	0.974	0.981
260 mμ O.D.	0.017	0.011	0.008
340 mμ I	0.984	0.989	0.992
340 mμ O.D.	0.007	0.005	0.003

(b)

	(a)	(b)	(c)
260 mμ I	0.226	0.371	0.474
260 mμ O.D.	0.645	0.430	0.322
340 mμ I	0.547	0.669	0.755
340 mμ O.D.	0.261	0.174	0.131

(c)	(a)	(b)	(c)
260 mμ I	0.070	0.170	0.264
260 mμ O.D.	1.155	0.770	0.577
340 mμ I	1.00	1.00	1.00
340 mμ O.D.	0	0	0

(d)	(a)	(b)	(c)
260 mμ I	0.419	0.560	0.648
260 mμ O.D.	0.377	0.251	0.188
340 mμ I	0.950	0.967	0.975
340 mμ O.D.	0.022	0.014	0.011

3. (a) [ATP] $= 3.43 \times 10^{-5} M$, [TPNH] $= 2.41 \times 10^{-5} M$; (b) [ATP] $= 4.87 \times 10^{-5} M$, [TPNH] $= 0$; (c) [ATP] $= 0$, [TPNH] $= 3.54 \times 10^{-5} M$

4. [A] $= 1.26 \times 10^{-5} M$, [B] $= 1.11 \times 10^{-5} M$

5. $5.98 \times 10^{-5} M$

6. To the unknown solution, add a solution containing excess DPN^+, P_i, and glyceraldehyde-3-phosphate dehydrogenase. The GAP present will be converted to 1,3-DiPGA. One mole of DPNH will be produced for every mole of GAP originally present. Then add triosephosphate isomerase to convert the DHAP to GAP which will then be converted to 1,3-DiPGA producing another mole DPNH for every mole of DHAP originally present. Finally, add aldolase to convert the FDP to GAP and DHAP which, upon conversion to 1,3-DiPGA, will yield 2 moles of DPNH per mole of FDP originally present.

7. Citric acid $= 1.36 \times 10^{-4} M$, isocitric acid $= 1.36 \times 10^{-4} M$

CHAPTER VII. ISOTOPES IN BIOCHEMISTRY

1. (a) 1.73×10^{-1} yr^{-1}, 4.74×10^{-4} days^{-1}, 1.99×10^{-5} hr^{-1}, 3.29×10^{-7} min^{-1}, 5.50×10^{-9} sec^{-1}; (b) 83.0%

2. (a) One atom out of every 11.5 radioactive atoms present decays per day, one atom out of every 16,600 radioactive atoms present decays per minute; (b) 27.65×10^{16} DPM/g, 12.45×10^4 curies/g, 950 curies/g-atom

3. 28.6%

4. (a) 8.93×10^3 curies/mole (8.93×10^3 mc/mmole), (b) 0.01%

5. 5.62×10^{-8} g

6. (a) $3.57 \times 10^{-3} M$, (b) 8.63×10^8 CPM/ml

7. (a) 1.56 mc/mg, (b) 228 mc/mmole, (c) 5.06×10^8 DPM/μmole, (d) 6.75×10^7 CPM/μmole of carbon

8. Take 10.6 μl of radioactive L-cysteine-S^{35} solution plus 0.1175 g solid unlabeled L-cysteine hydrochloride and dissolve in sufficient water to make 75 ml of solution

9. Take 3.38 ml of radioactive glucose-C^{14} stock solution plus 2.92 mg solid un-labeled glucose and dissolve in sufficient water to make 50 ml of solution

10. (a) 9.33×10^{-4} μmoles/min, (b) 0.622 μmoles/liter-min, (c) 1.29×10^{-3} μmoles/mg protein-min

11. 6 liters

12. 0.0128 M

13. 172.5 μg

14. 85.4 mg/50 ml or 5.56 × 10^{-3} M

15. 51.8 μmoles

16. 6 hr

17. (a) 7965 CPM/μmole, (b) 6765 CPM/μmole, (c) 4496 CPM/μmole

CHAPTER VIII. MISCELLANEOUS CALCULATIONS

1. (a) 0.25, (b) 1.08, (c) 0.90, (d) 1.20, (e) 0.03, (f) 1.20, (g) 5.29 × 10^{-4}, (h) 1.08, (i) 0.011

2. (a) −5.12°, (b) −794.6°

3. 0.226 g/ml

4. $\alpha = 67.5\%$, $\beta = 32.5\%$

5. (a) $C_6H_{13}NO_2$, (b) 131

6. 1980

7. 15,150

8. 158.5

9. 7105

Natural Numbers	0	1	2	3	4	5	6	7	8	9	Proportional Parts								
											1	2	3	4	5	6	7	8	9
10	0000	0043	0086	0128	0170	0212	0253	0294	0334	0374	4	8	12	17	21	25	29	33	37
11	0414	0453	0492	0531	0569	0607	0645	0682	0719	0755	4	8	11	15	19	23	26	30	34
12	0792	0828	0864	0899	0934	0969	1004	1038	1072	1106	3	7	10	14	17	21	24	28	31
13	1139	1173	1206	1239	1271	1303	1335	1367	1399	1430	3	6	10	13	16	19	23	26	29
14	1461	1492	1523	1553	1584	1614	1644	1673	1703	1732	3	6	9	12	15	18	21	24	27
15	1761	1790	1818	1847	1875	1903	1931	1959	1987	2014	3	6	8	11	14	17	20	22	25
16	2041	2068	2095	2122	2148	2175	2201	2227	2253	2279	3	5	8	11	13	16	18	21	24
17	2304	2330	2355	2380	2405	2430	2455	2480	2504	2529	2	5	7	10	12	15	17	20	22
18	2553	2577	2601	2625	2648	2672	2695	2718	2742	2765	2	5	7	9	12	14	16	19	21
19	2788	2810	2833	2856	2878	2900	2923	2945	2967	2989	2	4	7	9	11	13	16	18	20
20	3010	3032	3054	3075	3096	3118	3139	3160	3181	3201	2	4	6	8	11	13	15	17	19
21	3222	3243	3263	3284	3304	3324	3345	3365	3385	3404	2	4	6	8	10	12	14	16	18
22	3424	3444	3464	3483	3502	3522	3541	3560	3579	3598	2	4	6	8	10	12	14	15	17
23	3617	3636	3655	3674	3692	3711	3729	3747	3766	3784	2	4	6	7	9	11	13	15	17
24	3802	3820	3838	3856	3874	3892	3909	3927	3945	3962	2	4	5	7	9	11	12	14	16
25	3979	3997	4014	4031	4048	4065	4082	4099	4116	4133	2	3	5	7	9	10	12	14	15
26	4150	4166	4183	4200	4216	4232	4249	4265	4281	4298	2	3	5	7	8	10	11	13	15
27	4314	4330	4346	4362	4378	4393	4409	4425	4440	4456	2	3	5	6	8	9	11	13	14
28	4472	4487	4502	4518	4533	4548	4564	4579	4594	4609	2	3	5	6	8	9	11	12	14
29	4624	4639	4654	4669	4683	4698	4713	4728	4742	4757	1	3	4	6	7	9	10	12	13
30	4771	4786	4800	4814	4829	4843	4857	4871	4886	4900	1	3	4	6	7	9	10	11	13
31	4914	4928	4942	4955	4969	4983	4997	5011	5024	5038	1	3	4	6	7	8	10	11	12
32	5051	5065	5079	5092	5105	5119	5132	5145	5159	5172	1	3	4	5	7	8	9	11	12
33	5185	5198	5211	5224	5237	5250	5263	5276	5289	5302	1	3	4	5	6	8	9	10	12
34	5315	5328	5340	5353	5366	5378	5391	5403	5416	5428	1	3	4	5	6	8	9	10	11
35	5441	5453	5465	5478	5490	5502	5514	5527	5539	5551	1	2	4	5	6	7	9	10	11
36	5563	5575	5587	5599	5611	5623	5635	5647	5658	5670	1	2	4	5	6	7	8	10	11
37	5682	5694	5705	5717	5729	5740	5752	5763	5775	5786	1	2	3	5	6	7	8	9	10
38	5798	5809	5821	5832	5843	5855	5866	5877	5888	5899	1	2	3	5	6	7	8	9	10
39	5911	5922	5933	5944	5955	5966	5977	5988	5999	6010	1	2	3	4	5	7	8	9	10
40	6021	6031	6042	6053	6064	6075	6085	6096	6107	6117	1	2	3	4	5	6	8	9	10
41	6128	6138	6149	6160	6170	6180	6191	6201	6212	6222	1	2	3	4	5	6	7	8	9
42	6232	6243	6253	6263	6274	6284	6294	6304	6314	6325	1	2	3	4	5	6	7	8	9
43	6335	6345	6355	6365	6375	6385	6395	6405	6415	6425	1	2	3	4	5	6	7	8	9
44	6435	6444	6454	6464	6474	6484	6493	6503	6513	6522	1	2	3	4	5	6	7	8	9
45	6532	6542	6551	6561	6571	6580	6590	6599	6609	6618	1	2	3	4	5	6	7	8	9
46	6628	6637	6646	6656	6665	6675	6684	6693	6702	6712	1	2	3	4	5	6	7	7	8
47	6721	6730	6739	6749	6758	6767	6776	6785	6794	6803	1	2	3	4	5	5	6	7	8
48	6812	6821	6830	6839	6848	6857	6866	6875	6884	6893	1	2	3	4	4	5	6	7	8
49	6902	6911	6920	6928	6937	6946	6955	6964	6972	6981	1	2	3	4	4	5	6	7	8
50	6990	6998	7007	7016	7024	7033	7042	7050	7059	7067	1	2	3	3	4	5	6	7	8
51	7076	7084	7093	7101	7110	7118	7126	7135	7143	7152	1	2	3	3	4	5	6	7	8
52	7160	7168	7177	7185	7193	7202	7210	7218	7226	7235	1	2	2	3	4	5	6	7	7
53	7243	7251	7259	7267	7275	7284	7292	7300	7308	7316	1	2	2	3	4	5	6	6	7
54	7324	7332	7340	7348	7356	7364	7372	7380	7388	7396	1	2	2	3	4	5	6	6	7

Natural Numbers	0	1	2	3	4	5	6	7	8	9	Proportional Parts								
											1	2	3	4	5	6	7	8	9
55	7404	7412	7419	7427	7435	7443	7451	7459	7466	7474	1	2	2	3	4	5	5	6	7
56	7482	7490	7497	7505	7513	7520	7528	7536	7543	7551	1	2	2	3	4	5	5	6	7
57	7559	7566	7574	7582	7589	7597	7604	7612	7619	7627	1	2	2	3	4	5	5	6	7
58	7634	7642	7649	7657	7664	7672	7679	7686	7694	7701	1	1	2	3	4	4	5	6	7
59	7709	7716	7723	7731	7738	7745	7752	7760	7767	7774	1	1	2	3	4	4	5	6	7
60	7782	7789	7796	7803	7810	7818	7825	7832	7839	7846	1	1	2	3	4	4	5	6	6
61	7853	7860	7868	7875	7882	7889	7896	7903	7910	7917	1	1	2	3	4	4	5	6	6
62	7924	7931	7938	7945	7952	7959	7966	7973	7980	7987	1	1	2	3	3	4	5	6	6
63	7993	8000	8007	8014	8021	8028	8035	8041	8048	8055	1	1	2	3	3	4	5	5	6
64	8062	8069	8075	8082	8089	8096	8102	8109	8116	8122	1	1	2	3	3	4	5	5	6
65	8129	8136	8142	8149	8156	8162	8169	8176	8182	8189	1	1	2	3	3	4	5	5	6
66	8195	8202	8209	8215	8222	8228	8235	8241	8248	8254	1	1	2	3	3	4	5	5	6
67	8261	8267	8274	8280	8287	8293	8299	8306	8312	8319	1	1	2	3	3	4	5	5	6
68	8325	8331	8338	8344	8351	8357	8363	8370	8376	8382	1	1	2	3	3	4	4	5	6
69	8388	8395	8401	8407	8414	8420	8426	8432	8439	8445	1	1	2	2	3	4	4	5	6
70	8451	8457	8463	8470	8476	8482	8488	8494	8500	8506	1	1	2	2	3	4	4	5	6
71	8513	8519	8525	8531	8537	8543	8549	8555	8561	8567	1	1	2	2	3	4	4	5	5
72	8573	8579	8585	8591	8597	8603	8609	8615	8621	8627	1	1	2	2	3	4	4	5	5
73	8633	8639	8645	8651	8657	8663	8669	8675	8681	8686	1	1	2	2	3	4	4	5	5
74	8692	8698	8704	8710	8716	8722	8727	8733	8739	8745	1	1	2	2	3	4	4	5	5
75	8751	8756	8762	8768	8774	8779	8785	8791	8797	8802	1	1	2	2	3	3	4	5	5
76	8808	8814	8820	8825	8831	8837	8842	8848	8854	8859	1	1	2	2	3	3	4	5	5
77	8865	8871	8876	8882	8887	8893	8899	8904	8910	8915	1	1	2	2	3	3	4	4	5
78	8921	8927	8932	8938	8943	8949	8954	8960	8965	8971	1	1	2	2	3	3	4	4	5
79	8976	8982	8987	8993	8998	9004	9009	9015	9020	9026	1	1	2	2	3	3	4	4	5
80	9031	9036	9042	9047	9053	9058	9063	9069	9074	9079	1	1	2	2	3	3	4	4	5
81	9085	9090	9096	9101	9106	9112	9117	9122	9128	9133	1	1	2	2	3	3	4	4	5
82	9138	9143	9149	9154	9159	9165	9170	9175	9180	9186	1	1	2	2	3	3	4	4	5
83	9191	9196	9201	9206	9212	9217	9222	9227	9232	9238	1	1	2	2	3	3	4	4	5
84	9243	9248	9253	9258	9263	9269	9274	9279	9284	9289	1	1	2	2	3	3	4	4	5
85	9294	9299	9304	9309	9315	9320	9325	9330	9335	9340	1	1	2	2	3	3	4	4	5
86	9345	9350	9355	9360	9365	9370	9375	9380	9385	9390	1	1	2	2	3	3	4	4	5
87	9395	9400	9405	9410	9415	9420	9425	9430	9435	9440	0	1	1	2	2	3	3	4	4
88	9445	9450	9455	9460	9465	9469	9474	9479	9484	9489	0	1	1	2	2	3	3	4	4
89	9494	9499	9504	9509	9513	9518	9523	9528	9533	9538	0	1	1	2	2	3	3	4	4
90	9542	9547	9552	9557	9562	9566	9571	9576	9581	9586	0	1	1	2	2	3	3	4	4
91	9590	9595	9600	9605	9609	9614	9619	9624	9628	9633	0	1	1	2	2	3	3	4	4
92	9638	9643	9647	9652	9657	9661	9666	9671	9675	9680	0	1	1	2	2	3	3	4	4
93	9685	9689	9694	9699	9703	9708	9713	9717	9722	9727	0	1	1	2	2	3	3	4	4
94	9731	9736	9741	9745	9750	9754	9759	9763	9768	9773	0	1	1	2	2	3	3	4	4
95	9777	9782	9786	9791	9795	9800	9805	9809	9814	9818	0	1	1	2	2	3	3	4	4
96	9823	9827	9832	9836	9841	9845	9850	9854	9859	9863	0	1	1	2	2	3	3	4	4
97	9868	9872	9877	9881	9886	9890	9894	9899	9903	9908	0	1	1	2	2	3	3	4	4
98	9912	9917	9921	9926	9930	9934	9939	9943	9948	9952	0	1	1	2	2	3	3	4	4
99	9956	9961	9965	9969	9974	9978	9983	9987	9991	9996	0	1	1	2	2	3	3	4	4